ESSENTIAL DISCRETE

MATHEMATICS FOR

COMPUTER SCIENCE

ESSENTIAL DISCRETE
MATHEMATICS FOR
COMPUTER SCIENCE

Harry Lewis and Rachel Zax

PRINCETON UNIVERSITY PRESS ~ PRINCETON AND OXFORD

Requests for permission to reproduce material from this work
should be sent to permissions@press.princeton.edu

Published by Princeton University Press
41 William Street, Princeton, New Jersey 08540
6 Oxford Street, Woodstock, Oxfordshire OX20 1TR

press.princeton.edu

LCCN: 2018960770

ISBN 978-0-691-17929-2

British Library Cataloging-in-Publication Data is available

Editorial: Vickie Kearn, Susannah Shoemaker, and Arthur Werneck
Production Editorial: Kathleen Cioffi
Cover Design: Lorraine Doneker
Production: Erin Suydam
Publicity: Alyssa Sanford
Copyeditor: Alison S. Britton

This book has been composed in MinionPro

Printed on acid-free paper. ∞

Printed in the United States of America

10 9 8 7 6 5 4 3 2 1

To Alexandra, Stella, Elizabeth, and Annie
and
to David, Marcia, Ben, and Aryeh

An engineer is said to be a man who knows a great deal about a very little, and who goes around knowing more and more, about less and less, until finally, he practically knows everything about nothing; whereas, a salesman, on the other hand, is a man who knows a very little about a great deal, and keeps on knowing less and less about more and more until finally he knows practically nothing, about everything.

Van Nuys, California, *News,* June 26, 1933

CONTENTS

PREFACE

Τοῦ δὲ ποσοῦ τὸ μέν ἐστι διωρισμένον, τὸ δὲ συνεχες.

As to quantity, it can be either discrete or continuous.

—Aristotle, *Categories* (*ca.* 350 BCE)

This introductory text treats the discrete mathematics that computer scientists should know but generally do not learn in calculus and linear algebra courses. It aims to achieve breadth rather than depth and to teach reasoning as well as concepts and skills.

We stress the art of proof in the hope that computer scientists will learn to think formally and precisely. Almost every formula and theorem is proved in full. The text teaches the cumulative nature of mathematics; in spite of the breadth of topics covered, seemingly unrelated results in later chapters rest on concepts derived early on.

The text requires precalculus and occasionally uses a little bit of calculus. Chapter 21, on order notation, uses limits, but includes a quick summary of the needed basic facts. Proofs and exercises that use basic facts about derivatives and integrals, including l'Hôpital's rule, can be skipped without loss of continuity.

A fast-paced one-semester course at Harvard covers most of the material in this book. That course is typically taken by freshmen and sophomores as a prerequisite for courses on theory of computation (automata, computability, and algorithm analysis). The text is also suitable for use in secondary schools, for students of mathematics or computer science interested in topics that are mathematically accessible but off the beaten track of the standard curriculum.

The book is organized as a series of short chapters, each of which might be the subject of one or two class sessions. Each chapter ends with a brief summary and about ten problems, which can be used either as homework or as in-class exercises to be solved collaboratively in small groups.

Instructors who choose not to cover all topics can abridge the book in several ways. The spine of the book includes Chapters 1–8 on foundational concepts, Chapters 13–18 on digraphs and graphs, and Chapters 21–25 on order notation and counting. Four blocks of chapters are optional and can be included or omitted at the instructor's discretion and independently of each other:

- Chapters 9–12 on logic;
- Chapters 19–20 on automata and formal languages;

- Chapters 26–29 on discrete probability; and
- Chapters 30–31 on modular arithmetic and cryptography.

None of these blocks, if included at all, need be treated in full, since only later chapters in the same block rely on the content of chapters earlier in the block.

It has been our goal to provide a treatment that is generic in its tastes and therefore suitable for wide use, without the heft of an encyclopedic textbook. We have tried throughout to respect our students' eagerness to learn and also their limited budgets of time, attention, and money.

✳

With thanks to the CS20 team: Deborah Abel, Ben Adlam, Paul Bamberg, Hannah Blumberg, Crystal Chang, Corinne Curcie, Michelle Danoff, Jack Dent, Ruth Fong, Michael Gelbart, Kirk Goff, Gabriel Goldberg, Paul Handorff, Roger Huang, Steve Komarov, Abiola Laniyonu, Nicholas Longenbaugh, Erin Masatsugu, Keenan Monks, Anupa Murali, Eela Nagaraj, Rebecca Nesson, Jenny Nitishinskaya, Sparsh Sah, Maria Stoica, Tom Silver, Francisco Trujillo, Nathaniel Ver Steeg, Helen Wu, Yifan Wu, Charles Zhang, and Ben Zheng;

to Albert Meyer for his generous help at the start of CS20;

and to Michael Sobin, Scott Joseph, Alex Silverstein, and Noam Wolf for their critiques and support during the writing.

Harry Lewis and Rachel Zax

ESSENTIAL DISCRETE

MATHEMATICS FOR

COMPUTER SCIENCE

Chapter 1

The Pigeonhole Principle

How do we know that a computer program produces the right results? How do we know that a program will run to completion? If we know it will stop eventually, can we predict whether that will happen in a second, in an hour, or in a day? Intuition, testing, and "it has worked OK every time we tried it" should not be accepted as proof of a claim. Proving something requires formal reasoning, starting with things known to be true and connecting them together by incontestable logical inferences. This is a book about the mathematics that is used to reason about the behavior of computer programs.

The mathematics of computer science is not some special field. Computer scientists use almost every branch of mathematics, including some that were never thought to be useful until developments in computer science created applications for them. So this book includes sections on mathematical logic, graph theory, counting, number theory, and discrete probability theory, among other things. From the standpoint of a traditional mathematics curriculum, this list includes apples and oranges. One common feature of these topics is that all prove useful in computer science. Moreover, they are all *discrete mathematics*, which is to say that they involve quantities that change in steps, not continuously, or are expressed in symbols and structures rather than numbers. Of course, calculus is also important in computer science, because it assists in reasoning about continuous quantities. But in this book we will rarely use integrals and derivatives.

One of the most important skills of mathematical thinking is the art of *generalization*. For example, the proposition

> *There is no triangle with sides of lengths* 1, 2, *and* 6

is true, but very specific (see Figure 1.1). The sides of lengths 1 and 2 would have to join the side of length 6 at its two ends, but the two short sides together aren't long enough to meet up at the third corner.

Figure 1.1. Can there be a triangle with sides of lengths 1, 2 and 6?

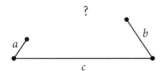

Figure 1.2. There is no triangle with sides of lengths a, b and c if $a + b \leq c$.

A more general statement might be (Figure 1.2)

There is no triangle with sides of lengths a, b, and c if a, b, c are any numbers such that a + b ≤ c.

The second form is more general because we can infer the first from the second by letting $a = 1$, $b = 2$, and $c = 6$. It also covers a case that the picture doesn't show—when $a + b = c$, so the three "corners" fall on a straight line. Finally, the general rule has the advantage of not just stating what is impossible, but explaining it. There is no $1 - 2 - 6$ triangle because $1 + 2 \leq 6$.

So we state propositions in general form for two reasons. First, a proposition becomes more useful if it is more general; it can be applied with confidence in a greater variety of circumstances. Second, a general proposition makes it easier to grasp what is really going on, because it leaves out irrelevant, distracting detail.

✳

As another example, let's consider a simple scenario.

Annie, Batul, Charlie, Deja, Evelyn, Fawwaz, Gregoire, and Hoon talk to each other and discover that Deja and Gregoire were both born on Tuesdays. (1.1)

Well, so what? Put two people together and they may or may not have been born on the same day of the week. Yet there is something going on here that can be generalized. As long as there are at least eight people, *some* two of them must have been born on the same day of the week, since a week has only seven days. *Some* statement like (1.1) must be true, perhaps with a different pair of names and a different day of the week. So here is a more general proposition.

In any group of eight people, some two of them were born on the same day of the week.

But even that isn't really general. The duplication has nothing to do with properties of people or days of the week, except how many there are of each. For the same reason, if we put eight cups on seven saucers, some saucer would have two cups on it. In fact there is nothing magic about "eight" and "seven," except that the one is larger than the other. If a hotel has 1000 rooms and 1001 guests, some room must contain at least two guests. How can we state a general principle that covers all these cases, without mentioning the irrelevant specifics of any of them?

First, we need a new concept. A *set* is a collection of things, or *elements*. The elements that belong to the set are called its *members*. The members of a set must be *distinct*, which is another way of saying they are all different from each other. So the people mentioned in (1.1) form a set, and the days of the week form another set. Sometimes we write out the members of a set explicitly, as a list within curly braces {}:

$P = \{$Annie, Batul, Charlie, Deja, Evelyn, Fawwaz, Gregoire, Hoon$\}$

$D = \{$Sunday, Monday, Tuesday, Wednesday, Thursday, Friday, Saturday$\}$.

When we write out the elements of a set, their order does not matter—in any order it is still the same set. We write $x \in X$ to indicate that the element x is a member of the set X. For example, Charlie $\in P$ and Thursday $\in D$.

We need some basic terminology about numbers in order to talk about sets. An *integer* is one of the numbers 0, 1, 2, …, or -1, -2, …. The *real* numbers are all the numbers on the number line, including all the integers and also all the numbers in between integers, such as $\frac{1}{2}$, $-\sqrt{2}$, and π. A number is *positive* if it is greater than 0, *negative* if it is less than 0, and *nonnegative* if it is greater than or equal to 0.

For the time being, we will be discussing finite sets. A *finite* set is a set that can (at least in principle) be listed in full. A finite set has a *size* or *cardinality*, which is a nonnegative integer. The cardinality of a set X is denoted $|X|$. For example, in the example of people and the days of the week on which they were born, $|P| = 8$ and $|D| = 7$, since eight people are listed and there are seven days in a week. A set that is not finite—the set of integers, for example—is said to be *infinite*. Infinite sets have sizes too—an interesting subject to which we will return in our discussion of infinite sets in Chapter 7.

Now, a *function* from one set to another is a rule that associates each member of the first set with exactly one member of the second set. If f is a function from X to Y and $x \in X$, then $f(x)$ is the member of Y that the function f associates with x. We refer to x as the *argument* of f and $f(x)$ as the *value* of f on that argument. We write $f : X \to Y$ to indicate that f is a function *from* set X *to* set Y. For example, we could write $b : P \to D$ to denote the function that associates each of the eight friends with the day of the week on which he or she was born; if Charlie was born on a Thursday, then $b(\text{Charlie}) = \text{Thursday}$.

A function $f : X \to Y$ is sometimes called a *mapping* from X to Y, and f is said to *map* an element $x \in X$ to the element $f(x) \in Y$. (In the same way, a real map associates a point on the surface of the earth with a point on a sheet of paper.)

Finally, we have a way to state the general principle that underlies the example of (1.1):

> *If* $f : X \to Y$ *and* $|X| > |Y|$, *then there are elements*
> $x_1, x_2 \in X$ *such that* $x_1 \neq x_2$ *and* $f(x_1) = f(x_2)$. (1.2)

The statement (1.2) is known as the *Pigeonhole Principle*, as it captures in mathematical form this commonsense idea: if there are more pigeons than

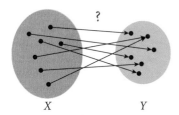

Figure 1.3. The Pigeonhole Principle. If $|X| > |Y|$ and f is any function from X to Y, then the values of f must be the same for some two distinct members of X.

pigeonholes and every pigeon goes into a pigeonhole, then some pigeonhole must have more than one pigeon in it. The pigeons are the members of X and the pigeonholes are the members of Y (Figure 1.3).

We will provide a formal proof of the Pigeonhole Principle on page 34, once we have developed some of the basic machinery for doing proofs. For now, let's scrutinize the statement of the Pigeonhole Principle with an eye toward understanding mathematical language. Here are some questions we might ask:

1. What are X and Y?

   They are finite sets. To be absolutely clear, we might have begun the statement with the phrase, "For any finite sets X and Y," but the assertion that f is a function from X to Y makes sense only if X and Y are sets, and it is understood from context that the sets under discussion are finite—and we therefore know how to compare their sizes.

2. Why did we choose "x_1" and "x_2" for the names of elements of X?

   We could in principle have chosen any variables, "x" and "y" for example. But using variations on "X" to name elements of the set X suggests that x_1 and x_2 are members of the set X rather than the set Y. So using "x_1" and "x_2" just makes our statement easier to read.

3. Was the phrase "such that $x_1 \neq x_2$" really necessary? The sentence is simpler without it, and seems to say the same thing.

   Yes, the "$x_1 \neq x_2$" is necessary, and no, the sentence doesn't say the same thing without it! If we didn't say "$x_1 \neq x_2$," then "x_1" and "x_2" could have been two names for the same element. If we did not stipulate that x_1 and x_2 had to be different, the proposition would not have been false—only trivial! Obviously if $x_1 = x_2$, then $f(x_1) = f(x_2)$. That is like saying that the mass of Earth is equal to the mass of the third planet from the sun. Another way to state the Pigeonhole Principle would be to say, "there are distinct elements $x_1, x_2 \in X$ such that $f(x_1) = f(x_2)$."

One more thing is worth emphasizing here. A statement like "there are distinct elements $x_1, x_2 \in X$ with property blah" does *not* mean that there are *exactly* two elements with that property. It just means that at least two such elements exist for sure—maybe more, but definitely not less.

✳

Mathematicians always search for the most general form of any principle, because it can then be used to explain more things. For example, it is equally obvious that we can't put 15 pigeons in 7 pigeonholes without putting at least 3 pigeons in some pigeonhole—but there is no way to derive that from the Pigeonhole Principle as we stated it. Here is a more general version:

Theorem 1.3. Extended Pigeonhole Principle. *For any finite sets X and Y and any positive integer k such that $|X| > k \cdot |Y|$, if $f : X \to Y$, then there are at least $k + 1$ distinct members $x_1, \ldots, x_{k+1} \in X$ such that $f(x_1) = \ldots = f(x_{k+1})$.*

The Pigeonhole Principle is the $k = 1$ case of the Extended Pigeonhole Principle.

We have used *sequence* notation here for the first time, using the same variable with numerical subscripts in a range. In this case the x_i, where $1 \le i \le k + 1$, form a sequence of length $k + 1$. This notation is very convenient since it makes it possible to use an algebraic expression such as $k + 1$ in a subscript. Similarly, we could refer to the $2i^{\text{th}}$ member of a sequence $y_1, y_2, \ldots$ as y_{2i}.

The minimum value of the parameter k in the Extended Pigeonhole Principle, as applied to particular sets X and Y, can be derived once the sizes of X and Y are known. It is helpful to introduce some notation to make this calculation precise.

An integer p *divides* another integer q, symbolically written as $p \mid q$, if the quotient $\frac{q}{p}$ is an integer—that is, dividing q by p leaves no remainder. We write $p \nmid q$ if p does not divide q—for example, $3 \nmid 7$. If x is any real number, we write $\lfloor x \rfloor$ for the greatest integer less than or equal to x (called the *floor* of x). For example, $\lfloor \frac{17}{3} \rfloor = 5$, and $\lfloor \frac{6}{2} \rfloor = 3$. We will also need the *ceiling* notation: $\lceil x \rceil$ is the smallest integer greater than or equal to x, so for example $\lceil 3.7 \rceil = 4$.

With the aid of these notations, we can restate the Extended Pigeonhole Principle in a way that determines the minimum size of the most heavily occupied pigeonhole for given numbers of pigeons and pigeonholes:

Theorem 1.4. Extended Pigeonhole Principle, Alternate Version. *Let X and Y be any finite sets and let $f : X \to Y$. Then there is some $y \in Y$ such that $f(x) = y$ for at least*

$$\left\lceil \frac{|X|}{|Y|} \right\rceil$$

values of x.

Proof. Let $m = |X|$ and $n = |Y|$. If $n \mid m$, then this is the Extended Pigeonhole Principle with $k = \frac{m}{n} - 1 = \lceil \frac{m}{n} \rceil - 1$. If $n \nmid m$, then again this is the Extended Pigeonhole Principle with $k = \lceil \frac{m}{n} \rceil - 1$, since that is the largest integer less than $\frac{|X|}{|Y|}$. ∎

❊

Once stated in their general form, these versions of the Pigeonhole Principle seem to be fancy ways of saying something obvious. In spite of that, we can use them to explain a variety of different phenomena—once we figure out what are the "pigeons" and the "pigeonholes." Let's close with an application to *number theory*—the study of the properties of the integers. A few basics first.

If $p \mid q$, then p is said to be a *factor* or *divisor* of q.

A *prime* number is an integer greater than 1 that is divisible only by itself and 1. For example, 7 is prime, because it is divisible only by 7 and 1, but 6 is not prime, because $6 = 2 \cdot 3$. Note that 1 itself is not prime.

Theorem 1.5. The Fundamental Theorem of Arithmetic. *There is one and only one way to express an integer greater than 1 as a product of distinct prime numbers in increasing order and with positive integer exponents.*

We'll prove this theorem in Chapter 4, but make some use of it right now. The *prime decomposition* of a number n is that unique product

$$n = p_1^{e_1} \cdot \ldots \cdot p_k^{e_k}, \tag{1.6}$$

where the p_i are primes in increasing order and the e_i are positive integers. For example, $180 = 2^2 \cdot 3^2 \cdot 5^1$, and there is no other product $p_1^{e_1} \cdot \ldots \cdot p_k^{e_k}$ equal to 180, where $p_1 < p_2 < \ldots < p_k$, all the p_i are prime, and the e_i are integer exponents.

The prime decomposition of the product of two integers m and n combines the prime decompositions of m and of n—every prime factor of $m \cdot n$ is a prime factor of one or the other.

Theorem 1.7. *If m, n, and p are integers greater than 1, p is prime, and $p \mid m \cdot n$, then either $p \mid m$ or $p \mid n$.*

Proof. By the Fundamental Theorem of Arithmetic (Theorem 1.5), there is one and only one way to write

$$m \cdot n = p_1^{e_1} \cdot \ldots \cdot p_k^{e_k},$$

where the p_i are prime. But then p must be one of the p_i, and each p_i must appear in the unique prime decomposition of either m or n. ∎

The exponent of a prime p in the prime decomposition of $m \cdot n$ is the sum of its exponents in the prime decompositions of m and n (counting the exponent as 0 if p does not appear in the decomposition). For example, consider the product $18 \cdot 10 = 180$. We have

$$
\begin{aligned}
18 &= 2^1 \cdot 3^2 && \text{(exponents of 2, 3, 5 are 1, 2, 0)} \\
10 &= 2^1 \cdot 5^1 && \text{(exponents of 2, 3, 5 are 1, 0, 1)} \\
180 &= 2^2 \cdot 3^2 \cdot 5^1 \\
&= 2^{1+1} \cdot 3^{2+0} \cdot 5^{0+1}.
\end{aligned}
$$

We have color-coded the exponents to show how the exponents of 2, 3, and 5 in the product 180 are the sums of the exponents of those primes in the decompositions of the two factors 18 and 10.

Another important fact about prime numbers is that there are infinitely many of them.

Theorem 1.8. *There are arbitrarily large prime numbers.*

"Arbitrarily large" means that for every $n > 0$, there is a prime number greater than n.

Proof. Pick some value of k for which we know there are at least k primes, and let $p_1, \ldots, p_k$ be the first k primes in increasing order. (Since $p_1 = 2$, $p_2 = 3, p_3 = 5$, we could certainly take $k = 3$.) We'll show how to find a prime number greater than p_k. Since this process could be repeated indefinitely, there must be infinitely many primes.

Consider the number N that is one more than the product of the first k primes:

$$N = (p_1 \cdot p_2 \cdot \ldots \cdot p_k) + 1. \tag{1.9}$$

Dividing N by any of $p_1, \ldots, p_k$ would leave a remainder of 1. So N has no prime divisors less than or equal to p_k. Therefore, either N is not prime but has a prime factor greater than p_k, or else N is prime itself.   ■

In the $k = 3$ case, for example, $N = 2 \cdot 3 \cdot 5 + 1 = 31$. Here N itself is prime; Problem 1.11 asks you to find an example of the case in which N is not prime.

A *common divisor* of two numbers is a number that divides both of them. For example, 21 and 36 have the common divisors 1 and 3, but 16 and 21 have no common divisor greater than 1.

With this by way of background, let's work a number theory example that uses the Pigeonhole Principle.

Example 1.10. *Choose m distinct numbers between 2 and 40 inclusive, where $m \geq 13$. Then at least two of the numbers have some common divisor greater than 1.*

"Between *a* and *b inclusive*" means including all numbers that are $\geq a$ and also $\leq b$—so including both 2 and 40 in this case.

Solution to example. Observe first that there are 12 prime numbers less than or equal to 40: 2, 3, 5, 7, 11, 13, 17, 19, 23, 29, 31, 37, no two of which share a factor greater than 1. Let's define P to be this set of 12 prime numbers. (We needed to specify that $m \geq 13$, because the claim would be false with

$m = 12$ instead: the set P would be a counterexample.) Now consider a set X of m numbers in the range from 2 to 40 inclusive. We can think of the members of X as pigeons and the members of P as pigeonholes. To place pigeons in pigeonholes, use the function $f : X \to P$, where $f(x)$ is the smallest prime that divides x. For example, $f(16) = 2$, $f(17) = 17$, and $f(21) = 3$. By the Pigeonhole Principle, since $m > 12$, the values of f must be equal for two distinct members of X, and therefore at least two members of X have a common prime divisor.   ∎

Chapter Summary

- Mathematical thinking focuses on general principles, abstracted from the details of specific examples.

- A *set* is an unordered collection of *distinct* things, or *elements*. The elements of a set are its *members*.

- A set is *finite* if its members can be listed in full one by one. The number of members of a finite set X is called its *cardinality* or *size* and is denoted $|X|$. A set's size is always a *nonnegative integer*.

- A *function* or *mapping* between two sets is a rule associating each member of the first set with a unique member of the second.

- The *Pigeonhole Principle* states that if X is a set of pigeons and Y a set of pigeonholes, and $|X| > |Y|$, then any function mapping pigeons to pigeonholes assigns more than one pigeon to some pigeonhole.

- The *Extended Pigeonhole Principle* states that if X is a set of pigeons and Y a set of pigeonholes, and $|X| > k|Y|$, then any function mapping pigeons to pigeonholes assigns more than k pigeons to some pigeonhole.

- A *sequence* of terms can be denoted by a repeated variable with different numerical subscripts, such as $x_1, \ldots, x_n$. The subscript of a term may be an algebraic expression.

- The *Fundamental Theorem of Arithmetic* states that every positive integer has exactly one *prime decomposition*.

Problems

1.1. What are each of the following?
- (a) $|\{0, 1, 2, 3, 4, 5, 6\}|$.
- (b) $\lceil \frac{111}{5} \rceil$.
- (c) $\lfloor \frac{5}{111} \rfloor$.
- (d) The set of divisors of 100.
- (e) The set of prime divisors of 100.

1.2. Let $f(n)$ be the largest prime divisor of n. Can it happen that $x < y$ but $f(x) > f(y)$? Give an example or explain why it is impossible.

1.3. Under what circumstances is $\lfloor x \rfloor = \lceil x \rceil - 1$?

1.4. Imagine a 9×9 square array of pigeonholes, with one pigeon in each pigeonhole. (So 81 pigeons in 81 pigeonholes—see Figure 1.4.) Suppose that all at once, all the pigeons move up, down, left, or right by one hole. (The pigeons on the edges are not allowed to move out of the array.) Show that some pigeonhole winds up with two pigeons in it. *Hint:* The number 9 is a distraction. Try some smaller numbers to see what is going on.

1.5. Show that in any group of people, two of them have the same number of friends in the group. (Some important assumptions here: no one is a friend of him- or herself, and friendship is *symmetrical*—if A is a friend of B, then B is a friend of A.)

1.6. Given any five points on a sphere, show that four of them must lie within a closed hemisphere, where "closed" means that the hemisphere includes the circle that divides it from the other half of the sphere. *Hint:* Given any two points on a sphere, one can always draw a "great circle" between them, which has the same circumference as the equator of the sphere.

1.7. Show that in any group of 25 people, some three of them must have birthdays in the same month.

1.8. A collection of coins contains six different denominations: pennies, nickels, dimes, quarters, half-dollars, and dollars. How many coins must the collection contain to guarantee that at least 100 of the coins are of the same denomination?

1.9. Twenty-five people go to daily yoga classes at the same gym, which offers eight classes every day. Each attendee wears either a blue, red, or green shirt to class. Show that on a given day, there is at least one class in which two people are wearing the same color shirt.

1.10. Show that if four distinct integers are chosen between 1 and 60 inclusive, some two of them must differ by at most 19.

1.11. Find a k such that the product of the first k primes, plus 1, is not prime, but has a prime factor larger than any of the first k primes. (There is no trick for solving this. You just have to try various possibilities!)

1.12. Show that in any set of 9 positive integers, some two of them share all of their prime factors that are less than or equal to 5.

1.13. A *hash function* from strings to numbers derives a numerical hash value $h(s)$ from a text string s; for example, by adding up the numerical codes for the characters in s, dividing by a prime number p, and keeping just the remainder. The point of a hash function is to yield a reproducible result (calculating $h(s)$ twice for the same string s yields the same numerical value) and to make it likely that the hash values for different strings will be spread out evenly across the possible hash values (from 0 to $p - 1$). If the hash function has identical hash values for two different strings, then these two strings are said to *collide* on that

Figure 1.4. Each pigeonhole in a 9×9 array has one pigeon. All simultaneously move to another pigeonhole that is immediately above, below, to the left, or to the right of its current hole. Must some pigeonhole wind up with two pigeons?

hash value. We count the number of *collisions* on a hash value as 1 less than the number of strings that have that hash value, so if 2 strings have the same hash value there is 1 collision on that hash value. If there are m strings and p possible hash values, what is the minimum number of collisions that must occur on the hash value with the most collisions? The maximum number of collisions that might occur on some hash value?

Chapter 2

Basic Proof Techniques

Here is an English-language restatement of the Pigeonhole Principle (page 3):

> *If there are more pigeons than pigeonholes and every pigeon goes*
> *into a pigeonhole, then some pigeonhole must contain more than*
> *one pigeon.*

But suppose your friend did not believe this statement. How could you convincingly argue that it was true?

You might try to persuade your friend that there is no way the opposite could be true. You could say, let's imagine that each pigeonhole has no more than one pigeon. Then we can count the number of pigeonholes, and since each pigeonhole contains zero or one pigeons, the number of pigeons can be at most equal to the number of pigeonholes. But we started with the assumption that there were more pigeons than pigeonholes, so this is impossible! Since there is no way that every pigeonhole can have at most one pigeon, some pigeonhole must contain more than one pigeon, and that is what we were trying to prove.

In this chapter, we'll discuss how to take informal, specific arguments like this and translate them into formal, general, mathematical proofs. A *proof* is an argument that begins with one or more premises (for example, "there are more pigeons than pigeonholes") and proceeds using logical rules to establish a conclusion (such as "some pigeonhole has more than one pigeon"). Although it may seem easier to write (and understand!) an argument in plain English, ordinary language can be imprecise or overly specific. So it is clearer, as well as more general, to describe a mathematical situation in more formal terms.

For example, what does the statement

$$\textit{Everybody loves somebody} \tag{2.1}$$

mean? It might mean that for every person in the world, there is someone whom that person loves—so different lovers might have different beloveds. In semi-mathematical language, we would state that interpretation as

For every person A, there is a person B such that A loves B.   (2.2)

But there is another interpretation of (2.1), namely that there is some special person whom everybody loves, or in other words,

There is a person B such that for every person A, A loves B.   (2.3)

There is a big difference between these interpretations, and one of the purposes of mathematical language is to resolve such ambiguities of natural language.

The phrases "for all," "for any," "for every," "for some," and "there exists" are called *quantifiers*, and their careful use is an important part of mathematical discourse. The symbol ∀ stands for "for all," "for any," or "for every," and the symbol ∃ stands for "there exists" or "for some." Using these symbols saves time, but in writing mathematical prose they can also make statements more confusing. So we will avoid them until we discuss the formalization of *quantificational logic*, in Chapter 12.

Quantifiers modify *predicates*, such as "*A* loves *B*." A predicate is a template for a proposition, taking one or more *arguments*, in this case *A* and *B*. On its own, a predicate has no truth value: without knowing the values of *A* and *B*, "*A* loves *B*" cannot be said to be either true or false. It takes on a truth value only when quantified (as in (2.2) and (2.3)), or when applied to specific arguments (for example, "Romeo loves Juliet"), and may be true for some arguments but false for others.

Let's continue with a simple example of a mathematical statement and its proof.

Theorem 2.4. Odd Integers. *Every odd integer is equal to the difference between the squares of two integers.*

First, let's make sure we understand the statement. An *odd* integer is any integer that can be written as $2k + 1$, where k is also an integer. The *square* of an integer n is $n^2 = n \cdot n$. For every value of k, Theorem 2.4 says that there are two integers—call them m and n—such that if we square them and subtract one result from the other, the resulting number is equal to $2k + 1$. (Note the quantifiers: for every k, there exist m and n, such that ....)

An integer m is said to be a *perfect square* if it is the square of some integer, so a compact way to state the theorem is to say that every odd integer is the difference of two perfect squares. It is typical in mathematics, as in this case, that defining just the right concepts can result in very simple phrasing of general truths.

The next step is to convince ourselves of why the statement is true. If the reason is not obvious, it may help to work out some examples. So let's start by listing out the first few squares:

$$0^2 = 0$$
$$1^2 = 1$$
$$2^2 = 4$$
$$3^2 = 9$$
$$4^2 = 16.$$

We can confirm that the statement is true for a few specific odd integers, say 1, 3, 5, and 7:

$$1 = 1 - 0 = 1^2 - 0^2$$
$$3 = 4 - 1 = 2^2 - 1^2$$
$$5 = 9 - 4 = 3^2 - 2^2$$
$$7 = 16 - 9 = 4^2 - 3^2.$$

After these examples we might notice a pattern: so far, all of the odd integers are the difference between the squares of two consecutive integers: 0 and 1, then 1 and 2, 2 and 3, and 3 and 4. Another observation—those consecutive integers add up to the target odd integer: $0 + 1 = 1$, $1 + 2 = 3$, $2 + 3 = 5$, and $3 + 4 = 7$. So we might conjecture that for the odd integer $2k + 1$, the integers that should be squared and subtracted are $k + 1$ and k. Let's try that:

$$(k + 1)^2 - k^2 = k^2 + 2k + 1 - k^2,$$

which simplifies to $2k + 1$.

So our guess was right! And by writing it out using the definition of an odd integer $(2k + 1)$ rather than looking at any specific odd integer, we've confirmed that it works for all odd integers. It even works for negative odd integers (since those are equal to $2k + 1$ for negative values of k), although the idea was inspired by trying examples of positive odd integers.

This chain of thought shows how we might arrive at the idea, but it's too meandering for a formal proof. The actual proof should include only the details that turned out to be relevant. For instance, the worked-out examples don't add anything to the argument—we need to show that the statement is true all the time, not just for the examples we tried—so they should be left out. Here is a formal proof of Theorem 2.4:

Proof. Any odd integer can be written as $2k + 1$ for some integer k. We can rewrite that expression:

$$2k + 1 = (k^2 + 2k + 1) - k^2 \quad \text{(adding and subtracting } k^2\text{)}$$
$$= (k + 1)^2 - k^2 \quad \text{(writing the first term as a square).}$$

Now let $m = k + 1$ and $n = k$. Then $2k + 1 = m^2 - n^2$, so we have identified integers m and n with the properties that we claimed. ∎

Before examining the substance of this proof, a few notes about its style. First, it is written in full sentences. Mathematical expressions are used for precision and clarity, but the argument itself is written in prose. Second, its structure is clear: it starts with the given assumption, that the integer is odd, and clearly identifies when we have reached the end, by noting that m and n are the two integers that we sought. Third, it is rigorous: it gives a mathematical definition for the relevant term ("odd integer"), which forces us to be precise and explicit; and each step of the proof follows logically and clearly from the previous steps.

Finally, it is convincing. The proof gives an appropriate amount of detail, enough that the reader can easily understand why each step is correct, but not so much that it distracts from the overall argument. For example, we might have skipped writing out some of the arithmetic, and stated just that

$$2k + 1 = (k + 1)^2 - k^2.$$

But this equality is not obvious, and a careful reader would be tempted to double-check it. When we include the intermediate step, the arithmetic is clearly correct. On the other hand, some assumptions don't need to be proven—for example, that it is valid to write

$$2k + 1 = (k^2 + 2k + 1) - k^2.$$

This relies on the fact that if we move the terms around and group them in different ways, they still add up to the same value. In this context, these rules seem rather basic and can be assumed; proving them would distract the reader from the main argument. But in a text on formal arithmetic, these properties might themselves be the subject of a proof. The amount of detail to include depends on the context and the proof's intended audience; as a rule of thumb, write as if you are trying to convince a peer.

Let's return now to the substance of the above proof. First, it is *constructive*. The statement to be proved merely asserts the existence of something: it says that for any odd integer, there exist two integers with the property that the difference of their squares is equal to the number we started with. A constructive proof not only demonstrates that the thing exists, but shows us exactly how to find it. Given a particular odd integer $2k + 1$, the proof of Theorem 2.4 shows us how to find the two integers that have the property asserted in the statement of the theorem—one is $k + 1$ and the other is k. For example, if the odd integer we wanted to express as the difference of two squares was 341, the proof shows that we can subtract 1 from 341 and divide by 2, and that integer and the next larger integer are the desired pair—170 and 171. This is easy to check:

$$171^2 - 170^2 = 29241 - 28900 = 341.$$

There was no real need to check this particular case, but doing so makes us more confident that we didn't make an algebraic mistake somewhere.

In general, a procedure for answering a question or solving a problem is said to be an *algorithm* if its description is sufficiently detailed and precise that it could, in principle, be carried out mechanically—by a machine, or by a human being mindlessly following instructions. A constructive proof implicitly describes an algorithm for finding the thing that the proof says exists. In the case of Theorem 2.4, the proof describes an algorithm that, given an odd integer $2k + 1$, finds integers m and n such that $m^2 - n^2 = 2k + 1$.

Not every proof is constructive—sometimes it is possible to show that something exists without showing how to find it. Such a proof is called *nonconstructive*. We will see some interesting examples of nonconstructive proofs—in fact, the proof of the Pigeonhole Principle will be nonconstructive, since it cannot identify *which* pigeonhole has more than one pigeon. But computer scientists love constructive arguments, because a constructive proof that something exists yields an algorithm to find it—one that a computer could be programmed to carry out.[1]

One final note about Theorem 2.4. The proof not only is constructive, but proves more than was asked for. It shows not just that every odd integer is the difference of two squares, but that every odd integer is the difference of the squares of *two consecutive integers*, as we noted while working through the examples. After finishing a proof, it is often worth looking back at it to see if it yields any interesting information beyond the statement it set out to prove.

✳

A common goal in mathematical proof is to establish that two statements are equivalent—that one statement is true in all the circumstances in which the second statement is true, and vice versa. For example, consider the statement

> *The square of an integer is odd if and only if the integer itself is odd.*

Or, to write the same thing in a more conventionally mathematical style:

Theorem 2.5. *For any integer n, n^2 is odd if and only if n is odd.*

This is a fairly typical mathematical statement. Several things about it are worth noting.

- It uses a variable n to refer to the thing that is under discussion, so the same name can be used in different parts of the statement to refer to the same thing.

[1] A stronger meaning of the term "constructive" is used in the school of *constructive mathematics*, which disallows any mathematical argument that does not lead to the construction of the thing that the argument says exists. In constructive mathematics it is impermissible to infer from the fact that a statement is demonstrably false that its negation is necessarily true; the truth of the negation has to be demonstrated directly. For example, constructive mathematics disallows proofs by contradiction (explained below) and arguments like that of Problem 2.14. Computer scientists prefer arguments that yield algorithms, but generally don't insist on "constructive proofs" in the strict sense used by constructive mathematicians. For us, to show something is true it suffices to show that its negation is not true.

- By convention, the use of the name "n" suggests that the thing is an integer. A different name, such as "x," might suggest that the thing is an arbitrary real number; "p" might suggest it is a prime number. Using suggestive variable names helps the reader understand the statement, though mathematically the choice of variable names is irrelevant.

- Though the variable name is suggestive, the statement does not rely on the name of the variable to carry meaning. It specifically states that n is an integer, since in other contexts "n" might be a positive integer, a nonnegative integer, or something else.

- The statement uses a quantifier ("for any") in order to be explicit that the property it describes holds for *all* integers.

- The statement "n^2 is odd if and only if n is odd" is really two statements in one:

    1. "n^2 is odd if n is odd," or in other words, "if n is odd then n^2 is odd"; and

    2. "n^2 is odd only if n is odd," or in other words, "if n^2 is odd then n is odd."

    Let's give names to the two components of this "if and only if" statement: p for "n^2 is odd," and q for "n is odd." Focus for a moment on the meaning of "only if": "p only if q" means the same thing as "if p, then q." More longwindedly, if we know that p is true, then the only way that could be is for q to be true as well; that is, "if p, then q."

    An "if and only if" statement is true just in case *both* of the statements "p if q" and "p only if q" are true; or as we shall say, just in case the two components p and q are *equivalent*. The phrase "if and only if" is commonly abbreviated as *iff*.

A proof of the equivalence of two statements often consists of two proofs, showing that each of the statements implies the other. In the case of Theorem 2.5, we will prove that if an integer is odd, its square is odd as well; and then we will also prove that if the square of an integer is odd, then the integer itself is odd. In general, the proofs in the two directions may look very different from each other. No matter how difficult one direction of the argument has been, an equivalence proof is incomplete until the other direction has been proved as well!

The first direction of Theorem 2.5 ("if n is odd then n^2 is odd") can be shown by a *direct proof*. This type of proof is straightforward: we assume that the premise is true and work from there to show that the conclusion is true. But for the second direction ("if n^2 is odd then n is odd"), it's easier if we think of the problem in a slightly different way. A direct proof would assume that n^2 is odd and attempt to prove, from this information, that n is odd; but there does not seem to be any simple way to do this. Instead, it is equivalent to show that whenever n is *not* odd—that is, n is even—then n^2 is not odd either.

Proof. First, we show that if n is odd then n^2 is odd. If n is odd, then we can write $n = 2k + 1$ where k is some other integer. Then

$$
\begin{aligned}
n^2 &= (2k+1)^2 && \text{(since } n = 2k + 1) \\
&= 4k^2 + 4k + 1 && \text{(expanding the binomial)} \\
&= 2(2k^2 + 2k) + 1 && \text{(factoring 2 from the first two terms).}
\end{aligned}
$$

Let j be the integer $2k^2 + 2k$; then $n^2 = 2j + 1$, so n^2 is odd.

Next, we show that if n^2 is odd, then n must be odd. Suppose this is not true for all n, and that n is a particular integer such that n^2 is odd but n is not odd. Since any integer that isn't odd is even, n must be even. To prove that no such n can exist, we need to show that if n is any even integer, then n^2 must also be even. Now if n is even, we can write $n = 2k$ where k is some other integer. Then

$$
\begin{aligned}
n^2 &= (2k)^2 \quad \text{(since } n = 2k) \\
&= 4k^2 \\
&= 2(2k^2).
\end{aligned}
$$

Let j be the integer $2k^2$; then $n^2 = 2j$, so n^2 is even. That is, the supposition ("n is an integer such that n^2 is odd but n is not odd") was false. So its negation is true: if n^2 is odd, then n is odd.   ∎

This proof has more steps than the proof of Theorem 2.4, and a bit more structure. In a longer proof, the steps must fit together coherently, so that a reader understands not just each individual step but also the overall thrust of the argument. To guide the reader through this proof, for example, we announced which direction we were proving before diving in to each part of the argument.

A more complicated proof may have multiple intermediate steps. When the proof of an intermediate step is especially long or difficult, it may make sense to first give a standalone proof of that step, so that the overall proof is not too long. An intermediate step can be its own theorem, but if it is not especially interesting outside the context of the original theorem, we call it a *lemma*. On the other hand, if there is an interesting result that is very easy to prove once the original theorem is proven, we sometimes call it a *corollary* rather than framing it as its own theorem. For example, here is a corollary to Theorem 2.5:

Corollary 2.6. *If n is odd, then n^4 is odd.*

Proof. Note that $n^4 = (n^2)^2$. Since n is odd, by Theorem 2.5, n^2 is odd. Then since n^2 is odd, again by Theorem 2.5, n^4 is odd.   ∎

In the second part of the proof of Theorem 2.5, we took an *implication*—a statement of the form "if p then q"—and turned it into the equivalent form

---

Proving an equivalence

To prove p if and only if q:
Prove p if q; that is,
> If q then p.

Prove p only if q; that is,
> If p then q.

[2] We are here for the first time using variables like "p" and "q" to represent *propositions*; that is, statements that can individually be true or false. Chapter 9 is devoted to calculations on such propositional variables, a system called the *propositional calculus*. For now we are just using propositional variables to help us describe how propositions can be combined.

Variations on statement

$s = $ "*if p, then q*"

q if p (equivalent to *s*)

p only if q (equivalent to *s*)

Contrapositive:

   if not q, then not p

(equivalent to *s*)

Inverse:

   if not p, then not q

(not equivalent to *s*!)

Converse:

   if q, then p

(not equivalent to *s*, but the contrapositive of the inverse, so equivalent to the inverse)

"if not q, then not p." In this part of the proof, p was "n^2 is odd," and q was "n is odd." There are several other variations on this theme, so it is important to name and identify them.[2]

Any implication "if p then q" can be flipped around in three different ways. The *inverse* is the statement we get if we simply negate both parts: "if not p then not q." The *converse* is the statement we get if we change the order: "if q then p." The *contrapositive* is the statement we get if we both flip and negate: "if not q then not p."

The contrapositive of any statement is logically equivalent to the statement itself, so we can always prove one by proving the other—this is like proving a statement by contradiction, as we saw above. This is why it was valid for us to prove "if n is not odd then n^2 is not odd" instead of proving "if n^2 is odd then n is odd": each is the other's contrapositive.

Note though that the inverse and the converse are *not* equivalent to the original statement. For example, in the proof above, we had to prove both that "if n is odd then n^2 is odd" and "if n^2 is odd then n is odd"—these statements are converses of each other, and are not logically equivalent.

We used one other transformation of a statement. The *negation* of a statement p is the statement that p is false—in other words, "not p." So the negation of "n is odd" is, in effect, "n is even." And the contrapositive of "if p, then q" is "if the negation of q, then the negation of p."

✳

Although mathematical language allows us to express ideas more generally and more precisely, greater abstraction can also create confusion. When expressing ideas in this way, we have to be extra careful to make sure that the statements we're writing down make sense. In a proof, every step has to follow from the previous step or steps; be skeptical! For example, consider the following bogus "proof" that $1 = 2$:

Let $a = b$. Then we can write

$$a^2 = ab \qquad \text{(multiplying both sides by } a\text{)}$$
$$a^2 - b^2 = ab - b^2 \qquad \text{(subtracting } b^2 \text{ from both sides)}$$
$$(a + b)(a - b) = b(a - b) \qquad \text{(factoring both sides)}$$
$$(a + b) = b \qquad \text{(dividing both sides by } a - b\text{)}$$
$$2b = b \qquad \text{(substituting } a \text{ for } b \text{ on the left side since } a = b\text{)}$$

and therefore $2 = 1$ (dividing both sides by b).

Where did we go wrong? We've come to a conclusion that is obviously impossible, starting from an assumption ($a = b$) that is certainly possible, so at some point in this proof, we must have taken a step that doesn't make logical sense.

The flaw occurred when we moved from the third line to the fourth. We stipulated that $a = b$, so when we divided by $a - b$, we actually divided by zero, and that isn't a valid arithmetic operation. So the logic went haywire at that point, and the rest is nonsense. When writing proofs, it's important to keep in mind the meaning behind the symbols in order to avoid such mistakes.

✳

In the bogus proof that $2 = 1$, we presented something that looked like a proof but really wasn't. We recognized that there was a mistake somewhere because we wound up with a contradiction, and that forced us to backtrack to figure out where the mistake was in our argument. This kind of reasoning—deriving a contradiction, which means there was an error in the argument—is actually a very useful proof technique. Indeed we have already seen an example of *proof by contradiction*, in the second part of the proof of Theorem 2.5.

A proof by contradiction starts off by assuming the negation of the statement we are hoping to prove. If we can derive a contradiction (without making any logical errors along the way), then the only thing that could have been wrong with the argument is the way it started. Since assuming the negation of the statement results in a contradiction, the statement must actually be true.

As another example, we'll use proof by contradiction to prove that $\sqrt{2}$ is irrational. As usual, the first thing is to be sure we understand the statement we are discussing. A *rational* number is one that can be expressed as the ratio of two integers. For example, 1.25 is rational, since it can be expressed as $\frac{5}{4}$ (or $\frac{125}{100}$ or in many other ways). A number is *irrational* if it is not rational; that is, if it cannot be expressed as the ratio of any two integers.

We could try to prove that $\sqrt{2}$ is irrational directly, but it's not at all clear how to proceed (or even where to start!). So let's instead consider what would happen if the statement weren't true, and see if the consequences of that assumption are logically contradictory.

> **Proof by contradiction**
> To prove p by contradiction:
> Assume p is *false*.
> Derive a contradiction.
> Conclude that p must be true.

Theorem 2.7. $\sqrt{2}$ *is irrational.*

Proof. In order to derive a contradiction, assume that the statement is false; that is, that $\sqrt{2}$ is not irrational. That means that $\sqrt{2}$ is rational, or in other words, is the ratio of two integers. So suppose that $\sqrt{2} = \frac{a}{b}$, for some integers a and b, where at most one of a and b is even. This additional assumption (at most one of a and b is even) is safe, since if both were even we could divide each by 2 and get a fraction with the same value but a smaller numerator and a smaller denominator:

$$\frac{a}{b} = \frac{a/2}{b/2}.$$

After repeating this step enough times, we would obtain a ratio in which either the numerator or the denominator is odd, or both are odd. So we can assume, *without loss of generality* as the saying goes, that $\sqrt{2} = \frac{a}{b}$ and at least one of a and b is odd. Now since $\sqrt{2} = \frac{a}{b}$, multiplying both sides by b, we get $b \cdot \sqrt{2} = a$. Squaring both sides we get

$$2b^2 = a^2. \tag{2.8}$$

So a^2 is divisible by 2. That is, a^2 is not odd, so by Theorem 2.5 a itself is not odd—in other words, a is divisible by 2. So we can write $a = 2k$, where k is some other integer, so (2.8) becomes

$$2b^2 = (2k)^2 = 4k^2.$$

Dividing both sides by 2, we get $b^2 = 2k^2$. So b^2 is even, and therefore b must itself be even. But now we've shown that both a and b are even, which contradicts our assumption that at most one of a and b was even! So $\sqrt{2}$ must be irrational, as we set out to prove. ∎

✳

Sometimes it becomes easier to develop a proof if the proposition to be proved can be broken into smaller chunks, which can be proved separately and reassembled to complete the argument. A simple version of this appeared in the proof of Theorem 2.5 on page 15, where we broke an "if and only if" statement into two implications and proved them separately. Another method of proof that uses this approach is called *case analysis*, in which a general statement about a class is broken down into statements about several subclasses. If the statement can be proven separately about entities of each possible kind, then the original, general statement must be true. We saw a simple case analysis on page 7, where we were trying to find a prime larger than the first k primes $p_1, \ldots, p_k$, and argued that $(p_1 \cdot \ldots \cdot p_k) + 1$ was either prime itself or had a prime factor larger than any of the p_i, and either way the desired result followed. The following example utilizes a more complicated case analysis.

Example 2.9. *Show that in any group of 6 people, there are either 3 who all know each other, or 3 who are all strangers to each other. ("Knowing" is symmetric—if A knows B, then B knows A.)*[3]

Solution to example. Start by picking an arbitrary person X from the 6. Of the remaining 5 people, there must be at least 3 whom X knows, or at least 3 whom X does not know (see Problem 2.15). Depending on which is true, the argument splits into two main cases, each of which has two subcases.

"Without loss of generality" means that some simple argument or symmetry implies that the more specific situation is equivalent to the more general.

[3]This is the first and simplest result in an entire subfield of mathematics known as *Ramsey theory*, named after the mathematician and philosopher Frank P. Ramsey (1903–30), who first proved this theorem while studying a problem in quantificational logic.

Case 1. X knows at least 3 people. Then consider the relations among those 3 people. If no 2 of them know each other, they form a set of 3 who mutually don't know each other. If some 2 of them do know each other, then those 2 plus X form a set of 3 who mutually know each other.

Case 2. There are at least 3 people whom X does not know. Consider 3 people whom X does not know. If those 3 mutually know each other, they form a set of 3 people who mutually know each other. Otherwise, some 2 of them do not know each other, and those 2 people plus X constitute a set of 3 people who mutually do not know each other. ∎

In this proof, the argument in case 2 sounds a lot like the argument in case 1, with the roles of "knowing" and "not knowing" reversed. In fact, the two arguments are identical except for that swap. This example illustrates another commonly used proof technique: *symmetry*.

The symmetry in the argument is illustrated in Figure 2.1, which shows a *graph*, a kind of diagram that will be studied extensively in Chapters 16 through 18. A, B, ..., F represent 6 people; a blue line connects two of them if they know each other, and a red line connects them if they do not. So case 1 applies if X is someone, such as E in the figure, who knows (is connected by blue lines to) 3 others, and case 2 applies if X is someone, such as A in the figure, who does not know (is connected by red lines to) 3 others. The arguments are completely symmetrical except for reversing the roles of blue and red. Anticipating language that will be defined later, we would say that *in a graph with 6 vertices, and edges between each pair of vertices, if the edges have one of two colors, then the graph includes a monochromatic (single-color) triangle*. (E, B, and D form such a triangle in Figure 2.1.)

Once the symmetry has been identified, there is no reason to give an essentially identical argument twice. Instead, in case 2, we might just say, "If there are at least 3 people whom X does not know, then the argument is symmetrical, reversing the roles of 'knowing' and 'not knowing.' " Arguments from symmetry can be both shorter and more revealing than ones not exploiting symmetry! And in this case even the statement of the result becomes simpler when the symmetry has been identified, since a monochromatic triangle could represent either 3 people who know each other or 3 people who don't.

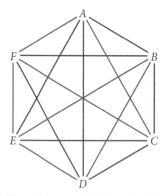

Figure 2.1. A graph illustrating the relations among 6 people, with blue lines connecting pairs that know each other and red lines connecting pairs that do not.

Chapter Summary

- A *proof* is a formal, general, precise mathematical argument that begins with one or more premises and proceeds using logical rules to establish a conclusion.

- *Quantifiers* such as "for all," "for any," "for every," "for some," and "there exists" specify the scope of a predicate.

- A *predicate* is a template for a proposition, taking one or more *arguments*. A predicate itself is neither true nor false, but takes on the values true or false when applied to specific arguments.

- A *constructive* proof shows how to find something that is claimed to exist. A *nonconstructive* proof demonstrates that something exists without showing how to find it.

- An *algorithm* is a detailed and precise procedure that can be carried out mechanically—for example, by a computer.

- To prove that two statements are *equivalent*, it is often easiest to prove each direction separately: to prove "*p* if and only if *q*," prove "if *p* then *q*" and "if *q* then *p*."

- A *direct proof* begins by assuming that the premises are true, and builds on that assumption to establish the conclusion.

- A *proof by contradiction* begins by assuming that the statement is false, and builds on that assumption to derive a contradiction.

- For the statement "if *p* then *q*," the *contrapositive* is "if not *q* then not *p*." The *inverse* is "if not *p* then not *q*." The *converse* is "if *q* then *p*." The *negation* of statement *p* is "not *p*."

- A statement and its contrapositive are equivalent (but the inverse and converse are not equivalent to the original statement). Therefore, to prove a statement, it suffices to prove its contrapositive.

- A *case analysis* proves a general statement by breaking it into restricted statements that together exhaust all possibilities, and then proving those various cases individually.

- An argument from *symmetry* can avoid repeating a nearly identical argument.

Problems

2.1. Is -1 an odd integer, as we have defined the term? Why or why not?

2.2. Write the inverse, converse, and contrapositive of the following statement: *If it is raining, then I have my umbrella.*

2.3. Prove that the product of two odd numbers is an odd number.

2.4. Prove that $\sqrt{3}$ is irrational.

2.5. Prove that $\sqrt[3]{2}$ is irrational.

2.6. Prove that for any positive integer n, $\sqrt{n}$ is either an integer or irrational.

2.7. Show that there is a fair seven-sided die; that is, a polyhedron with seven faces that is equally likely to fall on any one of its faces. *Hint:* The argument is not a strict mathematical proof, and relies on some intuitions about the properties of physical objects of similar geometric shapes, stretched in one of their dimensions.

2.8. Prove that all even square numbers are divisible by 4.

2.9. (a) Prove or provide a counterexample: if c and d are perfect squares, then cd is a perfect square.

(b) Prove or provide a counterexample: if cd is a perfect square and $c \neq d$, then c and d are perfect squares.

(c) Prove or provide a counterexample: if c and d are perfect squares such that $c > d$, and $x^2 = c$ and $y^2 = d$, then $x > y$. (Assume x, y are integers.)

2.10. Prove by contradiction that if $17n + 2$ is odd then n is odd.

2.11. Critique the following "proof":

$$x > y$$
$$x^2 > y^2$$
$$x^2 - y^2 > 0$$
$$(x + y)(x - y) > 0$$
$$x + y > 0$$
$$x > -y$$

2.12. What is the converse of the contrapositive of "if p, then q"? To what simpler statement is it equivalent?

2.13. Write the following statements in terms of quantifiers and implications. You may assume that the terms "positive," "real," and "prime" are understood, but "even" and "distinct" need to be expressed in the statement.

(a) Every positive real number has two distinct square roots.

(b) Every positive even number can be expressed as the sum of two prime numbers.[4]

2.14. Prove by a nonconstructive argument that there exist irrational numbers x and y such that x^y is rational. *Hint:* Consider the number $\sqrt{2}^{\sqrt{2}}$, and analyze the two cases, one in which this number is rational and the other in which it is irrational. In the latter case, try raising it to the $\sqrt{2}$ power.

2.15. Using concepts developed in Chapter 1, explain the step in the proof of Example 2.9 stating that once one individual X has been singled out, "of the remaining 5 people, there must be at least 3 whom X knows, or at least 3 whom X does not know."

[4]This is known as *Goldbach's conjecture*. It is not known to be true and not known to be false.

Chapter 3

Proof by Mathematical Induction

What is the sum of the first n powers of 2?

As usual, the first step in solving a problem like this is to make sure you understand what it says. What is a power of 2? It is a number such as 2^3—that is, 8—that could be the result of raising 2 to some integer. OK, then what is n? Nothing is said about that! So n could be anything for which the rest of the question makes sense. The question talks about adding things up, and n is the number of things that are added together. So "n" must be a variable representing a whole number, such as 10. Finally, we need to be sure which are the "first" n powers of 2. Is the "first" power of 2 equal to 2^0, or to 2^1, or maybe something else? In computer science we often start counting from 0.

If we knew the value of n, then we could calculate the answer. For example, in the $n = 10$ case, the question is asking for the value of

$$2^0 + 2^1 + 2^2 + 2^3 + 2^4 + 2^5 + 2^6 + 2^7 + 2^8 + 2^9$$
$$= 1 + 2 + 4 + 8 + 16 + 32 + 64 + 128 + 256 + 512,$$

which turns out to be 1023.

But we don't want the sum just for the $n = 10$ case. The question was stated in terms of a variable n, so we want an answer in some form that refers to that variable. In other words, we want the answer as a general formula that works for any value of n. Specifically, we want an expression for the value of

$$2^0 + 2^1 + 2^2 + \ldots + 2^{n-1}.$$

Each of the elements in a sum is called a *term*—so in the above sum, the terms are $2^0, 2^1, \ldots, 2^{n-1}$. Notice that the last term is 2^{n-1}, not 2^n. We started counting at 0, so the first n powers of 2 are 2 to the powers 0, 1, 2, ..., and $n - 1$, and so the last of the n terms is 2^{n-1}.

The three dots "$\ldots$" are called an *ellipsis*. They suggest a pattern that is supposed to be obvious to the reader. But obviousness is in the mind of the beholder, and may not align with the intentions of the author. If all we had was the formula

$$1 + 2 + 4 + \ldots + 512, \tag{3.1}$$

is it obvious what the omitted terms are supposed to be, or even how many there should be? Not really. We previously specified that we are discussing sums of powers of 2, but if all you had was the elliptical expression (3.1), there would be more than one way to extrapolate what the missing terms are. Maybe the idea is that the second term is 1 more than the first, the third term is 2 more than the second, and so on. In that case the fourth term would be 3 more than the third—that is, 7. Then the intention might have been

$$1 + 2 + 4 + 7 + 11 + 16 + \ldots + 512 \;??? \tag{3.2}$$

rather than

$$1 + 2 + 4 + 8 + 16 + 32 + \ldots + 512.$$

It would take some work to figure out whether 512 could really be the last number in a sequence like (3.2) (see Problem 3.10).

The way to avoid all ambiguity in situations like this is to find an expression for a typical term, and to use "sum notation" to show which terms should be added up. A typical power of 2 looks like 2^i—we need to pick a variable different from n as the exponent, since n is already being used to denote the number of terms to be added up. Then the sum of the first n terms of that form can be written unambiguously as

$$\sum_{i=0}^{n-1} 2^i. \tag{3.3}$$

The $\sum$ symbol is a Greek capital sigma, for "sum."

Unlike the elliptical notation, sum notation leaves nothing to the imagination. It makes sense even in case $n = 1$ or $n = 0$. When $n = 1$, (3.3) becomes

$$\sum_{i=0}^{1-1} 2^i = \sum_{i=0}^{0} 2^i = 2^0 = 1,$$

the sum of one term. When $n = 0$ the upper limit of the summation is smaller than the lower limit, so (3.3) becomes a sum of 0 terms, which by convention is 0:

$$\sum_{i=0}^{0-1} 2^i = \sum_{i=0}^{-1} 2^i = \text{the empty sum} = 0.$$

To take another example, how would we write the sum of the first n odd integers? Well, the odd integers are all of the form $2i + 1$, and the first odd integer is 1, which is $2i + 1$ when $i = 0$. So the sum of the first n odd integers can be written as[1]

$$\sum_{i=0}^{n-1} (2i + 1).$$

[1] We parenthesize "$(2i + 1)$" in this formula because it would otherwise be ambiguous whether the sum applies to the entire expression or just to the term $2i$. This expression has quite a different value from

$$\left(\sum_{i=0}^{n-1} 2i \right) + 1.$$

Now back to the question at the start of this chapter. Can we find a simple formula, without any dots or summation signs, equivalent to (3.3)?

The first thing to do when faced with a question like this (once you are sure you understand the question) is to try a few examples, which can help clarify why it might be true—the same strategy that we used when setting out to prove Theorem 2.4. Let's try plugging in $n = 1, 2, 3$:

$$\sum_{i=0}^{1-1} 2^i = 2^0 = 1$$

$$\sum_{i=0}^{2-1} 2^i = 2^0 + 2^1 = 1 + 2 = 3$$

$$\sum_{i=0}^{3-1} 2^i = 2^0 + 2^1 + 2^2 = 1 + 2 + 4 = 7. \tag{3.4}$$

The values 1, 3, and 7 are one less than 2, 4, and 8 respectively, which are successive powers of 2. Better try one more before framing a hypothesis:

$$\sum_{i=0}^{4-1} 2^i = 2^0 + 2^1 + 2^2 + 2^3 = 1 + 2 + 4 + 8 = 15. \tag{3.5}$$

And 15 is one less than 16, the next power of 2. We haven't proved anything yet, but the pattern seems too regular to be a coincidence. Let us conjecture

$$\sum_{i=0}^{n-1} 2^i = 2^n - 1. \tag{3.6}$$

Have we got that right, or should the right side be $2^{n-1} - 1$ or maybe $2^{n+1} - 1$? Plug in $n = 4$ to be sure. In the $n = 4$ case, the left side of (3.6) is the same as (3.5), namely 15, while the right side of (3.6) is $2^4 - 1 = 15$. So (3.6) looks like a good conjecture. It works even in the $n = 0$ case:

$$\sum_{i=0}^{0-1} 2^i = 0 = 2^0 - 1, \tag{3.7}$$

since the left side, with the ending value of i less than the beginning value of i, describes a sum of zero terms, which is 0.

Still this is not a proof. Now comes the need for some real insight. There is a *reason* why (3.5) follows from (3.4). Namely, if 2^3 is added to $2^3 - 1$, the result is $2^3 + 2^3 - 1 = 2 \cdot 2^3 - 1 = 2^4 - 1$.

And it holds true in general that adding the same power of 2 to itself yields the next power of 2. So suppose we knew that (3.6) held true for a given

Figure 3.1. The ladder metaphor for mathematical induction. If you can (**Base case**) get on the bottom rung of the ladder; and if, on the assumption (**Induction hypothesis**) that you have reached some arbitrary rung, you can (**Induction step**) climb to the next rung; then (**Conclusion**) you can reach any rung.

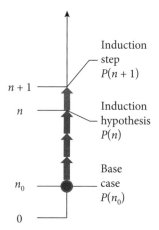

Figure 3.2. Principle of Mathematical Induction. Start by proving a base case ($P(n_0)$, in red). Then show that if $P(n)$ is true (the last green value), then $P(n+1)$ must also be true (the blue value).

value of n. In an inductive proof, this supposition is called the *induction hypothesis*. The crux of the proof, called the *induction step*, is to prove that if the induction hypothesis is true, then the same predicate also holds for the next value; that is, when n is replaced by $n+1$. Here's a proof of the induction step:

$$\sum_{i=0}^{(n+1)-1} 2^i = \sum_{i=0}^{n} 2^i$$

$$= \left(\sum_{i=0}^{n-1} 2^i\right) + 2^n \quad \text{(breaking off the last term)}$$

$$= 2^n - 1 + 2^n \quad \text{(by the induction hypothesis)}$$

$$= 2^{n+1} - 1 \quad \text{(since } 2^n + 2^n = 2^{n+1}\text{),}$$

which is exactly what (3.6) promises when $n+1$ is used in place of n.

This is a classic example of a *proof by mathematical induction*. The term "induction" refers to the idea that the truth of a statement about one value, say n, implies the truth of the same statement about the next value, $n+1$. So if we can establish the truth of the statement for *any* value, we know that it is true also for any larger value, in the same way that if we have a foolproof method for climbing from one rung of a ladder to the next higher rung, we can reach any rung above the first rung that we know we can reach (Figure 3.1).

Here, then, is the general form of a proof by mathematical induction (Figure 3.2).

To prove that a predicate $P(n)$ holds for every number n greater than or equal to some particular number n_0:

Base case. Prove $P(n_0)$.

Induction hypothesis. Let n be an arbitrary but fixed number greater than or equal to n_0, and assume $P(n)$.

Induction step. Assuming the induction hypothesis, prove $P(n+1)$.

Let's apply that schema to give a proper proof of the conjecture (3.6), hiding all the special cases and guesses we made to figure out what we thought was the general predicate.

Example 3.8. *For any $n \geq 0$,*

$$\sum_{i=0}^{n-1} 2^i = 2^n - 1.$$

Solution to example. In terms of the induction schema, $P(n)$ is the predicate that

$$\sum_{i=0}^{n-1} 2^i = 2^n - 1,$$

and $n_0 = 0$.

Base case. $P(0)$ is the statement that

$$\sum_{i=0}^{0-1} 2^i = 2^0 - 1,$$

which is true since both the left and right sides of the equation are equal to 0 (see (3.7)).

Induction hypothesis. The induction hypothesis is $P(n)$; that is,

$$\sum_{i=0}^{n-1} 2^i = 2^n - 1,$$

when $n \geq 0$ is some fixed but arbitrary value.

Induction step. We need to show that $P(n+1)$ holds; that is,

$$\sum_{i=0}^{n} 2^i = 2^{n+1} - 1. \tag{3.9}$$

Breaking off the last of the terms being added up, the left side of (3.9) is

$$\sum_{i=0}^{n} 2^i = \left(\sum_{i=0}^{n-1} 2^i \right) + 2^n.$$

But $\sum_{i=0}^{n-1} 2^i$ is exactly what the induction hypothesis says is equal to $2^n - 1$, so to complete the proof of (3.9) we just have to show that

$$(2^n - 1) + 2^n = 2^{n+1} - 1,$$

which is true since $2^n + 2^n = 2^{n+1}$. ∎

✻

The technique of proof by mathematical induction is very versatile and powerful, as the following examples illustrate.

Example 3.10. *For any $n \geq 0$,*

$$\sum_{i=1}^{n} i = \frac{n \cdot (n+1)}{2}. \tag{3.11}$$

Solution to example. Let $P(n)$ denote the predicate that

$$\sum_{i=1}^{n} i = \frac{n \cdot (n+1)}{2}. \tag{3.12}$$

Base case. $n_0 = 0$. For $P(0)$, the sum on the left is empty and therefore has the value 0, and the expression on the right is $\frac{0 \cdot (0+1)}{2}$, which also has the value 0.

Induction hypothesis. Assume $P(n)$; that is, that

$$\sum_{i=1}^{n} i = \frac{n \cdot (n+1)}{2}$$

when $n \geq 0$ is fixed but arbitrary.

Induction step. We need to prove $P(n+1)$; that is,

$$\sum_{i=1}^{n+1} i = \frac{(n+1) \cdot (n+2)}{2}.$$

Once again, breaking off the last term of the left side, we can utilize the induction hypothesis:

$$\sum_{i=1}^{n+1} i = \left(\sum_{i=1}^{n} i \right) + (n+1)$$

$$= \frac{n \cdot (n+1)}{2} + (n+1) \quad \text{(by the induction hypothesis)}$$

$$= (n+1) \cdot \left(\frac{n}{2} + 1 \right)$$

$$= \frac{(n+1) \cdot (n+2)}{2}. \qquad \blacksquare$$

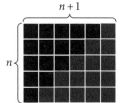

Figure 3.3. How many black tiles are there? We can count in two ways: it is the sum of the numbers $1, \ldots, n$, and also one-half the area of the grid—so these values must be equal.

Sometimes, a formal proof isn't quite enough to convey the intuition: after manipulating symbols to get the desired result, it may still not be clear why the statement is true. A concrete interpretation may be more satisfying. For (3.12) (and many other problems with sums), a good geometric interpretation is to imagine tiling a grid.

Figure 3.3 is a geometric representation of (3.12) for the case $n = 5$. The first row has n black tiles, the next has $n-1$, ..., all the way down to the last, which has 1. So the number of black tiles is $\sum_{i=1}^{n} i$. But we can calculate this value another way: the grid has length $n+1$ and height n, and half the tiles are black (compare the black shape to the red shape—they are the same, just rotated), so the number of black tiles is one-half the area of the rectangle; that is, $\frac{n \cdot (n+1)}{2}$.

We can also give a geometric interpretation to the induction step. The induction hypothesis says that $\sum_{i=1}^{n} i = \frac{n\cdot(n+1)}{2}$. For the induction step, we'd like to add an additional $n+1$ squares to get $\frac{(n+1)\cdot(n+2)}{2}$ in total. Figure 3.4 shows an $(n+1) \times (n+2)$ grid, with the top left half in black and gray and the bottom right half in red. The black squares in Figure 3.4 exactly match those in Figure 3.3; together with the additional $n+1$ gray squares on the diagonal, they cover one-half of the new $(n+1) \times (n+2)$ grid.

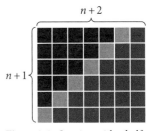

Figure 3.4. Starting with a half-covered grid of width $n+1$ and height n, we can add $n+1$ gray squares to cover half the area of a new grid that has width $n+2$ and height $n+1$.

Sometimes we want to express the product, rather than the sum, of a sequence of values. There is a notation for products, using the capital Greek pi, $\prod$, analogous to $\sum$ for sums. For example, (1.6) on page 6 can be rewritten as

$$p_1^{e_1} \cdot \ldots \cdot p_k^{e_k} = \prod_{i=1}^{k} p_i^{e_i}.$$

The elements that are multiplied together in a product are called *factors*, in the same way that the elements of a sum are called its terms. In the above product the factors are $p_1^{e_1}, \ldots, p_k^{e_k}$.

Let's try the induction schema on a statement about a product.

Example 3.13. *For any $n \geq 1$,*

$$\prod_{i=1}^{n} \left(1 + \frac{1}{i}\right) = n+1. \tag{3.14}$$

To work through an example: in the $n=3$ case, (3.14) states that

$$\left(1 + \frac{1}{1}\right) \cdot \left(1 + \frac{1}{2}\right) \cdot \left(1 + \frac{1}{3}\right) = 4.$$

Writing this in a different form suggests why it is true more generally:

$$\frac{1+1}{1} \cdot \frac{2+1}{2} \cdot \frac{3+1}{3} = 4.$$

The numerator of each fraction is equal to the denominator of the next, so all but the last numerator cancel out. We can use induction to make that argument solid.

Solution to example. Let $P(n)$ denote the predicate that

$$\prod_{i=1}^{n} \left(1 + \frac{1}{i}\right) = n+1. \tag{3.15}$$

Base case. $n_0 = 1$; (3.15) states that

$$\prod_{i=1}^{1} \left(1 + \frac{1}{i}\right) = 1 + 1,$$

which is true since both sides are equal to 2.

Induction hypothesis. Suppose that for some fixed but arbitrary $n \geq 1$, $P(n)$ is true:

$$\prod_{i=1}^{n} \left(1 + \frac{1}{i}\right) = n + 1.$$

Induction step. Now consider $P(n + 1)$. The left side of (3.15) becomes

$$\prod_{i=1}^{n+1} \left(1 + \frac{1}{i}\right) = \left(\prod_{i=1}^{n} \left(1 + \frac{1}{i}\right)\right) \cdot \left(1 + \frac{1}{n+1}\right) \quad \text{(breaking off the last factor)}$$

$$= (n+1) \cdot \left(1 + \frac{1}{n+1}\right) \quad \text{(by the induction hypothesis)}$$

$$= (n+1) \cdot \frac{1+n+1}{n+1} = n + 2,$$

as was to be shown. ∎

Equation (3.14) makes sense even in the $n = 0$ case, if we adopt the convention that *a product of zero factors is* 1:

$$\prod_{i=1}^{0} \left(1 + \frac{1}{i}\right) = \text{the empty product} = 1 = 0 + 1.$$

Why does this convention make sense? We adopted a similar convention for the sum of zero terms—namely, that the sum is 0. Intuitively, if we add some terms together, and then add zero more terms, then "adding zero more terms" is equivalent to adding 0. Similarly, if we multiply some factors together, and then multiply by zero more factors, then "multiplying by zero more factors" is equivalent to multiplying by 1.

⁎

Induction can be used to prove facts about things other than numbers. Sequences of 0s and 1s are often encountered in computer science. We'll call 0 and 1 the two *bits*,[2] and a sequence of bits, such as 10001001, a *binary string*, *bit string*, or *string of bits*—a string of *length* 8, in this case. The *complement* of a string of bits is the result of replacing all the 0s by 1s and vice versa. For example, the complement of 10001001 is

[2] For BInary digiTS, and also because they are small pieces of information.

`01110110`. The *concatenation* of two bit strings is just the result of writing one after the other. For example, the concatenation of `10001001` and `111` is `10001001111`. The length of the concatenation of two bit strings is the sum of their lengths—$8 + 3 = 11$, in this case.

There is an interesting sequence of bit strings called the *Thue sequence*,[3] which is defined as follows:

[3] Named after the Norwegian mathematician Axel Thue (1863–1922), whose name is pronounced "Too-eh." Also sometimes called the *Thue-Morse sequence* or *Thue-Morse-Prouhet sequence*, after the American mathematician Marston Morse (1892–1977), and the French mathematician Eugène Prouhet (1819–67), who implicitly used the sequence as early as 1851.

$$T_0 = 0, \text{ and for any } n \geq 0,$$

$$T_{n+1} = \text{ the concatenation of } T_n \text{ and the complement of } T_n.$$

This is an example of an *inductive definition* or a *recursive definition*. The definition of T_0 is the *base case*, and the definition of T_{n+1} in terms of T_n is the *constructor case*. Let's work through the first few strings of the Thue sequence.

0. $T_0 = 0$; this is the base case.

1. T_1 is the result of concatenating T_0 (that is, 0) with its complement (that is, 1). So $T_1 = 01$.

2. T_2 is the result of concatenating T_1 (that is, 01) with its complement (that is, 10). So $T_2 = 0110$.

3. T_3 is the result of concatenating T_2 (that is, 0110) with its complement (that is, 1001). So $T_3 = 01101001$.

4. T_4 is the result of concatenating T_3 (that is, 01101001) with its complement (that is, 10010110). So $T_4 = 0110100110010110$.

Since T_n is the first half of T_{n+1} for every $n \geq 0$, we can define an infinite bit string $t_0 t_1 t_2 \ldots$ simply by saying that t_i is the ith bit of T_n for all n that are large enough for T_n to have at least i bits. For example, if $i = 4$, then T_0, T_1, and T_2 have fewer than five bits, but for all $n \geq 3$, T_n has length at least five, and they all have the same fifth bit. That bit is t_4, which happens to be 1.

The Thue sequence has some very interesting properties. It looks repetitive, but it isn't repetitive at all. It also looks random, but it isn't random at all. Let's use mathematical induction to prove a few simple properties.

The first few strings of the Thue sequence:

$$T_0 = 0$$

$$T_1 = 01$$

$$T_2 = 0110$$

$$T_3 = 01101001$$

$$T_4 = 0110100110010110$$

Example 3.16. *For every $n \geq 0$, the length of T_n is 2^n.*

Solution to example. We prove this by induction.

Base case. $n_0 = 0$. $T_{n_0} = T_0 = 0$, which has length $1 = 2^0 = 2^{n_0}$.

Induction hypothesis. Assume that T_n has length 2^n.

Induction step. T_{n+1} is the result of concatenating T_n with its complement. Since these are both strings of length 2^n, the length of T_{n+1} is $2^n + 2^n = 2^{n+1}$. ∎

Example 3.17. *For every $n \geq 1$, T_n begins with 01, and ends with 01 if n is odd or with 10 if n is even.*

Solution to example. $T_1 = 01$, so those are the first two bits of T_n for every $n \geq 1$. The proof of how T_n ends proceeds by induction.

Base case. $n_0 = 1$. The number 1 is odd, and $T_{n_0} = T_1 = 01$, which ends in 01.

Induction hypothesis. Fix $n \geq 1$, and assume that T_n ends with 01 if n is odd or 10 if n is even.

Induction step. T_{n+1} ends with the complement of the last two bits of T_n, and $n + 1$ is even if n is odd and odd if n is even. So T_{n+1} ends with 01 if $n + 1$ is odd or 10 if $n + 1$ is even. ∎

Example 3.18. *For every $n \geq 0$, T_n never has more than two 0s or two 1s in a row.*

Solution to example. T_0 has only one bit, so the statement is true of T_0. We prove this by induction starting with $n_0 = 1$.

Base case. $n_0 = 1$. Then $T_{n_0} = T_1 = 01$, which has no consecutive identical bits, so it certainly has no more than two identical bits in a row.

Induction hypothesis. Fix $n \geq 1$ and assume that T_n has no more than two 0s or two 1s in a row.

Induction step. T_{n+1} includes bits inherited from T_n in the first half, and bits inherited from the complement of T_n in the second half. Neither has 000 or 111 by the induction hypothesis. So if one of these patterns occurs, it must occur at the junction, with two bits from one of these bit strings and one bit from the other. But by Example 3.17, the four bits at the junction can only be 0110 or 1010, neither of which contains a string of three identical bits (see Figure 3.5). ∎

We consider some other properties of the Thue sequence in Problems 3.11 and 11.11.

✳

We can use mathematical induction to prove the Pigeonhole Principle (page 3)! This requires a little more thought, since the Pigeonhole Principle does not mention a particular number n on which to "do the induction." In general, that variable is known as the *induction variable*—it has been denoted by "n" in the previous examples of induction.

Example 3.19. *If $f : X \to Y$ and $|X| > |Y|$, then there are elements $x_1, x_2 \in X$ such that $x_1 \neq x_2$ and $f(x_1) = f(x_2)$.*

T_2 0 1 1 0

T_3 0 1 1 0 1 0 0 1

T_4 0 1 1 0 1 0 0 1 1 0 0 1 0 1 1 0

Figure 3.5. Forming T_2 from T_1 and its complement, T_3 from T_2 and its complement, and T_4 from T_3 and its complement. The red bits are the complement of the black bits. Neither the black bits nor the red bits contain a run 000 or 111, so if such a run arises as a result of the concatenation, it must happen in the four-bit window at their junction. But those four bits are always either 0110 or 1010.

Solution to example. To set this up as a proof by induction, we have to identify the induction variable. There are two obvious possibilities: the size of the set X and the size of the set Y. Either can be used, but the proofs are different. Let's do the induction on $|X|$, and restate the principle as the claim that $P(n)$ is true for every $n \geq 2$, where $P(n)$ states that

> For any finite sets X and Y such that $|X| > |Y|$ and $|X| = n$, if $f : X \to Y$, then there are distinct elements $x_1, x_2 \in X$ such that $f(x_1) = f(x_2)$.

Problem 3.2 carries out the induction using the size of Y as the induction variable, rather than the size of X.

Base case. $n_0 = 2$. If $|X| > |Y|$ and $|X| = 2$, and there is a function from X to Y, then $|Y|$ must be 1: Y must contain at least one element since $f(x) \in Y$ for each $x \in X$, and Y cannot contain more than one element since it is smaller than X. But then the two elements of X must both map to the unique element of Y, as was to be shown.

Induction hypothesis. Suppose that for some fixed but arbitrary $n \geq 2$, and for any finite sets X and Y such that $|X| > |Y|$ and $|X| = n$, if $f : X \to Y$, there are distinct elements $x_1, x_2 \in X$ such that $f(x_1) = f(x_2)$.

Induction step. We want to prove that if $|X| > |Y|$, $|X| = n+1$, and $f : X \to Y$, then there are distinct elements $x_1, x_2 \in X$ such that $f(x_1) = f(x_2)$.

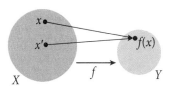

Figure 3.6. Proof by induction of the Pigeonhole Principle, case 1: x and x' are distinct elements of X that f maps to $f(x)$.

Pick an arbitrary element $x \in X$. Now there are two possibilities (what follows is a case analysis). Either there is another element $x' \in X$ such that $x' \neq x$ but $f(x') = f(x)$, or there is not. In the first case (Figure 3.6), the induction step is proved; we have found two distinct elements of X that map to the same element of Y. In the second case, x is the only element of X for which the value of f is equal to $f(x)$ (Figure 3.7). Let X' be the result of removing x from X and let Y' be the result of removing $f(x)$ from Y. Now $|X'| = n$ and $|Y'| = |Y| - 1 < n$, so the induction hypothesis applies. The function $f' : X' \to Y'$ that is identical to f on the elements of X' is then a function from a set of size n to a smaller set. By the induction hypothesis there are distinct $x_1, x_2 \in X'$ such that $f'(x_1) = f'(x_2) \in Y'$. But then x_1 and x_2 are also distinct members of X for which f has identical values. ∎

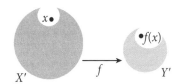

Figure 3.7. Case 2: x is the sole element of X that f maps to $f(x)$. Then removing x from X and $f(x)$ from Y leaves a function $f' : X' \to Y'$ to which the induction hypothesis applies, since $|X'| = |X| - 1$.

Chapter Summary

- Sum notation is used to denote the sum of a sequence of *terms*. For example, the expression

$$\sum_{i=0}^{n-1} 2^i$$

denotes the sum of the terms 2^i, for values of i ranging from 0 to $n - 1$.

■ By convention, the sum of zero terms is 0.

■ A *proof by induction* consists of three parts:

Base case. Prove that the predicate holds for a particular value n_0.

Induction hypothesis. Suppose that the predicate holds for some fixed value $n \geq n_0$.

Induction step. Prove that if the induction hypothesis is true, then the predicate holds for $n + 1$ as well.

This establishes that the predicate holds for all values greater than or equal to n_0.

■ Product notation is used to denote the product of a sequence of *factors*. For example, the expression

$$\prod_{i=1}^{k} p_i^{e_i}$$

denotes the product of the factors $p_i^{e_i}$, for values of i ranging from 1 to k.

■ By convention, the product of zero factors is 1.

■ An *inductive definition*, or *recursive definition*, defines some sequence of elements in terms of a base case and a constructor case. The *Thue sequence* is an example.

Problems

3.1. Find a closed-form expression for the expression on page 26,

$$\sum_{i=0}^{n-1} (2i + 1).$$

3.2. Prove the Pigeonhole Principle (Example 3.19) by induction on $|Y|$ instead of on $|X|$.

3.3. Prove the Extended Pigeonhole Principle (page 5) by induction.

3.4. (a) Using $\sum$ notation, write an expression for the sum of the first n odd powers of 2 (that is, the sum of 2^1, 2^3, and so on). Prove by induction that the value of this sum is $\frac{2}{3} \cdot (4^n - 1)$.

(b) Using $\prod$ notation, write an expression for the product of the first n negative powers of 2 (that is, the product of 2^{-1}, 2^{-2}, and so on). What is its value?

3.5. Prove by induction that for any $n \geq 0$,

$$\sum_{i=0}^{n} i^2 = \frac{n(n+1)(2n+1)}{6}.$$

3.6. For any $n \geq 0$, let

$$S(n) = \sum_{i=0}^{n} 2^{-i}.$$

We want to show that $S(n)$ is always less than 2, but becomes as close as we wish to 2 if n is sufficiently large.

(a) What are $S(0)$, $S(1)$, $S(2)$, and $S(3)$?

(b) Conjecture a general formula for $S(n)$ of the form

$$S(n) = 2 - \ldots.$$

(c) Prove by induction that the formula is correct for all $n \geq 0$.

(d) Now let ε be a small positive real number. How big does n have to be for $S(n)$ to be within ε of 2?

3.7. Prove by induction that for any $n \geq 0$,

$$\sum_{i=0}^{n} i^3 = \left(\sum_{i=0}^{n} i\right)^2.$$

3.8. Prove by induction that for any $n \geq 1$,

(a)
$$\sum_{i=1}^{n} 2^{i-1} \cdot i = 2^n(n-1) + 1.$$

(b)
$$\sum_{i=1}^{n} 2^{i-1} \cdot i^2 = 2^n(n^2 - 2n + 3) - 3.$$

3.9. What is the flaw in the following "proof"?

All horses are the same color.

Base case. Consider a set of horses of size 1. There is only one horse, which is the same color as itself, so the statement holds.

Induction hypothesis. Suppose that for all sets of horses of size $n \geq 1$, all horses in a set are the same color.

Induction step. Prove that for all groups of horses of size $n + 1$, all horses in a group are the same color: Consider a set of horses

$$H = \{h_1, h_2, \ldots, h_n, h_{n+1}\}.$$

Now we can consider two different subsets of H:

$$A = \{h_1, h_2, \ldots, h_n\} \text{ and}$$

$$B = \{h_2, h_3, \ldots, h_{n+1}\}.$$

Since A and B are both of size n, all horses in each group are the same color. But then h_{n+1} is the same color as h_2 (since both are in B), and h_2 (which is in set A) is the same color as every other horse in set A. So h_{n+1} is the same color as every horse in set A, so all horses in set H are actually the same color.

Thus for a set of horses of any size, all horses in the set are the same color as each other.

3.10. Could 512 actually appear as the last number in the sum (3.2), if the differences between successive elements increase by 1 at each step?

3.11. Prove the following about the Thue sequence.

(a) For every $n \geq 1$, T_{2n} is a *palindrome*; that is, a string that reads the same in both directions.

(b) For every n, if 0 is replaced by 01 and 1 is replaced by 10 simultaneously everywhere in T_n, the result is T_{n+1}. This means that in the infinite Thue bit string $t_0 t_1 \ldots$, if we imagine these replacements happening everywhere simultaneously, the result is the infinite Thue bit string again!

3.12. Show that the Principle of Mathematical Induction is a bit more general than is really necessary. That is, any proof that can be done by mathematical induction can be done using the narrower Weak Principle of Mathematical Induction:

If $P(0)$ is true, and for every n, $P(n+1)$ is true if $P(n)$ is true, then $P(n)$ is true for all n.

The Weak Principle is the Principle of Mathematical Induction with n_0 fixed to have the value 0.

Chapter 4

Strong Induction

Mathematical induction is a powerful, general proof method, but sometimes it seems not to be powerful enough. Here is a simple example that almost fits into the induction paradigm on page 28, but doesn't quite. To answer the question, we will need a stronger version of the induction principle.

Example 4.1. *A simple game involves two players and a pile of coins. Let's call the players Ali and Brad. Ali moves first, and then the two players alternate. There are only two other rules.*

1. *On each move, a player picks up one or two coins.*

2. *The player who picks up the last coin loses.*

Who wins, if each player plays as well as possible?

The answer depends on how many coins were in the pile at the beginning. Let's call that number n, and work through some small values of n to see if we can detect a pattern.

- If $n = 1$ then Ali loses, because she has to pick up at least one coin, but there *is* only one coin. So she has to pick up the "last" coin. It makes sense to say that $n = 1$ is a *losing position* for Ali, since she can only lose from that position (Figure 4.1).

Figure 4.1. Ali loses when $n = 1$.

- What if $n = 2$? Of course Ali would lose if she foolishly picked up both coins, but we are analyzing the situation in which Ali and Brad both play as well as possible. If Ali picks up 1 coin, then she leaves 1 behind, thus forcing Brad into a losing position. (Note the important concept here: if n is any losing position for Ali, then it would also be a losing position for Brad if he were the one left with n coins.) So starting with 2 coins is a *winning position* for Ali, since she can force Brad into a losing position (Figure 4.2).

Figure 4.2. Ali wins when $n = 2$.

- If the pile has 3 coins to begin with, Ali again can force Brad into a losing position by picking up 2 of them. So starting with 3 coins is a winning position for Ali (Figure 4.3).

Figure 4.3. Ali wins when $n = 3$.

Figure 4.4. Ali loses when $n = 4$, if she picks up 1 coin.

Figure 4.5. Ali loses when $n = 4$, if she picks up 2 coins.

Figure 4.6. Ali wins when $n = 5$.

- If the pile has 4 coins, Ali has a choice of what to do, but she can't win in either case. She can pick up one coin and leave 3, or she can pick up 2 coins and leave 2. But then the game goes over to Brad with a pile of either 3 or 2 coins. Those are both winning positions for Brad. So starting with 4 coins is a losing position for Ali (Figures 4.4–4.5).

- Let's try just one more situation—when Ali starts the game with $n = 5$ coins. Now she can pick up one and leave Brad with 4 coins, which is a losing position for Brad just as it was for Ali when she had to start with 4 coins. So 5 coins is a winning position for Ali (Figure 4.6).

Now what is really going on here? Whether a position with n coins is winning or losing for the next player seems to depend on whether the remainder is 0, 1, or 2 when n is divided by 3:

1. If n divided by 3 has a remainder of 0 or 2, n coins is a winning position.

2. If n divided by 3 has a remainder of 1, n coins is a losing position.

Facts (1) and (2) depend on each other. If the remainder is 1, then picking up 1 or 2 coins creates a pile with a remainder of 0 or 2, respectively, which are winning positions for the other side. That is, if there is some integer k such that $n = 3k + 1$, then $n - 1$ is divisible by 3, and

$$n - 2 = 3 \cdot (k - 1) + 2$$

leaves a remainder of 2 when divided by 3. If the remainder is 0 or 2, then picking up 2 coins or 1 coin, respectively, creates a pile with a remainder of 1, which is a losing position for the other side.

So let's make the following conjecture:

Conjecture. For any $n \geq 1$, starting position with n coins is a losing position for Ali if and only if n is one more than a multiple of 3; that is, $n = 3k + 1$ for some integer k.

Because of the way facts (1) and (2) interlock, this seems to be a perfect candidate for mathematical induction except for one glitch: the argument for n coins depends on an induction hypothesis not just for $n - 1$ but also for $n - 2$ coins. Let's write out the proof and then formulate the more general version of mathematical induction needed in order to prove it.

Solution to example. We are going to prove the conjecture for every n by induction on n.

Base case. The conjecture is true for $n_0 = 1$ and for $n_1 = 2$, since the former ($n_0 = 3 \cdot 0 + 1$) is a losing position for Ali and the latter ($n_1 = 3 \cdot 0 + 2$) is a winning position, as argued on the previous page.

Induction hypothesis. Let us assume that for a fixed but arbitrary $n \geq 2$, and for any m such that $1 \leq m \leq n$, m coins is a losing position for Ali if and only if m leaves a remainder of 1 when divided by 3.

Induction step. Suppose Ali has $n + 1$ coins where $n \geq 2$. Ali can pick up either one or two coins, leaving either n or $n - 1$ coins for Brad. Because $n \geq 2$, both n and $n - 1$ are ≥ 1 and $\leq n$.

If $n + 1$ leaves a remainder of 1 when divided by 3, then n and $n - 1$ leave remainders of 0 and 2, respectively, and since both n and $n - 1$ are less than or equal to n, the induction hypothesis applies, and both are winning positions for Brad. Therefore, $n + 1$ is a losing position for Ali.

If $n + 1$ leaves a remainder of 0 when divided by 3, say $n + 1 = 3k$, then Ali can pick up two coins and leave Brad with $n - 1 = 3 \cdot (k - 1) + 1$ coins, and since $n - 1 < n$, the induction hypothesis applies and $n - 1$ is a losing position for Brad.

Similarly, if $n + 1$ leaves a remainder of 2 when divided by 3, Ali can pick up one coin and leave Brad in a losing position with 1 more than a multiple of 3. ∎

✳

Let $P(n)$ denote the predicate

> *The starting position with n coins is a losing position for Ali if and only if n is 1 more than a multiple of 3.*

This argument involved two special features that were not relevant to the proofs by induction in the previous chapter. First is the point already mentioned: to prove $P(n + 1)$, we used as the induction hypothesis not just that we knew $P(n)$ was true but also that we knew that $P(m)$ was true for multiple values $m \leq n$. (In this particular argument, for both $m = n$ and $m = n - 1$.) This makes sense. Using our ladder metaphor, it is like imagining that to reach the next rung of the ladder we may need to prop ourselves not just on the rung immediately below but also (like a centipede!) on one or more other rungs that are lower down (Figure 4.7). Once we have gotten past all the lower rungs and reached a certain height, it is fair to use any of the lower rungs as part of the foundation for ascending one rung higher.

The other important detail is that we needed multiple base cases—in this argument, we needed to establish both the $n = 1$ and $n = 2$ cases before we could proceed to the induction step. This is a subtle point, and it is easy to overlook. In the induction step, we took a number $n + 1$ and subtracted either 1 or 2 from it, and then argued that it was safe to assume by induction the truth of both $P(n)$ and $P(n - 1)$. But to do that we needed to note both that n and $n - 1$ were $\leq n$, which is obvious, and also that they were ≥ 1. For $n + 1 - 2$ to be ≥ 1 we needed to start the induction with the assumption that $n \geq 2$; that is, that both the $n = 1$ and $n = 2$ cases were dealt with separately in arguing the base case.

Figure 4.7. Strong induction: To climb to the next rung of the ladder, you can push off from as many of the lower rungs as may be needed.

Putting all these considerations together, we can present the *Strong Principle of Mathematical Induction*:

To prove that a predicate $P(n)$ holds for every number n greater than or equal to some particular number n_0:

Base case. Prove $P(m)$ for all m such that $n_0 \le m \le n_1$, where n_1 is some number $\ge n_0$. (That is, n_0 is the lowest rung we can reach, and we establish $P(m)$ for each m between n_0 and n_1 inclusive, each by a separate argument. In the coin game, $n_0 = 1$ and $n_1 = 2$.)

Induction hypothesis. Let $n \ge n_1$ be arbitrary but fixed, and assume that $P(m)$ holds whenever $n_0 \le m \le n$.

Induction step. Assuming the induction hypothesis, show that $P(n+1)$ holds.

Strong induction is illustrated in Figure 4.8, which shows the numbers $0, 1, \ldots$. For numbers $m < n_0$, $P(m)$ may not be true. The base cases are the numbers $n_0, \ldots, n_1$; some special arguments must be made to show that $P(m)$ is true for each m in that range. At an arbitrary stage of the induction argument, we have proved $P(m)$ for all m between n_0 and n inclusive, the values in green; at the first step of the induction, $m = n_1$ and the values in that range are just the base cases. The induction step (in blue) is to prove $P(n+1)$ on the basis of knowing that $P(m)$ is true for all the green values, between n_0 and n.

✳

This example shows how strong induction can be used to prove properties of a series of numbers.

Example 4.2. *Consider the sequence $a_1 = 3$, $a_2 = 5$, and for all $n > 2$, $a_n = 3a_{n-1} - 2a_{n-2}$. Prove by strong induction that $a_n = 2^n + 1$ for all integers $n \ge 1$.*

Solution to example. Let $P(n)$ be the predicate that $a_n = 2^n + 1$. Our aim is to prove $P(n)$ for all $n \ge 1$.

Base cases.
$$P(1): a_1 = 2^1 + 1 = 3$$
$$P(2): a_2 = 2^2 + 1 = 4 + 1 = 5.$$

Induction hypothesis. Fix $n \ge 2$, and assume that for all m, $1 \le m \le n$, $P(m)$ holds; that is, $a_m = 2^m + 1$.

Induction step. Show $P(n+1)$; that is, $a_{n+1} = 2^{n+1} + 1$:
$$a_{n+1} = 3a_{(n+1)-1} - 2a_{(n+1)-2}$$
$$= 3a_n - 2a_{n-1}$$

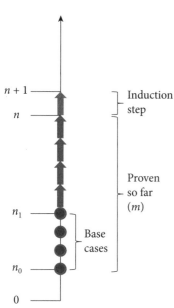

Figure 4.8. The strong induction schema. The base cases are in red, and the induction hypothesis is that $P(m)$ is true for all m in green or red. The induction step is to prove $P(n+1)$, the blue value. If that argument can be made regardless of the value of $n \ge n_1$, then we can conclude that $P(n)$ holds for all $n \ge n_0$.

$$= 3(2^n + 1) - 2(2^{n-1} + 1) \quad \text{(by the induction hypothesis)}$$
$$= 3 \cdot 2^n + 3 - 2^n - 2$$
$$= 2 \cdot 2^n + 1$$
$$= 2^{n+1} + 1. \qquad \blacksquare$$

✳

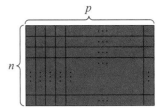

Figure 4.9. An n-by-p chocolate bar.

Let's work another example involving strategy for a game. In this example, the induction has to go back not just one or two values of the induction variable, but through all smaller values. Again, we'll call the players Ali and Brad.

The game involves a chocolate bar that is a rectangular grid of n by p individual pieces (Figure 4.9). The bar can be broken along any of the seams that run the length or width of the bar to break it into two smaller rectangles. We use the term "piece" just for the individual 1×1 squares.

Ali gets the first move. She breaks the bar either horizontally or vertically, and now has two separate rectangles. (The players can rotate the bar 90°, so "horizontal" and "vertical" have meaning only in our illustrations.) After breaking the bar, Ali eats one rectangle and hands the other to Brad. Brad does the same, and hands one of his two rectangles back to Ali. Both Ali and Brad keep going until one hands the other a rectangle that is just a single piece, so that the other can't break it any further. The player who receives this 1×1 rectangle loses.

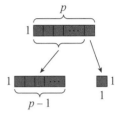

Figure 4.10. A $1 \times p$ or $p \times 1$ rectangle is a winning position for whoever holds it, as long as $p > 1$, since the holder can break off a single piece and hand it to the other player.

What is the best strategy for the players to use?

Let's look at the game from the perspective of Ali, the first player to move, and work backward from the end of the game.

Ali wants to hand Brad a 1×1 rectangle. If that is what she starts with ($n = p = 1$), she loses. But if she starts with a $1 \times p$ rectangle for any $p > 1$, she can break off $p - 1$ of the pieces, hand Brad the single leftover piece, and win. So a 1×1 rectangle is a losing position for whoever holds it, and a $1 \times p$ (or $p \times 1$) rectangle is a winning position if $p > 1$ (Figure 4.10).

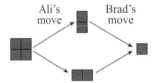

Figure 4.11. Ali loses if she starts with a 2×2 square.

If Ali starts with a 2×2 square, she can break it in two ways. But either way she does it, she will hand Brad a 2×1 rectangle (or a 1×2 rectangle, which is equivalent). That is as much a winning situation for Brad as it would have been for Ali if she started with the same rectangle. So Ali can't win if she starts with a 2×2 square (Figure 4.11), and a 2×2 square is a losing position for whoever holds it.

What if Ali starts with a 3×3? She has a few more options: either she can give Brad a 2×3 or 3×2—which are effectively the same—or she can give Brad a 1×3 or a 3×1—again essentially the same. No matter which way Ali breaks the 3×3 bar, Brad can find a way to give Ali a smaller square back (Figure 4.12)—and we already know that Ali will lose if she receives a 1×1 or 2×2 square.

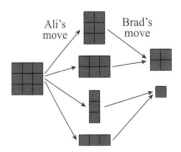

Figure 4.12. Ali also loses if she starts with a 3×3 square.

So let's conjecture that the first player has a winning strategy if that player starts with something other than a square bar, and the second player has a winning strategy if the first player starts with a square bar.

Example 4.3. *A player who moves starting with a rectangle of unequal dimensions ($n \times p$, where without loss of generality, $p > n$) has a winning strategy of breaking off $n \times (p - n)$ squares and returning an $n \times n$ square to the other player. A player who starts with an $n \times n$ square cannot win if the other player uses the same strategy.*

Solution to example. Say the dimensions of the bar are $n \times p$, where $p \geq n$. The proof is by induction on n, the smaller of the dimensions (unless p is equal to n). As usual, Ali moves first.

Base case. $n_0 = 1$. If also $p = 1$ then Ali has a single piece and loses. If $p > 1$, then Ali can break off $p - 1$ pieces (a $1 \times (p - 1)$ rectangle) and hand Brad a single piece. Brad then loses, and Ali wins.

Induction hypothesis. Let $n \geq 1$ be some fixed number, and assume that for all $m \leq n$, when $p > m$, a player who moves first with an $m \times p$ bar has a winning strategy of breaking off $m \times (p - m)$ squares and returning an $m \times m$ square to the other player, but a player who starts with an $m \times m$ square cannot win if the other player uses the same strategy.

Induction step. Suppose Ali starts with a bar of dimensions $(n + 1) \times p$, where $p \geq n + 1 \geq 2$. If $p = n + 1$, Ali is stuck; no matter how she breaks the bar, she has to hand to Brad a bar of dimension $(n + 1) \times m$, where $m < n + 1$. That is, $m \leq n$, and by the induction hypothesis Brad has a winning strategy if he starts with a bar of unequal dimensions for which the smaller dimension is $\leq n$. On the other hand, if $p > n + 1$, then Ali can break off an $(n + 1) \times (p - (n + 1))$ rectangle and hand Brad an $(n + 1) \times (n + 1)$ rectangle, which is a losing position for Brad in exactly the same way it would have been for Ali.   ∎

Note the critical use of strong induction in this proof. When one player has an $n \times p$ rectangle, where $p \geq n$, that player's move returns to the other player a rectangle, the smaller dimension of which might be any number m less than or equal to n.

*

In our previous examples, when trying to prove that $P(n)$ is always true, we have started with a base case of $n_0 = 0$ or $n_0 = 1$. But sometimes, a predicate does not hold for some finite number of cases, and we instead want to prove that it is true for "large enough" values. That is, there is some number n_0 such that $P(n)$ may be false when $n < n_0$, but is true for any $n \geq n_0$. In these situations we start with this value n_0 as the base case (and we must be careful that our induction step does not rely on any smaller value of n for which $P(n)$ is false).

Example 4.4. *A set of measuring cups includes a 4-cup, 9-cup, 11-cup, and 14-cup measure. Show that this set can be used to measure out any number of cups greater than or equal to* 11.

Solution to example. The general proposition to be proved is that $P(n)$ holds for any $n \geq n_0 = 11$.

Base cases. Let's go through the first few quantities starting with 11. (It is important not to try to start the induction with any number smaller than 11, since there is no way to measure out 10 with cups of the given sizes.) Since there is an 11-cup measure, using it once measures out 11 cups. We can measure 12 cups by using the 4-cup measure three times. For 13, we can use the 4-cup measure combined with the 9-cup. And for 14, we can use the 14-cup. So if we let $P(n)$ be the predicate that n cups can be measured using the available cup sizes, we have established that $P(11)$, $P(12)$, $P(13)$, and $P(14)$ are all true—four base cases.

Induction hypothesis. For some fixed but arbitrary $n \geq n_1 = 14$, $P(m)$ holds for all m such that $n_0 \leq m \leq n$.

Induction step. Prove $P(n+1)$ on the assumption that $P(m)$ is true for all m between $n_0 = 11$ and n inclusive: Since $n + 1 \geq 15$, we can measure $(n+1) - 4$ cups by the induction hypothesis, since $(n+1) - 4$ is greater than or equal to 11 and is therefore a case already proven. But then we can measure $n + 1$ cups by measuring $(n+1) - 4$ cups and then adding a single additional scoop using the 4-cup measure. ∎

＊

Though we have presented strong induction as though it were a new and more general principle than ordinary mathematical induction, it really isn't. In fact, anything that can be proved using strong induction could also be proved using ordinary induction, perhaps less elegantly (see Problem 4.6).

Since strong induction is a convenience, not really a new mathematical principle, we won't distinguish sharply, when we say "proof by induction," whether ordinary or strong induction is being used.

There is yet another variation on the principle of mathematical induction, which at first doesn't look like induction at all:

Well-Ordering Principle. Any nonempty set of nonnegative integers has a smallest element.

Now this looks pretty obvious; if S is a set of integers, each greater than or equal to 0, and S contains at least one element, then of course one of the elements has to be the smallest. It turns out that from the perspective of the foundations of mathematics, even this principle needs a proof! It is actually equivalent to the Principle of Mathematical Induction. Proving either would require digging deeper into the metamathematical question of what we *can* assume, if not these principles, but it is not hard to show that either implies the other (Problem 4.8).

With the aid of the Well-Ordering Principle, we can prove Theorem 1.5, the Fundamental Theorem of Arithmetic, that every integer n greater than 1 has a unique prime decomposition (page 6).

Proof. The proof is by induction on n.

Base case. $n_0 = 2$. Clearly $2 = 2^1$, and no other product of primes with positive exponents can equal 2.

Induction hypothesis. Fix $n \geq 2$, and suppose we know that every m, $2 \leq m \leq n$, has a unique prime decomposition.

Induction step. Consider now $n + 1$. If $n + 1$ is prime, then $n + 1 = (n + 1)^1$ is the unique prime decomposition of $n + 1$. If $n + 1$ is not prime, then let S be the set of all its factors greater than 1. By the Well-Ordering Principle, S has a smallest element p, which must be prime, since otherwise any factor of p would be a smaller factor of $n + 1$. Let $q = (n + 1)/p$. Then $q \leq n$ (in fact, $q < n$), so by the induction hypothesis q has a unique prime decomposition

$$q = p_1^{e_1} \cdot \ldots \cdot p_k^{e_k}.$$

Now p is either p_1 or is a prime smaller than p_1, which we can call p_0. In the first case,

$$n + 1 = p_1^{e_1+1} \cdot p_2^{e_2} \cdot \ldots \cdot p_k^{e_k},$$

and in the second case,

$$n + 1 = p_0^1 \cdot p_1^{e_1} \cdot \ldots \cdot p_k^{e_k}.$$

We leave it as an exercise to show that no number can have more than one prime decomposition (Problem 4.7). ■

Chapter Summary

- A proof by strong induction differs from a proof by induction in two ways:

  1. The base case may prove that the predicate holds for multiple values, n_0 through n_1 (whereas in ordinary induction there is just one value, n_0).

  2. The induction hypothesis is the assertion that the predicate holds for all values up to a fixed value n, where $n \geq n_1$ (whereas in ordinary induction the hypothesis applies just to n, and $n \geq n_0$).

- A proof by induction may be used even if the proposition is false in a finite number of cases. In this situation, the smallest base case n_0 is chosen such that $P(n)$ is true whenever $n \geq n_0$, and the induction step may not rely on any case where the proposition is false.

- Strong induction is a convenience: anything that can be proven using strong induction can also be proven with ordinary induction.

- The Well-Ordering Principle, which states that any nonempty set of nonnegative integers has a smallest element, is equivalent to the principle of induction.

Problems

4.1. Let $a_1 = 0, a_2 = 6$, and $a_3 = 9$. For $n > 3$, $a_n = a_{n-1} + a_{n-3}$. Show that for all n, a_n is divisible by 3.

4.2. Show that for all $n \geq 8$, there are integers a and b for which $n = 3a + 5b$; that is, n is the sum of some number of 3s and 5s.

4.3. Prove that for all $n > 1$, it is possible to tile a square of dimensions $2^n \times 2^n$ using just L-shaped pieces made up of three squares (Figure 4.13), leaving just a single corner square blank (Figure 4.14).
  The L-shaped pieces can be rotated and used in any orientation.

Figure 4.13. An L-shaped tile.

4.4. Suppose you are given a real number x such that $x + \frac{1}{x}$ is an integer. Use strong induction to show that $x^n + \frac{1}{x^n}$ is an integer for all integers $n \geq 0$.
  Hint: Multiply out $\left(x + \frac{1}{x}\right)\left(x^n + \frac{1}{x^n}\right)$ to find an equality that may be helpful.

4.5. Let S be the sequence $a_1, a_2, a_3, \ldots$ where $a_1 = 1$, $a_2 = 2$, $a_3 = 3$, and $a_n = a_{n-1} + a_{n-2} + a_{n-3}$ for all $n \geq 4$. Use strong induction to prove that $a_n < 2^n$ for any positive integer n.

4.6. Show that any proposition that can be proved using strong induction can also be proved using ordinary induction.

4.7. Show that the prime decomposition is unique; that is, that if

$$n = \prod_{i=1}^{k} p_i^{e_i} = \prod_{i=1}^{\ell} r_i^{f_i}$$

where the p_i are increasing and are all prime, the r_i are increasing and are all prime, and all the exponents e_i and f_i are positive integers, then $k = \ell$, $p_i = r_i$ for each i, and $e_i = f_i$ for each i.

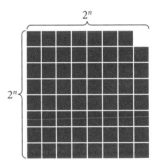

Figure 4.14. The L-shaped tiles fit together to cover every square except for one corner (this is the $n = 3$ case).

4.8. (a) Prove that the Principle of Mathematical Induction implies the Well-Ordering Principle. *Hint:* Proceed by induction on the size of the nonempty set of nonnegative integers. You may assume that if a set is nonempty, it is possible to find a member of it.

  (b) Prove that the Well-Ordering Principle implies the Principle of Mathematical Induction. *Hint:* Suppose $P(n_0)$ is true and for every n, we know that $P(n+1)$ is true if $P(n)$ is true, but for some reason it is *not* true that $P(n)$ is true for all $n \geq n_0$. Reason about the set of nonnegative integers of which P is false.

4.9. Use the method presented in the proof of the Fundamental Theorem of Arithmetic to find the prime decomposition of 100. Show the values of S, p, and q at each stage.

4.10. Suppose that the only postage stamps available to you are for 2 cents and 3 cents. Show that as long as $n \geq 2$, n cents in postage can be made up using at most $\frac{n}{2}$ postage stamps.

Chapter 5

Sets

A *set* is a collection of distinct objects, which are called its *members*. We defined sets already on page 2, but will review the terminology here. As an example, a set might consist of the three numbers 2, 5, and 7. We can write out sets like this one explicitly, using curly braces. The set containing just 2, 5, and 7 is written $\{2, 5, 7\}$. The members of a set might themselves be sets. For example, $\{\{2, 5, 7\}\}$ is a set containing just one member, and that member is the set $\{2, 5, 7\}$. These two sets are different:

$$\{2, 5, 7\} \neq \{\{2, 5, 7\}\}.$$

The first set has three members and the second has one. Both are different from the set $\{\{2\}, \{5\}, \{7\}\}$, which is a set with three members, each of which is itself a set, with one member.

When we say that the members of a set are *distinct*, we mean that no object is repeated. For example, if five students take an exam and get the scores 83, 90, 90, 100, and 100, then the set of exam scores for the five students is just $\{83, 90, 100\}$.[1] The order in which the members of a set are presented is also of no significance; the set $\{90, 100, 83\}$ is equal to the set $\{83, 90, 100\}$, since they have the same members.

Here are a few sets that we'll use often enough that they have special names:

$$\mathbb{Z} = \{\ldots, -3, -2, -1, 0, 1, 2, 3, \ldots\} = \text{the set of integers}$$

$$\mathbb{N} = \{0, 1, 2, 3, \ldots\} = \text{the set of nonnegative integers, also known}$$
$$\text{as the } natural \text{ numbers}$$

$$\mathbb{R} = \text{the set of } real \text{ numbers}$$

$$\mathbb{Q} = \text{the set of } rational \text{ numbers}$$

$$\emptyset = \{\} = \text{the } empty \text{ set, containing no objects.}$$

A *subset* of a set is a collection of elements drawn from the original set, but possibly missing some elements of the original. For example, $\mathbb{N}$ is a subset of $\mathbb{Z}$, and $\mathbb{Z}$ is a subset of $\mathbb{Q}$, and $\emptyset$ is a subset of every set. We write $A \subseteq B$ to mean that A is a subset of B, possibly equal to B. If we want to specify that A

[1] On page 238 we will define a more general structure called a *multiset*, which correctly accounts for the multiplicity of duplicated elements. For example, in this case, there would be two occurrences of 90 and two of 100, but only one of 83.

is a subset of B, but is definitely *not* equal to B, then we write $A \subsetneq B$. If $A \subsetneq B$, then A is said to be a *proper* subset of B. (The symbol $\subset$ is sometimes used instead. It can mean either $\subseteq$ or $\subsetneq$, and in most recent mathematical writing is avoided because of that ambiguity.)

We accept both sets and numbers (integer and real) as given—what computer scientists would term primitive or basic *data types*. From the standpoint of the foundations of mathematics, however, there is no need to treat numbers as primitive if we can form sets, since numbers can be defined in terms of sets by means of what computer scientists would call a coding trick. This coding is explored later on in this book, in Problem 8.7, once a more sophisticated form of inductive definition and proof is available to us.

If A is a subset of B, then B is a *superset* of A. If A is a proper subset of B, then B is said to be a *proper superset* of A.

We said that the members of a set can themselves be sets. An important example is the *power set* of a set A, represented by $\mathcal{P}(A)$, which is the set of all subsets of A. For example, if $A = \{3, 17\}$, then

$$\mathcal{P}(A) = \{\emptyset, \{3\}, \{17\}, \{3, 17\}\}.$$

We use the symbol $\in$ to express set membership. For example, $a \in S$ says that a is a member of the set S, and the statement

$$\text{"}2 \in \{2, 5, 7\}\text{" is true,}$$
$$\text{while "}3 \in \{2, 5, 7\}\text{" is false.}$$

Similarly $\mathbb{Z} \in \mathcal{P}(\mathbb{Q})$, since the set of all integers is one of the subsets of the set of rational numbers. The symbol $\notin$ means that the given element is not in the given set. For example, $3 \notin \{2, 5, 7\}$. The empty set is the unique set with no elements. That is, there is no object x such that $x \in \emptyset$.

Do not confuse

- 1 and $\{1\}$: 1 is a number, while $\{1\}$ is a set, containing a single object, the number 1.

- 0 and $\emptyset$: 0 is a number, while $\emptyset$ is a set, specifically the empty set. $\{\emptyset\}$ is yet a third thing—a set containing one element, that one element being the empty set.

- $\in$ and $\subseteq$: $1 \in \{1, 2\}$, since 1 is one of the two elements of $\{1, 2\}$. But it is not true that $1 \subseteq \{1, 2\}$, since 1 is not a set and so cannot be a subset. (Computer scientists would say that "$1 \subseteq \{1, 2\}$" has a type mismatch, since the entities on both sides of "$\subseteq$" must be sets.) On the other hand, $\{1\} \subseteq \{1, 2\}$ and indeed $\{1\} \subsetneq \{1, 2\}$, but $\{1\} \notin \{1, 2\}$, since the elements of $\{1, 2\}$ are not sets but numbers.[2]

[2]Note in particular that extra curly braces can *not* be sprinkled with abandon, like parentheses in algebra or curly braces in certain programming languages. Throwing in an extra pair of { } changes the meaning entirely.

✳

A set can be either *finite* or *infinite*. It is finite if the number of members it has is equal to some nonnegative integer. For example, Ø has 0 members, so it is finite, and if $A = \{3, 17\}$, then $\mathcal{P}(A) = \{\emptyset, \{3\}, \{17\}, \{3, 17\}\}$ has 4 members, so it is also finite. Remember when counting $\mathcal{P}(A)$ that it is a set of sets, not a set of numbers, so we count the number of sets that it contains rather than the number of numbers in those sets.

We denote the size of a set S, also called its *cardinality*, by $|S|$. So $|\emptyset| = 0$ and if $A = \{3, 17\}$, then $|A| = 2$ and $|\mathcal{P}(A)| = |\{\emptyset, \{3\}, \{17\}, \{3, 17\}\}| = 4$.

The set of integers is infinite, as is the set $\{\ldots, -4, -2, 0, 2, 4, \ldots\}$ of all even integers. For now, it will suffice to say that any set that isn't finite is infinite, though it turns out that two infinite sets might or might not be the same size as each other. Some distinct infinite sets are the same size—for example, the set of integers and the set of even integers. And some infinite sets are of different sizes—for example, the set of integers and the set of real numbers. We'll return to the question of how "big" an infinite set is in Chapter 6.

What is $|\{\mathbb{Z}\}|$? Like $\mathcal{P}(A)$, this is a set of sets, so we don't care about the sizes of the internal sets—even if they happen to be infinite! So $|\{\mathbb{Z}\}| = 1$, since $\{\mathbb{Z}\}$ is the set containing the single object $\mathbb{Z}$, even though $\mathbb{Z}$ itself is a set of infinite size.

✳

Sometimes we want to write down a set with more specifications than just "the integers" or "the rational numbers." To write down a set like "the even integers" without just listing out sample members as we did above, we use the following notation:[3]

$$\{n \in \mathbb{Z} : n \text{ is even}\}$$

or equivalently

$$\{n \in \mathbb{Z} : n = 2m \text{ for some } m \in \mathbb{Z}\},$$

or even

$$\{2m : m \in \mathbb{Z}\}.$$

In general, if we want to talk about the set of elements of A that satisfy predicate P, we write

$$\{x \in A : P(x)\}.$$

Another way to paraphrase $P(x)$ is to say that "x has property P."

There are a number of other ways in which we can construct new sets out of old ones. If we have two sets A and B, we can take the *union* of those

[3] The vertical bar "|" is commonly used in place of the colon ":" in mathematical writing; for example,

$$\mathcal{P}(S) = \{T \mid T \subseteq S\}.$$

We prefer the colon because vertical bars are used for the size of a set.

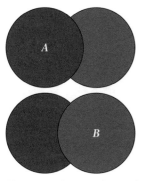

Figure 5.1. Sets A and B overlap.

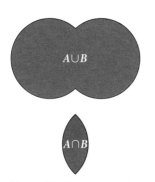

Figure 5.2. The union and intersection of the overlapping sets A and B of Figure 5.1.

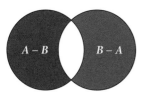

Figure 5.3. The two set differences between A and B, namely $A - B$ and $B - A$.

sets—that is, the elements that are contained in either A or B—or the *intersection*—the elements contained in both A and B. These concepts are often illustrated using a *Venn diagram* like the ones in Figures 5.1–5.3, which show two overlapping sets A and B, and their union and intersection.

For the union of A and B, we write

$$A \cup B = \{x : x \in A \text{ or } x \in B\}$$

and for the intersection we write

$$A \cap B = \{x : x \in A \text{ and } x \in B\}.$$

Like addition and multiplication of numbers, union and intersection of sets are *associative* operations. That is,

$$(A \cup B) \cup C = A \cup (B \cup C)$$

for any sets A, B, and C, and similarly

$$(A \cap B) \cap C = A \cap (B \cap C).$$

We'll return to this subject in Chapter 9, but for now it suffices to observe that "or" and "and" are associative in ordinary English. For example, to say that you are over 17 years old and female, and are also American, is exactly the same thing as to say that you are over 17 years old, and are also both female and American. You can form the intersection of the three sets of over-17-year-olds, females, and Americans either by grouping the first two together and taking the intersection with the third, or by grouping the last two together and taking the intersection of the first with the result. Similarly, set union and intersection are *commutative*: $A \cup B = B \cup A$ for any sets A and B, and $A \cap B = B \cap A$.

We can also take the *complement* of a set, $\overline{B} = \{x : x \notin B\}$. Often the *universe* U—the set of things x could possibly be—is left unspecified, since it is implied by the context. But to be completely unambiguous, if $B = \{x \in U : P(x)\}$ then we could write $\overline{B} = \{x \in U : x \notin B\}$. For example, if $B = \{\text{Saturday, Sunday}\}$ it's probably safe to assume that Wednesday $\in \overline{B}$. But is January $\in \overline{B}$? That depends on whether U is just the days of the week, or all words in the English language, or something else entirely.

Once we have the complement, we can take the *difference* of A and B, the elements that are in A but not in B:

$$A - B = A \cap \overline{B} = \{x : x \in A \text{ and } x \notin B\}.$$

Figure 5.3 shows $A - B$ and $B - A$ for the sets A and B of Figure 5.1. These expressions mean very different things; in the example, both are nonempty. In general,

$$A \cup B = (A - B) \cup (B - A) \cup (A \cap B).$$

That is, $A \cup B$ is the union of the two set differences shown in the margin, with the set intersection—the lens-shaped "missing piece" in the middle.

The set difference $A - B$ is also sometimes denoted $A \setminus B$.

The distributive laws for set union and intersection resemble the distributive law for addition and multiplication in arithmetic, that $a \cdot (b + c) = a \cdot b + a \cdot c$. But there is an important difference: the operations of set union and intersection distribute over each other symmetrically.

Theorem 5.1. Distributive Laws.

$$A \cap (B \cup C) = (A \cap B) \cup (A \cap C)$$
$$A \cup (B \cap C) = (A \cup B) \cap (A \cup C).$$

We'll prove just the first, leaving the second for Problem 5.2.

Proof. Consider an arbitrary element x. First suppose that $x \in A \cap (B \cup C)$. By the definition of $\cap$ and $\cup$, this means that both

  (a) x is a member of the set A, and

  (b) x is a member either

    (b1) of the set B or

    (b2) of the set C.

Then depending on which subcase of (b) is true, either x is a member of both A and B, or x is a member of both A and C—which is to say that $x \in (A \cap B) \cup (A \cap C)$.

Now suppose $x \in (A \cap B) \cup (A \cap C)$. We can simply follow the same chain of reasoning in reverse. Whether $x \in (A \cap B)$ or $x \in (A \cap C)$, x must be a member of A, and must also be a member of either B or C—that is, $x \in A \cap (B \cup C)$. ∎

    This argument alternates between reasoning about sets and the operations $\cup$ and $\cap$, and commonsense reasoning about statements and the connectives "or" and "and." Formalizing this commonsense reasoning about how the truth and falsity of compound statements depend on the truth and falsity of their component statements is the subject of the *propositional calculus*, which is the subject of Chapter 9. In fact, there are associative, commutative, and distributive laws in propositional logic that parallel those for sets (page 95).

    The union and intersection symbols $\cup$ and $\cap$ can be used with limits like those in the notations $\sum$ and $\prod$ for extended sums and products. For example, to say that $\mathbb{N}$ is the union of all the one-element sets $\{n\}$ for various

$n \in \mathbb{N}$, we could write

$$\mathbb{N} = \bigcup_{n=0}^{\infty} \{n\},$$

and to say that a set S is the intersection of all subsets of the universe that include the entire universe except for single elements not in S, we could write

$$S = \bigcap_{x \notin S} (U - \{x\}).$$

✳

An *ordered pair* $\langle x, y \rangle$ is a mathematical structure that assembles the *components*, x and y, into a single structure in which the order matters. That is, $\langle x, y \rangle$ is a different thing from $\langle y, x \rangle$, unless x happens to be the same thing as y. In general, $\langle x, y \rangle = \langle z, w \rangle$ just in case $x = z$ and $y = w$.

Also, $\langle x, y \rangle$ is a very different thing from $\{x, y\}$ (which is, of course, the same thing as $\{y, x\}$). We will treat the ordered pair $\langle x, y \rangle$ as another primitive data type, something different from any set.

In fact, it is possible to define ordered pairs in terms of sets, as we noted on page 50 could be done to define numbers in terms of sets. Mathematical purists and fundamentalists, who consider it important to get away with as few primitive concepts as possible, sometimes define the ordered pair $\langle x, y \rangle$ as $\{x, \{x, y\}\}$. With this definition, ordered pairs have the essential property that two ordered pairs are equal if and only if their first components are equal and their second components are also equal (Problem 5.11).

The concept of ordered pair is extended to ordered triples, et cetera. We write $\langle x, y, z \rangle$ for the ordered triple with components x, y, and z, and in general use the term *ordered n-tuple*, where n is a nonnegative integer, to refer to a sequence of n elements. Two ordered n-tuples are equal if and only if their i^{th} components are equal for each i, $1 \leq i \leq n$.

The set of all ordered pairs with the first component from set A and the second component from set B is called the *Cartesian product* or *cross product* of A and B, written $A \times B$. For example, if $A = \{1, 2, 3\}$ and $B = \{-1, -2\}$, then

$$A \times B = \{\langle 1, -1 \rangle, \langle 1, -2 \rangle, \langle 2, -1 \rangle, \langle 2, -2 \rangle, \langle 3, -1 \rangle, \langle 3, -2 \rangle\}.$$

Plainly $A \times B$ is generally different from $B \times A$. If A and B are finite, then $|A \times B| = |A| \cdot |B|$. In the example just shown, $|A| = 3$, $|B| = 2$, and $|A \times B| = 6$. But cross products can also be constructed from infinite sets. For example, $\mathbb{N} \times \mathbb{Z}$ is the set of all ordered pairs of integers in which the first component is nonnegative, and we could write

$$\{1, 2, 3\} \times \{-1, -2\} \subsetneq \mathbb{N} \times \mathbb{Z}.$$

Chapter Summary

- A *set* is an unordered collection of *distinct* things, or *elements*. The elements of a set are its *members*.

- The *empty set* is the set that contains no objects, denoted ∅ or {}.

- Some numerical sets have names: the integers $\mathbb{Z}$, the *natural* numbers $\mathbb{N}$, the *real* numbers $\mathbb{R}$, and the *rational* numbers $\mathbb{Q}$.

- A *subset* of a set consists of some of the members of the original set (possibly none, or all).

- The notation $A \subseteq B$ means that A is a subset of B, and is possibly equal to B. The notation $A \subsetneq B$ means that A is a *proper* subset of B (definitely not equal to B).

- If A is a (proper) subset of B then B is a (proper) *superset* of A.

- Set membership is denoted by $\in$; its negation is $\notin$.

- A set may itself be regarded as an element in a different set. It is important to distinguish when an object is being treated as an element, and when as a set. The membership symbols are used to indicate whether an element is a member of a set, while the subset symbols are used to indicate whether all the elements of one set are members of another set.

- The *power set* of a set S, denoted $\mathcal{P}(S)$, is the set of all subsets of S.

- The size of a set S, also called its *cardinality*, is denoted $|S|$. A set may be *finite*, if its size is a nonnegative integer, or *infinite*, if not.

- The *complement* of a set S, denoted $\bar{S}$, is the set of all elements in the *universe* of S that are not in S.

- Sets may be defined based on other sets via *union*, denoted $\cup$; *intersection*, denoted $\cap$; and *difference*, denoted $-$ or $\setminus$.

- Set union and intersection obey associative, commutative, and distributive laws:

$$(A \cup B) \cup C = A \cup (B \cup C)$$
$$(A \cap B) \cap C = A \cap (B \cap C)$$
$$A \cup B = B \cup A$$
$$A \cap B = B \cap A$$
$$A \cap (B \cup C) = (A \cap B) \cup (A \cap C)$$
$$A \cup (B \cap C) = (A \cup B) \cap (A \cup C).$$

- An *ordered pair* consists of two components, the order of which matters. The ordered pair with first component x and second component y is denoted $\langle x, y \rangle$. An *ordered n-tuple* is the extension of this definition to n components.

■ The set of ordered pairs with the first component taken from a set A and the second taken from a set B is called the *Cartesian product* or *cross product* of A and B, denoted $A \times B$.

Problems

5.1. What are these sets? Write them explicitly, listing their members.

(a) $\{\{2, 4, 6\} \cup \{6, 4\}\} \cap \{4, 6, 8\}$

(b) $\mathcal{P}(\{7, 8, 9\}) - \mathcal{P}(\{7, 9\})$

(c) $\mathcal{P}(\emptyset)$

(d) $\{1, 3, 5\} \times \{0\}$

(e) $\{2, 4, 6\} \times \emptyset$

(f) $\mathcal{P}(\{0\}) \times \mathcal{P}(\{1\})$

(g) $\mathcal{P}(\mathcal{P}(\{2\}))$

5.2. Prove the second version of the distributive law:
$$A \cup (B \cap C) = (A \cup B) \cap (A \cup C).$$

5.3. Show that if A is a finite set with $|A| = n$, then $|\mathcal{P}(A)| = 2^{|A|}$.

5.4. If $|A| = n$, what is $|\mathcal{P}(A) - \{\{x\} : x \in A\}|$?

5.5. (a) Suppose A and B are finite sets. Compare the two quantities $|\mathcal{P}(A \times B)|$ and $|\mathcal{P}(A)| \cdot |\mathcal{P}(B)|$. Under what circumstances is one larger than the other, and what is their ratio?

(b) Is it inevitably true that $(A - B) \cap (B - A) = \emptyset$? Prove it or give a counterexample.

5.6. Describe the following sets using formal set notation:

(a) The set of irrational numbers.

(b) The set of all integers divisible by either 3 or 5.

(c) The power set of a set X.

(d) The set of 3-digit numbers.

5.7. Decide whether each statement is true or false and why:

(a) $\emptyset = \{\emptyset\}$

(b) $\emptyset = \{0\}$

(c) $|\emptyset| = 0$

(d) $|\mathcal{P}(\emptyset)| = 0$

(e) $\emptyset \in \{\}$

(f) $\emptyset = \{x \in \mathbb{N} : x \leq 0 \text{ and } x > 0\}$

5.8. Prove that if A, B, C, and D are finite sets such that $A \subseteq B$ and $C \subseteq D$, then $A \times C \subseteq B \times D$.

5.9. Prove the following.

(a) $A \cap (A \cup B) = A$

(b) $A - (B \cap C) = (A - B) \cup (A - C)$

5.10. There are 100 students each enrolled in at least one of three science classes. Of those students, 60 are enrolled in chemistry, 45 in physics, and 30 in biology. Some students are enrolled in two science classes, and 10 students are enrolled in all three.

(a) How many students are enrolled in exactly two science classes?

(b) There are 9 students taking both chemistry and physics (but not biology), and 4 students taking both physics and biology (but not chemistry). How many are taking both chemistry and biology (but not physics)?

5.11. Suppose we do not accept ordered pairs as primitive, and instead define $\langle x, y \rangle$ to be $\{x, \{x, y\}\}$. Prove that $\langle x, y \rangle = \langle u, v \rangle$ if and only if $x = u$ and $y = v$.

Chapter 6

Relations and Functions

A relation is a connection between things. For instance, we know what it means for one number to be greater than another. We can think of the "greater-than" relation as the property that is shared by all such pairs of numbers. Relations can be between any kind of things, not just numbers; for example, the relation of "parenthood" is whatever all parent-child pairs share. That is, we identify a relation with the set of pairs of things (or people) that bear that relation to each other.

One of the triumphs of modern mathematics is to pin down such metaphysical abstractions, replacing them with concrete definitions based in set theory. The basic idea is to replace an abstract property with the set of all things that have it—its *extension*, as we would say. So a *relation* is a set of ordered n-tuples, a subset of the cross product of sets. We'll be most interested in *binary* relations; that is, relations between two things, which we represent as ordered pairs. As a concrete example, let's consider the people who live at various addresses in a town (Figure 6.1).

This is a relation between people and addresses: that is, a subset of the set of ordered pairs from $P \times A$, where P is the set of people including Alan, David, Grace, and Mary, and A is the set of all addresses, including 33 Turing Terrace, 66 Hilbert Hill, 77 Hopper Hollow, and 22 Jackson Junction. Specifically, the relation is this set of four ordered pairs:

Name	Address
Alan	33 Turing Terrace
David	66 Hilbert Hill
Grace	77 Hopper Hollow
Mary	22 Jackson Junction

Figure 6.1. A simple relation, consisting of names paired with addresses.

$$\{\langle Alan, 33\ Turing\ Terrace\rangle, \langle David, 66\ Hilbert\ Hill\rangle,$$
$$\langle Grace, 77\ Hopper\ Hollow\rangle, \langle Mary, 22\ Jackson\ Junction\rangle\}. \qquad (6.1)$$

This is a particularly simple kind of binary relation, in which each first component is paired with one and only one second component, and vice versa. A more general example would be the relation between students and the courses they are taking. A student takes more than one course, and courses enroll more than one student. For example, the relation E might represent the enrollments in a college:

$$E = \{\langle Aisha, CS20\rangle, \langle Aisha, Ec10\rangle, \langle Aisha, Lit26\rangle,$$
$$\langle Ben, CS20\rangle, \langle Ben, Psych15\rangle, \langle Ben, Anthro29\rangle,$$

$$\langle \text{Carlos, CS1}\rangle, \langle \text{Carlos, Lit60}\rangle, \langle \text{Carlos, Ethics22}\rangle,$$
$$\langle \text{Daria, CS50}\rangle, \langle \text{Daria, Ethics22}\rangle), \langle \text{Daria, Anthro80}\rangle\}. \qquad (6.2)$$

In this relation, each student is taking three courses. That is, for each person x, there are three different y such that E contains the pair $\langle x, y\rangle$. Also, Aisha and Ben have a course in common; that is, when $y = \text{CS20}$, there are two distinct values of x such that $\langle x, y\rangle \in E$. As shown in Figure 6.2, a relation R on disjoint sets A and B (Students and Courses in the example) can be illustrated by representing the two sets as blobs and representing elements of A and B as points, with an arrow from $x \in A$ to $y \in B$ if the pair $\langle x, y\rangle \in R$. To keep the picture simple, we have left out most of the arrows. In general, A and B need not be disjoint, as they are in this example.

Generally we won't write out relations explicitly. Instead, we can describe a relation by giving a rule that describes the pairs, in the same way we described sets in Chapter 5. For example, we can define the birthday relation between P, the set of all the people in the world, and D, all the dates, by saying

$$B = \{\langle p, d\rangle : \text{person } p \text{ was born on date } d\}.$$

To take a numerical example, a two-dimensional graph is a useful way to visualize the relation between two real variables. The circle shown in Figure 6.3 is the set of points at distance 1 from the origin; that is, the relation $\{\langle x, y\rangle : x^2 + y^2 = 1\} \subseteq \mathbb{R} \times \mathbb{R}$.

The *inverse* of a binary relation $R \subseteq A \times B$ is the relation

$$R^{-1} \subseteq B \times A, \text{ where } R^{-1} = \{\langle y, x\rangle : \langle x, y\rangle \in R\}.$$

That is, R^{-1} is simply the relation obtained by reversing all the arrows in the blob-and-arrow diagram for R. The inverse of the part of the students and courses relation shown in Figure 6.2 is shown in Figure 6.4.

A function is a special kind of binary relation. On page 3 we described a function as a rule that associates, with each member of one set, a member of a second set. We are now ready to formalize that description.

A *function f from* a set A *to* a set B is a relation $f \subseteq A \times B$ with the property that for each $x \in A$, there is one and only one $y \in B$ such that $\langle x, y\rangle \in f$. Since the value of y corresponding to a given x is uniquely determined if f is a function, we can refer to that unique y as $f(x)$. We say that $y = f(x)$ is the *value* of f on *argument* x. If A and B are disjoint and we illustrate a function as a blob diagram, a function differs from an arbitrary relation in that there is one and only one arrow out of each point in the left blob (Figure 6.5).

The relation between people and addresses in (6.1) is a function, since in this example each person lives at just one address. For example, we could write

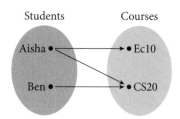

Students   Courses

Figure 6.2. Part of the course enrollment relation. Aisha and Ben are both enrolled in CS20, and Aisha is also enrolled in Ec10. A is the set of students and B is the set of courses.

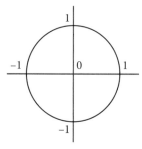

Figure 6.3. A graph drawn in two-dimensional space represents a relation. In this case, each point $\langle x, y\rangle$ on the circle satisfies the equation $x^2 + y^2 = 1$.

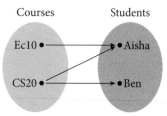

Courses   Students

Figure 6.4. The inverse of the relation shown in Figure 6.2.

$$f(\text{Alan}) = 33 \text{ Turing Terrace.}$$

On the other hand, the enrollment relation (6.2) is not a function, since students take multiple classes. (It would fail to be a function if there were even just one student who took two classes.) Nor is the circle relation a function, since for any value of x such that $-1 < x < 1$, there are two values of y such that $x^2 + y^2 = 1$, namely $y = \pm\sqrt{1 - x^2}$. But the birthday relation is a function, since every person is born on one and only one day.

If $f \subseteq A \times B$ is a function from A to B, we write $f : A \to B$. The set A is called the *domain* and the set B is called the *codomain* (Figure 6.5). For example, the domain of the birthday function is the set of all people, and the codomain is the set of days of the year. If we consider the function $f(x) = x^2$ where the arguments x are integers, then the domain is $\mathbb{Z}$. Since the square of an integer is always nonnegative, the codomain can be considered either $\mathbb{N}$ or $\mathbb{Z}$. (Based on the way we've defined functions, the domain must be exactly the set of elements for which the function gives a value. The codomain, on the other hand, must include all values of the function, but it can also include extra elements.) In the same way, the codomain of the birthday function is the set of all the days of the year, even if there is some day on which no person was born.

The domain and codomain can be any sets. They do not need to be disjoint or even distinct. For example, the function $f : \mathbb{Z} \to \mathbb{Z}$ such that $f(n) = n + 1$ for every $n \in \mathbb{Z}$ has the same domain and codomain.

The domain or codomain could be a set of sets. As a simple example in which the value of a function is a set, let's look again at the course enrollment relation (6.2). This is not a function, since one student takes several courses. But for each student, it is possible to stipulate precisely the *set* of courses the student is taking. So the information in (6.2) can be presented as a function from students to sets of courses, as follows:

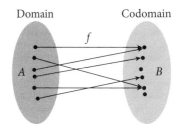

Figure 6.5. In a function, there is one and only one arrow coming out of each element of A. The set A is called the domain, and the set B is the codomain.

Student	Courses
Aisha	{CS20, Ec10, Lit26}
Ben	{CS20, Psych15, Anthro29}
Carlos	{CS1, Lit60, Ethics22}
Daria	{CS50, Ethics22, Anthro80}

(6.3)

If S is the set of students and C is the set of courses, then (6.2) is a relation on $S \times C$, while (6.3) is a function $e : S \to \mathcal{P}(C)$. For example, $e(\text{Aisha}) = \{\text{CS20, Ec10, Lit26}\}$.

As explained on page 3, a function *maps* an argument *to* a value, and a function is sometimes called a *mapping*. The value of a function for a given argument is also called the *image* of that argument, and if A is a subset of the

Domain   Codomain

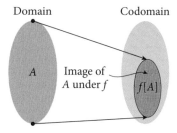

Figure 6.6. The image of a function is the subset of the codomain consisting of all values of the function on arguments from the domain; that is, $\{f(x) : x \in A\}$.

[1] Another traditional term, now falling into disuse, is the *range* of a function. This term is ambiguous: sometimes it means what we are calling the codomain, sometimes what we are calling the image. So we won't use it at all.

domain of f, then $f[A]$, called the *image* of the set A, is the set of all images of the elements of A (Figure 6.6). That is, if $f : S \to T$ and $A \subseteq S$, then

$$f[A] = \{f(x) : x \in A\} \subseteq T.$$

To get the visual metaphor, think of the domain as a picture and the function as a projector: the image of a portion of the picture is the set of points on the screen corresponding to that part of the picture.

For example, for the function $f : \mathbb{Z} \to \mathbb{Z}$ where $f(n) = n^2$ for every n, $f[\mathbb{Z}]$, the image of $\mathbb{Z}$, is the set of perfect squares $\{0, 1, 4, 9, \ldots\}$. If $g(n) = 2n$ for every n, then $g[\mathbb{Z}]$ is the set of all even integers. If we let $\mathbb{E} \subseteq \mathbb{Z}$ be the set of even integers, then $f[\mathbb{E}] = \{0, 4, 16, 36, \ldots\}$ is the set of even perfect squares.[1]

What are we to make of the real-valued function $f(x) = \frac{1}{x}$? We are tempted to say $f : \mathbb{R} \to \mathbb{R}$; that is, f takes a real argument and has a real value. But that is not quite right; $\frac{1}{0}$ is undefined, so $f(0)$ is undefined. For every number $x \in \mathbb{R}$ except 0, however, $f(x)$ is defined. So f is properly described as a function with domain $\mathbb{R} - \{0\}$; that is, $f : \mathbb{R} - \{0\} \to \mathbb{R}$.

An alternative terminology distinguishes between *partial* and *total* functions, where a total function is what we call a function, and a partial function from A to B is a total function for which the domain is a subset of A. Using this alternative terminology, $\frac{1}{x}$ could be described as a partial function from $\mathbb{R}$ to $\mathbb{R}$.

Certain kinds of functions have commonly used descriptors.

An *injective* function is one for which each element of the codomain is the value of the function for *at most one* argument. To put it another way, a function is injective if no two different arguments map to the same value. For example, the function that maps every integer n to its successor $n + 1$ is injective, since for any integer value m, there is only one value of n such that $m = n + 1$, namely $n = m - 1$. On the other hand, the birthday function is *not* injective, since sometimes two people have the same birthday.

Whether a function is injective can depend on the domain. Even the birthday function might be injective if the set of people was sufficiently restricted. The function $f : \mathbb{N} \to \mathbb{N}$ such that $f(n) = n^2$ for every $n \in \mathbb{N}$ is injective, since every natural number is the square of at most one natural number. But the similar function $g : \mathbb{Z} \to \mathbb{N}$ where $g(n) = n^2$ for every $n \in \mathbb{Z}$ is not injective, since, for example, $g(-1) = g(+1) = 1$.

A blob-and-arrow diagram represents an injective function just in case (see Figure 6.7)

(1) it represents a function (there is exactly one arrow out of each point in the domain) and

(2) no two different arrows end at the same point in the codomain.

Domain   Codomain

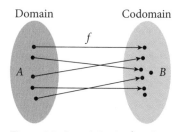

Figure 6.7. In an injective function, there is at most one arrow ending at each element of B.

Since a function is a special kind of binary relation, every function has an inverse, which is a relation. But the inverse of a function need not be a function. For example, the function $f : \{0, 1\} \to \{0\}$ such that $f(0) = f(1) = 0$ has the inverse $\{\langle 0, 0 \rangle, \langle 0, 1 \rangle\}$, but this relation is not a function.

Injective functions, however, are *invertible*. That is, the inverse of a injective function is itself a function. If $f : A \to B$ is injective, then the function $f^{-1} : f[A] \to A$, called the *inverse* of f, has the property that

$$f(f^{-1}(y)) = y \text{ for any } y \in f[A].$$

The inverse of f is defined by the rule that for any $y \in f[A]$, $f^{-1}(y)$ is that unique element $x \in A$ such that $f(x) = y$. We know that there cannot be two such values of x since f is injective, and we know there is at least one such x since $y \in f[A]$. In general, however, the domain of the inverse need not be all of B, since there may be some elements of B that are not the value of f for any argument $x \in A$.

The set (6.1) is an injective function, since no two people live at the same address. If we denote this function by f, so that $f(\text{Alan}) = 22$ Turing Terrace for example, then $f^{-1}(22 \text{ Turing Terrace}) = \text{Alan}$.

With the aid of the concept of injective functions, the Pigeonhole Principle (page 3) can be stated succinctly as follows: *if A and B are finite sets and $|A| > |B|$, then there is no injective function from A to B.* Or alternatively in the contrapositive: *if $f : A \to B$ is injective where A and B are finite, then $|A| \leq |B|$.*

✳

A function is *surjective* if every element of the codomain is the value of the function for some argument in the domain, or in other words, if the image is equal to the codomain. That is, $f : A \to B$ is surjective just in case for every $y \in B$, there is at least one $x \in A$ such that $f(x) = y$. A blob-and-arrow diagram represents a surjective function if every element of the codomain has at least one arrow pointing to it (Figure 6.8).

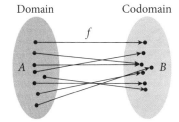

Figure 6.8. In a surjective function, there is at least one arrow ending at each element of B.

The birthday function is surjective in reality if the domain is the set of all people, since births occur on every day of the year. The simple address-book function (6.1) is surjective, given that the four addresses constitute the entire codomain. On the other hand, consider the square function on the natural numbers; that is, $f : \mathbb{N} \to \mathbb{N}$, where $f(x) = x^2$ for each $x \in \mathbb{N}$. Then f is not surjective, since there is, for example, no natural number x such that $x^2 = 2$.

A *bijection* is a function that is both injective and surjective. That is, $f : A \to B$ is a bijection if and only if, for every element $y \in B$, there is one and only one element $x \in A$ such that $f(x) = y$. Or in terms of blobs and arrows, every element of B has exactly one arrow pointing to it (Figure 6.9).

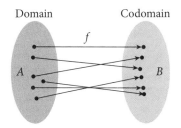

Figure 6.9. In a bijection, there is exactly one arrow ending at each element of B.

For the *successor* function f over the integers such that $f(n) = n + 1$ for every n, f^{-1} is that function from $\mathbb{Z}$ to $\mathbb{Z}$ such that $f(m) = m - 1$ for any *m*—commonly known as the *predecessor* function. Both the successor and the predecessor functions are bijective on the integers, since for every integer there is both a unique next integer and a unique previous integer.

A more interesting example of a bijection is the function $f : \mathbb{Z} \to \mathbb{E}$, where $\mathbb{E}$ is the set of even integers, given by $f(x) = 2x$ for every x. For every $y \in \mathbb{E}$, there is one and only one integer $x \in \mathbb{Z}$ such that $y = 2x$, namely $x = \frac{y}{2}$ (which is an integer since y is even). So the function (Figure 6.10) is a bijection.

If $f : A \to B$ is a bijection, then it has an inverse f^{-1}, since any injective function has an inverse. In the case of a bijection, the domain of the inverse is the entire codomain of f—in other words, $f[A] = B$. So the inverse of f is $f^{-1} : B \to A$, where $f^{-1}(y)$ is for any $y \in B$ the unique $x \in A$ such that $f(x) = y$. This inverse f^{-1} is also a bijection: in the corresponding blob-and-arrow diagram we can just reverse the arrows, so now every element of A has exactly one arrow pointing to it.

$\mathbb{Z}$	$\leftrightarrow$	$\mathbb{E}$
$\vdots$		$\vdots$
-2	$\leftrightarrow$	-4
-1	$\leftrightarrow$	-2
0	$\leftrightarrow$	0
1	$\leftrightarrow$	2
2	$\leftrightarrow$	4
$\vdots$		$\vdots$

Figure 6.10. A bijection between $\mathbb{Z}$ and $\mathbb{E}$.

If A and B are finite and f is an injective function from A to B, then $|A| \leq |B|$ since all of the values $f(x)$, where $x \in A$, are distinct for different x. And if f is surjective, then $|A| \geq |B|$ since B cannot contain any element that is not the image of an argument in A. So if f is a bijection, that is, both injective and surjective, then $|A|$ and $|B|$ must be equal. In other words, *if there is a bijection between finite sets, the sets are the same size.*

The converse is certainly true as well. Suppose A and B are sets with n members each, say

$$A = \{a_1, \ldots, a_n\}$$
$$B = \{b_1, \ldots, b_n\},$$

where if $i \neq j$ ($1 \leq i, j \leq n$), then $a_i \neq a_j$ and $b_i \neq b_j$. Then the function $f : A \to B$ such that $f(a_i) = b_i$ for $i = 1, \ldots, n$ is a bijection.

So finite sets have the same size if and only if there is a bijection between them. This may seem too obvious to dress up in fancy language, but there is a reason to do so. We will use the existence of a bijection between sets as the *definition* of what it means for two *infinite* sets to be of equal size. Using this definition, we shall see that not all infinite sets are the same size, and important consequences will follow.

To lay the groundwork, we need to establish one more fact about bijectively related sets: if there exist bijections from set A both to set B and to set C, then there is a bijection between B and C. So all the sets that stand in a bijective relationship to a given set have that same "family resemblance" to each other. Let's prove the technical fact first and then pause to consider what it means.

Theorem 6.4. *Let A, B, and C be any sets. Suppose there exist bijections $f : A \to B$ and $g : A \to C$. Then there is a bijection $h : B \to C$.*

Proof. Because f is a bijection, its inverse $f^{-1}: B \to A$ exists and is a bijection from B to A. Define the function $h: B \to C$ as follows: for any $y \in B$, $h(y) = g(f^{-1}(y))$. That is, given $y \in B$, follow the f arrow backward to find the corresponding element of A, and then follow the g arrow forward to an element of C (Figure 6.11). Distinct elements of B correspond to distinct elements of A under f^{-1}, which are mapped by g to distinct elements of C. Therefore h is an injection. It is also a surjection, since for any element $z \in C$, z is the value of h on the argument $f(g^{-1}(z)) \in B$:

$$h(f(g^{-1}(z)) = g(f^{-1}(f(g^{-1}(z))))$$
$$= g(g^{-1}(z))$$
$$= z.$$

So h is a bijection from B to C.   ∎

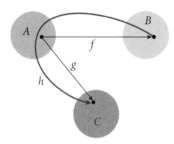

Figure 6.11. Given bijections f from A to B and g from A to C, a bijection from B to C can be constructed: for any element of B, follow the f arrow backward into A and then the g arrow forward into C.

In this proof we chained together f^{-1} and g to define the function h by the rule that $h(y) = g(f^{-1}(y))$. We have, in other words, applied a certain operation on two functions to produce a third. This operation is called *composition* and denoted by $\circ$:

$$h = g \circ f^{-1}.$$

In general, if $f: A \to B$, $g: C \to D$, and $f[A] \subseteq C$, then the function $g \circ f: A \to D$ is defined by the rule that $(g \circ f)(a) = g(f(a))$ for any $a \in A$.

The proof of Theorem 6.4 suggests a way of comparing the size of arbitrary sets: say that they are the same size if there is a bijection between them. That produces the natural result for finite sets, that they are the same size if they have the same number of elements. This idea produces counterintuitive results for infinite sets, however. For example, since there is a bijection between the integers and the even integers, we would have to accept that the set of integers and the set of even integers have the same size, though the even integers are a proper subset of the integers. Strange as that sounds at first, it is exactly what we want, as we shall see in the next chapter.

✳

A function may have more than one argument. Multiplication can be regarded as a two-argument function $M: \mathbb{Z} \times \mathbb{Z} \to \mathbb{Z}$, where $M(m, n) = m \cdot n$ for any $m, n \in \mathbb{Z}$. For example, we would write $M(3, -5) = -15$. Of course it is conventional to use a symbol such as "$\cdot$" written *between* the arguments rather than a letter such as M written *before* the arguments. When a function has two arguments and its name is written between the arguments, the notation is said to be *infix*; writing the name before the arguments is *prefix* notation.

Functions with more than one argument are not really a new idea. A two-argument function $f: A \times B \to C$ is really a one-argument function with domain $A \times B$ and codomain C. As a matter of notational simplification, we write $f(a, b) = c$ instead of $f(\langle a, b \rangle) = c$; that is, without explicitly naming the arguments as an ordered pair. In the same way, a k-argument function, or as we would usually say a k-ary function, is simply a function with a domain that is a Cartesian product of k sets, and we write $f(x_1, \ldots, x_k) = y$ rather than $f(\langle x_1, \ldots, x_k \rangle) = y$.

Chapter Summary

- A *relation* is a connection between things, and can be identified with the set of tuples of things that have that relation to each other; for instance, $R = \{\langle x, y \rangle : P(x, y)\}$ for some property P. A *binary relation* is a relation between two things.

- A binary relation may associate each first component x with any number of second components y, and each second component y with any number of first components x. The components x and y may come from different sets, or from the same set.

- The *inverse* of a binary relation consists of all pairs of the original relation with their components reversed: that is, all pairs $\langle y, x \rangle$ where the original relation contains $\langle x, y \rangle$.

- A *function* (or *mapping*) f *from* a set A to a set B, denoted $f : A \to B$, is a relation that associates each $x \in A$ with exactly one $y \in B$. That unique y is called the *value* of the function on the *argument* x.

- For a function $f : A \to B$, A is the *domain* and B is the *codomain*.

- The value of a function f on an argument x is the *image* of x. The set of values of f for all arguments in a set X is the image of X, denoted $f[X]$.

- An *injective* function $f : A \to B$ is a function where each element of A maps to a different element of B.

- Every function $f : A \to B$ has an inverse f^{-1} that is a relation, but f^{-1} is a function only if f is injective. Injective functions are therefore called *invertible*. The domain of f^{-1} is a subset of (possibly equal to) B.

- A *surjective* function $f : A \to B$ is a function where each element of B is the value of f for some argument in A.

- A *bijection* $f : A \to B$ is a function that is both injective and surjective; that is, every element of B is the value of exactly one argument in A.

- Two sets (finite or infinite) are the same size if and only if there is a bijection between them.

- Two functions may be chained together by *composition*, denoted by $\circ$.

- A function may have multiple arguments; we can view such a function as a single-argument function, the domain of which is the Cartesian product of multiple sets.

Problems

6.1. Let f be any function. Suppose that the inverse relation
$$f^{-1} = \{\langle y, x \rangle : y = f(x)\}$$
is a function. Is f^{-1} a bijection? Explain.

6.2. For each of the following functions, decide whether it is injective, surjective, and/or bijective. If the function is a bijection, what is its inverse? If it is injective but not surjective, what is its inverse on the image of its domain?

(a) $f : \mathbb{Z} \to \mathbb{Z}$, where $f(n) = 2n$.

(b) $f : \mathbb{R} \to \{x \in \mathbb{R} : 0 \le x < 1\}$, where $f(x) = x - \lfloor x \rfloor$.

(c) $f : \mathbb{N} \times \mathbb{N} \to \mathbb{N}$, where $f(n, m)$ is the larger of m and n.

(d) $f : \mathbb{Z} \to \mathbb{R}$, where $f(n) = \frac{n}{3}$.

(e) $f : \mathbb{R} \to \mathbb{R}$, where $f(x) = \frac{x}{3}$.

(f) $f : \mathbb{N} \to \mathbb{Z}$, where $f(n) = \frac{-n}{2}$ if n is even and $f(n) = \frac{n+1}{2}$ if n is odd.

6.3. (a) Show that if two finite sets A and B are the same size, and r is an injective function from A to B, then r is also surjective; that is, r is a bijection.

(b) Give a counterexample showing that the conclusion of part (a) does not necessarily hold if A and B are two bijectively related infinite sets.

6.4. What is the inverse of the circle relation of Figure 6.3?

6.5. Suppose $f : A \to B$, $g : C \to D$, and $A \subseteq D$. Explain when $(f \circ g)^{-1}$ exists as a function from a subset of B to C, and express it in terms of f^{-1} and g^{-1}.

6.6. Fifteen people use the same computer at different times, with no two people using the computer simultaneously. Each reserves a one-hour time slot to use every day, beginning and ending on the hour. So one might take the 3am–4am time slot every day, and another might take the 11pm–midnight slot. Show that there is some continuous time span of seven hours in which five different people are using the computer. *Hint:* Define a function s taking two arguments, a person and an integer between 0 and 6 inclusive, such that $s(p, i)$ is the seven-hour block beginning i hours before the one-hour time block of person p. Apply the Extended Pigeonhole Principle.

6.7. The function $f(n) = 2n$ is a bijection from $\mathbb{Z}$ to the even integers and the function $g(n) = 2n + 1$ is a bijection from $\mathbb{Z}$ to the odd integers. What are f^{-1}, g^{-1}, and the function h of Theorem 6.4?

Chapter 7

Countable and Uncountable Sets

Does it make sense to talk about the sizes of infinite sets? It does, but we have to be prepared for some counterintuitive results, such as the one mentioned in the last chapter—that the set of even integers is the same size as the set of all integers.

Let's work with an analogy. Suppose you run a hotel with 67 rooms.[1] One day the hotel is empty and 67 people show up wanting rooms. If you issue one room key to each individual, you will run out of keys just as you give a key to the last person. If the keys are sequentially numbered from 0 to 66, you can just tell guests to look at their keys and go to the room bearing the number on the key.

Now the hotel is full. A 68th person who showed up looking for a room would be out of luck.

Next imagine instead that you own a hotel with an infinite number of rooms, numbered 0, 1, 2, 3, …. If the hotel starts out empty, and an infinite number of people, p_0, p_1, …, show up simultaneously, you can again give each guest a numbered key and the instruction to go to the room identified by the number on the key.

Now the hotel is full—every room has an occupant. Suppose that another traveler now arrives and asks for a room. If you were to say "you can have room n," for any particular room number n, your new arrival would find that room already occupied. And yet in the case of an infinite hotel, you can squeeze in one more person!

Tell each of your current guests to move up one room, so that the person in room 0 moves to room 1, at the same time the person displaced from room 1 moves to room 2, and so on (Figure 7.1). Then your newly arrived traveler can settle into room 0, and everyone is happy. The same thing could be done if you had five extra guests, or ten, or k for any finite number k—just have everyone move up k rooms, and give the first k rooms, now empty, to the new guests.

You can do more than that. Imagine you have two infinite hotels G and J, with rooms numbered with the nonnegative integers, and both of them are full. And now suppose you have to move everyone into another hotel H, where the rooms are also numbered 0, 1, 2, …. How can you make space for all your guests from both hotels in H?

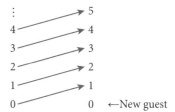

Figure 7.1. Always room for one more in an infinite hotel. Just tell everyone to move up one room, and put the new arrival in room 0.

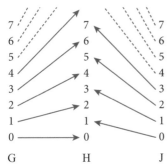

Figure 7.2. Two infinite hotels G and J can be merged into one, H, if the rooms are numbered 0, 1, 2, ... The arrows on the left represent the function f_G and the arrows on the right, f_J.

Tell occupants of hotel G to double their room number and move into the room in H with that number—so the guest in room G0 (as we will refer to room 0 of hotel G) goes to H0, the guest in room G1 moves to H2, G2 moves to H4, and so on (Figure 7.2). Now all the even-numbered rooms of H are full and the odd-numbered rooms are empty—so you can fill the odd-numbered rooms with all the guests from J. Guests from hotel J double their room number and add 1, and move to that room in H: so the person from room J0 moves to room H1, the person from room J1 moves to room H3, J2 moves to H5, and so on. Something similar could be done to merge together three infinite hotels, or k hotels for any finite number k.

Each of these tricks is based on the fact that there is a bijection between a proper subset of the natural numbers and the set of all the natural numbers, like the bijection between $\mathbb{Z}$ and $\mathbb{E}$ on page 64. For example, in the last scenario, we split $\mathbb{N}$ into two disjoint proper subsets—the even numbers and the odd numbers—and constructed bijections between $\mathbb{N}$ and each of these subsets, the first $f_G(n) = 2n$ and the second $f_J(n) = 2n + 1$. We then moved the party from room n of hotel G into room $f_G(n)$ of hotel H, and the party from room n of hotel J into room $f_J(n)$ of hotel H. The image $f_G[\mathbb{N}]$ is the set of even numbers and the image $f_J[\mathbb{N}]$ is the set of odd numbers, so $f_G[\mathbb{N}] \cap f_J[\mathbb{N}] = \emptyset$ and $f_G[\mathbb{N}] \cup f_J[\mathbb{N}] = \mathbb{N}$: everyone wound up accommodated with no double-booking, and every room of H winds up occupied!

We say that a set is *countably infinite* if there is a bijection between that set and the natural numbers. The metaphor is that a bijection $f : \mathbb{N} \to S$ for some set S counts off the members of S, in an unending sequence $f(0), f(1)$, ..., in much the same way that a bijection from $\{0, \ldots, n-1\}$ to a finite set S demonstrates that S has exactly n elements. In both cases, it is essential that the "counting" enumerate all of S; that is, for each $x \in S$, there is some n for which $f(n) = x$.

By Theorem 6.4, any two countably infinite sets have a bijection between them, since each has a bijection with $\mathbb{N}$. So it makes sense to say that all countably infinite sets are the same size; namely, they are the size of $\mathbb{N}$. (That "size" has a name, $\aleph_0$, pronounced "aleph zero" or "aleph null." Naming the other infinite numbers, and doing arithmetic with them, is fascinating, but beyond the scope of this book.)

We have already seen several subsets of the natural numbers that are countably infinite; for example, the set $\mathbb{N} - \{0\}$, and the set of nonnegative even numbers. It is also true that some sets are countably infinite even though they are proper supersets of the natural numbers—$\mathbb{Z}$, for example.

Theorem 7.1. $\mathbb{Z}$ *is countably infinite.*

Proof. We need to "count" all the integers; that is, to assign numerical tags 0, 1, ... to the integers, without missing any. The first thing to try might be to assign 0 to 0, 1 to 1, and so on, but then we would use up all the natural numbers without ever getting around to tagging the negative numbers. So instead we will start at zero, but then alternate, counting a positive integer and then a negative integer, and so on, in this order: $0, +1, -1, +2, -2, \ldots$. The bijection is shown in Figure 7.3.

That is, $f : \mathbb{N} \to \mathbb{Z}$ is a bijection, such that for any $n \in \mathbb{N}$,

$$f(n) = \begin{cases} -n/2 & \text{if } n \text{ is even} \\ (n+1)/2 & \text{if } n \text{ is odd.} \end{cases} \quad \blacksquare$$

Even $\mathbb{N} \times \mathbb{N}$, the set of ordered pairs of natural numbers, is no larger than $\mathbb{N}$.

Theorem 7.2. $\mathbb{N} \times \mathbb{N}$ *is countably infinite.*

Proof. We need to find a way of listing all the ordered pairs $\langle x, y \rangle$ such that $x, y \in \mathbb{N}$, in some order, without missing any. We can't use perhaps the most obvious order, listing all the pairs in which $x = 0$, and then those in which $x = 1$, and so on, because we would use up all the natural numbers on the infinitely many pairs in which $x = 0$, and never get to $x = 1$ or beyond.

Instead we will list the pairs in order of increasing value of the sum $x + y$, and among pairs with the same sum, list pairs with smaller x first. For any given value z, there are only finitely many pairs $\langle x, y \rangle$ such that $x + y = z$. So this ordering does eventually get to every value z—and therefore every pair $\langle x, y \rangle$—in a finite number of steps, unlike our first proposed ordering. If we think of $\mathbb{N} \times \mathbb{N}$ as the points in the plane with nonnegative integer coordinates, we are marching through the diagonals one at a time, as shown in Figure 7.4 along with the first few values of the bijection. We leave it as an exercise (Problem 7.8) to give the function in algebraic terms. $\quad \blacksquare$

Theorem 7.2 is quite remarkable, given that we started with the hotelier trying to squeeze one more person into an infinite but fully occupied hotel. It implies that a countable infinity of countably infinite hotels can be merged into one! In mathematical language (see Problem 7.11),

Theorem 7.3. *The union of countably many countably infinite sets is countably infinite.*

A set is *countable* if it is finite or countably infinite. And a set is *uncountable* if it is not countable.

Now uncountability is a disturbing notion. We have already seen that some sets that seem to be "bigger" than the set of natural numbers actually are not. What sort of set could be uncountably infinite? We do not have any examples at hand, nor any obvious reason to think that uncountable sets even exist.

$n \in \mathbb{N}$	$\leftrightarrow$	$f(n) \in \mathbb{Z}$
0	$\leftrightarrow$	0
1	$\leftrightarrow$	1
2	$\leftrightarrow$	-1
3	$\leftrightarrow$	2
4	$\leftrightarrow$	-2
5	$\leftrightarrow$	3
6	$\leftrightarrow$	-3
$\vdots$		$\vdots$

Figure 7.3. A bijection between $\mathbb{N}$ and $\mathbb{Z}$.

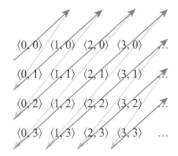

$n \in \mathbb{N}$	$\leftrightarrow$	$f(n) \in \mathbb{N} \times \mathbb{N}$
0	$\leftrightarrow$	$\langle 0, 0 \rangle$
1	$\leftrightarrow$	$\langle 0, 1 \rangle$
2	$\leftrightarrow$	$\langle 1, 0 \rangle$
3	$\leftrightarrow$	$\langle 0, 2 \rangle$
4	$\leftrightarrow$	$\langle 1, 1 \rangle$
5	$\leftrightarrow$	$\langle 2, 0 \rangle$
6	$\leftrightarrow$	$\langle 0, 3 \rangle$
$\vdots$		$\vdots$

Figure 7.4. To enumerate $\mathbb{N} \times \mathbb{N}$, go through the pairs one diagonal at a time, in order of increasing distance from the $\langle 0, 0 \rangle$ pair in the upper left corner. Every pair will be reached eventually.

But indeed they do. The canonical example of a set that turns out to be uncountable is $\mathcal{P}(\mathbb{N})$, the set of all sets of natural numbers. Certainly this set $\mathcal{P}(\mathbb{N})$ is infinite—each of the one-element sets $\{0\}, \{1\}, \{2\}, \ldots$ is a member of $\mathcal{P}(\mathbb{N})$. So the countably infinite set $\{\{n\} : n \in \mathbb{N}\}$ is a subset of $\mathcal{P}(\mathbb{N})$.

But those members of $\mathcal{P}(\mathbb{N})$ are all finite sets, indeed sets of cardinality 1. The full power set $\mathcal{P}(\mathbb{N})$ also contains infinite sets, such as the set of positive even numbers $\{2, 4, 6, 8, 10, \ldots\}$ and $\mathbb{N}$ itself. It is not at all obvious that there is no clever way to enumerate all the sets of natural numbers—some way to list them all in order as we did with $\mathbb{N} \times \mathbb{N}$, so that every set gets listed eventually.

But there isn't. Here is why.

Let's consider an arbitrary subset of $\mathbb{N}$—let's call it S. Since a set is determined by its members, S could be described by specifying, for each element of $\mathbb{N}$, whether or not that element is in S. This is a yes-or-no question for each natural number, so we can represent a "yes" with the bit 1 and a "no" with 0. Then every subset S can be represented as an infinite-length bit string—a *countably* infinite bit string—where the nth bit of the representation of S is the answer to the question, "is the natural number n in S?" For example, the set $\{1, 2, 3\}$ would be represented as

$$01110000000\ldots$$

with nothing but 0s after the first four digits. (The first bit is 0 since $0 \notin \{1, 2, 3\}$.) The set of all even numbers $\{0, 2, 4, 6, 8, 10, \ldots\}$ would be represented as

$$10101010101\ldots$$

with its digits alternating forever. What we are calling an infinite bit string representing S is just a function χ_S from the natural numbers to $\{0, 1\}$, where $\chi_S(n)$ is the bit value in position n, and has the value 1 if and only if $n \in S$:

$$\chi_S(n) = \begin{cases} 0 & \text{if } n \notin S \\ 1 & \text{if } n \in S. \end{cases}$$

The function χ_S is called the *characteristic function* of S. (The symbol χ is the Greek letter chi, pronounced "kiye" with a hard "ch" like that in "characteristic.")

Now we will prove that the set $\mathcal{P}(\mathbb{N})$ is uncountable, using what is known as a *diagonalization* argument. Diagonalization is a particular kind of proof by contradiction. It shows that a certain kind of object—a bijection between $\mathbb{N}$ and $\mathcal{P}(\mathbb{N})$—cannot exist. The proof proceeds by assuming such a bijection does exist and proving that it is not a bijection after all, because it "missed" a set—that is, we construct from the supposed bijection $b : \mathbb{N} \to \mathcal{P}(\mathbb{N})$ a set

$S \subseteq \mathbb{N}$ that is not the value of b for any argument. Since the assumption that b exists and was a bijection led to the conclusion that it wasn't a bijection after all, no such b could have existed.

Theorem 7.4. $\mathcal{P}(\mathbb{N})$ *is uncountable.*

Proof. Let's assume, in order to derive a contradiction, that $\mathcal{P}(\mathbb{N})$ is countable. Then we can number the sets in $\mathcal{P}(\mathbb{N})$ via a bijection $b : \mathbb{N} \to \mathcal{P}(\mathbb{N})$. That is, every set of natural numbers is $b(n)$ for some natural number n. To give these sets convenient names, let's write $S_n = b(n)$ for each n, so $\mathcal{P}(\mathbb{N}) = \{S_0, S_1, S_2, \ldots\}$.

For illustrative purposes, let's represent these sets by their characteristic functions. So that we can draw a concrete picture of the way a "missing" subset of $\mathbb{N}$ is constructed, let's say $S_0 = 0101010101\ldots$ and $S_1 = 1110000000\ldots$, and the next is the set of all even numbers, $S_2 = 1010101010\ldots$. Every possible set $S \subseteq \mathbb{N}$ is in some spot in this ordering. So if we list the characteristic functions as rows in an infinite matrix, we get the following picture:

$$
\begin{array}{cccccccccccc}
S_0 & 0 & 1 & 0 & 1 & 0 & 1 & 0 & 1 & 0 & 1 & \cdots \\
S_1 & 1 & 1 & 1 & 0 & 0 & 0 & 0 & 0 & 0 & 0 & \cdots \\
S_2 & 1 & 0 & 1 & 0 & 1 & 0 & 1 & 0 & 1 & 0 & \cdots \\
S_3 & \cdot & \cdot & \cdot & \cdot & \cdot & \cdot & \cdot & \cdot & \cdot & \cdot & \cdot \\
S_4 & \cdot & \cdot & \cdot & \cdot & \cdot & \cdot & \cdot & \cdot & \cdot & \cdot & \cdot \\
S_5 & \cdot & \cdot & \cdot & \cdot & \cdot & \cdot & \cdot & \cdot & \cdot & \cdot & \cdot \\
\end{array}
$$

Now we'll construct a new set D, the characteristic function of which is the *complement of the diagonal* of this matrix. We've written the diagonal in red so it stands out. Bit 0 of D is the complement of bit 0 of S_0. Since S_0 starts with 0, the first bit of D will be 1. Bit 1 of S_1 is 1, so bit 1 of D will be 0. Bit 2 of S_2 is a 1, so that bit of D will be 0. We continue in this way, always looking at the ith digit of the set S_i and then taking the opposite bit to be the ith digit of D. So D in our example begins $100\ldots$, the complement of $011\ldots$.

Now D was constructed in such a way that it cannot possibly be the same as S_0, because the bit in position 0 of D and S_0 are complementary. In exactly the same way, D cannot be equal to S_i for *any* natural number i, since D has a different bit at the ith position! That is, D differs from each row at the position where the diagonal and the row intersect. So D is none of the S_i; it appears nowhere in our supposedly exhaustive list of all the members of $\mathcal{P}(\mathbb{N})$.

That is a contradiction. To recap, without reference to the picture: Given that we have an enumeration $S_0, S_1, S_2, \ldots$ of all the subsets of $\mathbb{N}$, the set D is a perfectly good set of natural numbers, namely, $D = \{i : i \notin S_i\}$. Then for each $d \in \mathbb{N}$, it cannot be the case that $D = S_d$. For suppose d were a natural number such that $D = S_d$. Then

$$d \in D \text{ if and only if } d \in S_d \text{ (since } D = S_d)$$
$$\text{if and only if } d \notin D \text{ (since } D = \{i : i \notin S_i\}).$$

So $d \in D$ if and only if $d \notin D$. That is a contradiction. So D is none of the S_i, contradicting the assumption that every set of integers was one of the S_i. So there is no bijection $\mathbb{N} \to \mathcal{P}(\mathbb{N})$, and $\mathcal{P}(\mathbb{N})$ is uncountable.[2] ∎

[2]This is called Cantor's diagonal argument (or diagonalization argument), after the mathematician Georg Cantor (1845–1918), who first used this technique in his study of infinite numbers.

We said that two sets are the same size if there is a bijection between them. It now makes sense to say that one set is *smaller* than another (or that the second is *larger* than the first) if there is an injection from the first to the second but no surjection from the first to the second (and therefore no bijection between them). For finite sets these definitions align with our commonsense understanding of relative size—that A is smaller than B if $|A| < |B|$. By Theorem 7.4, we know there is no surjection from $\mathbb{N}$ to $\mathcal{P}(\mathbb{N})$, but there certainly is an injection; for example, the function $f : \mathbb{N} \to \mathcal{P}(\mathbb{N})$ such that $f(n) = \{n\}$ for every $n \in \mathbb{N}$. So we can now say that $\mathcal{P}(\mathbb{N})$ is larger than $\mathbb{N}$.

✳

So there are uncountably many sets of natural numbers. Likewise, there are uncountably many subsets of any countably infinite set—there are uncountably many sets of even integers, for example. As long as there is a bijection between $\mathbb{N}$ and a set S, there are uncountably many subsets of S.

However, there are only countably many *finite* sets of natural numbers!

Theorem 7.5. *The set of all finite subsets of $\mathbb{N}$ is countable.*

Proof. Notice that the bit string representing the characteristic function of a finite set has only finitely many 1s; from some point on, all the bits are 0. So we can truncate such a string after the last 1, and list these finite bit strings in order of increasing length, using the standard dictionary ordering to sequence bit strings of the same length. The empty set is finite, and since its characteristic function would be represented as $000\ldots$, in this one case there is no last 1; we'll represent it as the "empty string of bits," denoted by λ (the Greek letter lambda). So our enumeration of all finite sets begins as shown in Figure 7.5. ∎

i	χ_{S_i}	S_i
0	λ	$\emptyset$
1	1	$\{0\}$
2	01	$\{1\}$
3	11	$\{0, 1\}$
4	001	$\{2\}$
5	011	$\{1, 2\}$
6	101	$\{0, 2\}$
7	111	$\{0, 1, 2\}$
8	0001	$\{3\}$
⋮	⋮	⋮

Figure 7.5. A bijection between $\mathbb{N}$ and the set of finite subsets of $\mathbb{N}$. The second column shows the values of $\chi_{S_i}(n)$ for $n = 0, 1, \ldots$ up to $n =$ the largest member of S_i.

This way of ordering bit strings—in groups by increasing length, and alphabetically among strings of the same length—is known as *lexicographic order*.

If there are countably many finite sets of natural numbers, and uncountably many sets of natural numbers, how many *sets of sets of natural numbers* are there? There are more sets of sets of natural numbers than there are sets of natural numbers, as we shall see. But pursuing the general theory of *transfinite cardinals*, as it is known, would take us too far afield. Instead, let's just generalize Theorem 7.4.

We might have noticed that $\mathcal{P}(S)$ is larger than S not just when S is countably infinite, but also when S is finite. A set with two elements has four subsets, a set with three elements has eight subsets, and so on. This pattern holds even for the smallest sets: a set with one element has two subsets, and the set with zero elements has one subset (the empty set has no elements, but one subset, namely the empty set).

So we have a general pattern: *every* set has more subsets than elements. The diagonalization argument generalizes nicely to show that $\mathcal{P}(A)$ is always larger than A, for any set A—whether A is finite, countably infinite, or uncountably infinite.

Theorem 7.6. *There is an injection but no bijection between A and $\mathcal{P}(A)$ for any set A. Therefore $\mathcal{P}(A)$ is larger than A for any set A.*

Proof. The mapping $g : A \to \mathcal{P}(A)$ such that $g(x) = \{x\}$ for any $x \in A$ is an injection from A to $\mathcal{P}(A)$. To show that there is no bijection, again the proof is by contradiction. Assume there exists such a bijection $f : A \to \mathcal{P}(A)$. Define the set $D = \{a \in A : a \notin f(a)\}$—that is, D is an element of $\mathcal{P}(A)$, consisting of all elements a of A such that a is *not* in the corresponding set $f(a)$. D is exactly the same set as we defined in the proof of Theorem 7.4 in the case when A was the set of natural numbers $\mathbb{N}$.

Since f is a bijection, and D is a member of $\mathcal{P}(A)$, there is some member of A that gets mapped by this bijection to S. Let's call this element d, so $f(d) = D$.

Because $f(d) = D = \{a : a \notin f(a)\}$, we know that for any $a \in A$, $a \in f(d)$ if and only if $a \notin f(a)$.

But now let's apply this statement to the specific case of $a = d$: it says that $d \in f(d)$ if and only if $d \notin f(d)$.

This is impossible: we've derived a contradiction. Therefore no such bijection can exist. ∎

So $\mathcal{P}(A)$ is always larger than A, and *any set has more subsets than elements*.

✳

The idea of countable and uncountable infinities is important to computer scientists because the set of things that computers can possibly compute is countable. Any algorithm must be written using only a finite number of characters, drawn from a finite character set, so the set of all algorithms can be ordered lexicographically, making it countable.

However, the set of all possible functions is uncountable (see Problem 7.12 for an example of a particular set of functions that is uncountable). So it follows that some functions—in fact, uncountably many functions—do not correspond to any algorithm. A function that cannot be computed by any algorithm is called an *uncomputable function*.

There are many interesting problems in computer science that one might hope to solve by an algorithm, but that turn out to be uncomputable. One example is the *halting problem*: whether a given algorithm will halt on a given input after a finite number of steps, or whether it will instead run forever without halting. Identifying the halting problem as computationally unsolvable was the seminal contribution of Alan Turing, and arguably the result that gave birth to the field of computer science.[3] But showing that *some* problems must be uncomputable follows immediately once it is realized that the set of all computer programs (and therefore the set of all algorithms) is countable, but the set of all functions is uncountable.

[3] Alan Mathison Turing (1912–54), "On Computable Numbers, with an Application to the Entscheidungs-problem," *Proceedings of the London Mathematical Society* 2–42, no. 1 (1936, published 1937): 230–65.

Chapter Summary

- Infinite sets may have the same size even when one is a proper subset of the other.

- A set is *countably infinite* if it is the same size as the natural numbers; that is, there is a bijection between the set and $\mathbb{N}$.

- All countably infinite sets are the same size.

- A set is *countable* if it is either finite or countably infinite, and is *uncountable* if not.

- The set $\mathbb{N} \times \mathbb{N}$ is countably infinite; equivalently, the union of countably many countably infinite sets is countably infinite.

- The power set of the natural numbers, $\mathcal{P}(\mathbb{N})$, is uncountable. This can be proven by contradiction, using a *diagonalization* argument applied to the *characteristic functions* of the sets.

- The power set of any countably infinite set is uncountable.

- Set A is *smaller* than set B, and B is *larger* than A, if there is an injection from A to B but no surjection (and hence no bijection) from A to B.

- The string ordering that orders strings first by length, and then alphabetically among strings of the same length, is called *lexicographic order*. The first few bit strings in lexicographic order are $\lambda, 0, 1, 00, 01, 10, 11, 000, \ldots$.

- For every set S, whether finite, countable, or uncountable, $\mathcal{P}(S)$ is larger than S; that is, every set has more subsets than elements.

Problems

7.1. Two sets have the same size if there is a bijection between them. Prove that there are infinitely many different sizes of infinite sets; that is, that there are at least a countably infinite number of infinite sets, no two of which have the same size.

7.2. Is the set of finite strings of letters from the Roman alphabet $\{a, \ldots, z\}$ countable or uncountable? Why?

7.3. Johnny is skeptical about the proof of Theorem 7.4. He agrees that the diagonalization process yields a set of natural numbers D that is not in the original list $S_0, S_1, \ldots$. He claims, however, that the new set can be accommodated by moving all the indices up by 1, and sliding the new set in at the beginning. That is, Johnny wants to set $T_{i+1} = S_i$ for each $i \in \mathbb{N}$, and to set $T_0 = D$, the newly constructed set. Now all the sets of natural numbers have been enumerated, he claims, in the list $T_0, T_1, \ldots$. What is wrong with his argument?

7.4. Give examples of infinite sets A and B for which $|A - B|$ is equal to:

(a) 0.

(b) n, where $n > 0$ is an integer.

(c) $|A|$, where $|A| = |B|$.

(d) $|A|$, where $|A| \neq |B|$.

7.5. Which of the following are possible? Give examples or explain why it is not possible.

(a) The set difference of two uncountable sets is countable.

(b) The set difference of two countably infinite sets is countably infinite.

(c) The power set of a countable set is countable.

(d) The union of a collection of finite sets is countably infinite.

(e) The union of a collection of finite sets is uncountable.

(f) The intersection of two uncountable sets is empty.

7.6. Prove using diagonalization that the set of real numbers between 0 and 1 is uncountable. *Hint:* Any real number can be represented by an infinite decimal, between $0 = 0.000\ldots$ and $1 = 0.999\ldots$. But be careful—some numbers have more than one decimal representation! Use the method of the proof of Theorem 7.4, without simply relying on that result.

7.7. (a) Show that there are as many ordered pairs of reals between 0 and 1 as there are reals in that interval. That is, exhibit a bijection $f : [0, 1] \times [0, 1] \leftrightarrow [0, 1]$. *Hint:* Represent reals as decimals, as in Problem 7.6.

(b) Extend the result of part (a) to give a bijection between pairs of nonnegative real numbers and nonnegative real numbers.

7.8. Figure 7.4 illustrates a bijection $f : \mathbb{N} \to \mathbb{N} \times \mathbb{N}$. Give the inverse function $f^{-1} : \mathbb{N} \times \mathbb{N} \to \mathbb{N}$, which is also a bijection, in algebraic terms. (We have asked for $f^{-1} : \mathbb{N} \times \mathbb{N} \to \mathbb{N}$ rather than $f : \mathbb{N} \to \mathbb{N} \times \mathbb{N}$ because it is the simpler of the two to write down.)

7.9. In each case, state whether the set is finite, countably infinite, or uncountable, and explain why.

(a) The set of all books, where a "book" is a finite sequence of uppercase and lowercase Roman letters, Arabic numerals, the space symbol, and these 11 punctuation marks:

$$; \quad , \quad . \quad ' \quad : \quad - \quad (\quad) \quad ! \quad ? \quad ''$$

(b) The set of all books of less than $500,000$ symbols.

(c) The set of all finite sets of books.

(d) The set of all irrational numbers greater than 0 and less than 1.

(e) The set of all sets of numbers that are divisible by 17.

(f) The set of all sets of even prime numbers.

(g) The set of all sets of powers of 2.

(h) The set of all functions from $\mathbb{Q}$ to $\{0, 1\}$.

7.10. This problem refers to Theorem 7.5 and its proof.

(a) Why can't the proof proceed by enumerating finite subsets of $\mathbb{N}$ in order of their size?

(b) Give a different proof by enumerating finite subsets of $\mathbb{N}$ in order of the sum of their elements, and specifying an ordering among the sets for which this sum is the same.

(c) Why doesn't this new proof generalize to show that the set of *all* subsets of $\mathbb{N}$ is countable?

7.11. Prove Theorem 7.3.

7.12. (a) Prove that the set of all functions $f : \{0, 1\} \to \mathbb{N}$ is countable.

(b) Prove that the set of all functions $f : \mathbb{N} \to \{0, 1\}$ is uncountable.

Chapter 8

Structural Induction

Proof by induction is a fundamental tool of computer science. Using induction, we can establish the truth of infinitely many different propositions in a single, finite argument. For example, on page 27 we proved by induction that for any $n \geq 0$,

$$\sum_{i=0}^{n-1} 2^i = 2^n - 1. \tag{8.1}$$

The proof took a page or two, but it proved infinitely many facts, including

$$\sum_{i=0}^{0-1} 2^i = 2^0 - 1 \text{ (when } n = 0),$$

$$\sum_{i=0}^{1-1} 2^i = 2^1 - 1 \text{ (when } n = 1),$$

$$\sum_{i=0}^{2-1} 2^i = 2^2 - 1 \text{ (when } n = 2),$$

and a similar statement for each value of n.

Proofs by induction are important to computer science because they parallel the way a single computer program can perform different but similar calculations by repeatedly executing the same piece of code. For example, a simple "for loop" that adds 2^i to a running total, looping as the value of i increases from 0 to n, could compute the sum on the left side of (8.1) for any value n. The inductive proof of this equation establishes that the value computed by such a loop is indeed $2^n - 1$.

Computers can manipulate objects other than numbers—we have already spent some time reasoning about the properties of binary strings; for example, the Thue sequence on page 33. This chapter generalizes the notion of proof by induction to facilitate reasoning about nonnumeric mathematical objects.

First, we need to set a general framework. Objects are built up from smaller objects, or are irreducible and atomic. This idea is familiar to programmers using higher-level languages. Such languages support atomic *data types* such as integers, reals, and characters, and provide structuring capabilities such as arrays and strings so that larger, more complex objects can be treated as single entities. The same holds in mathematics.

We could think of a string of symbols, for example, as constructed from the empty string by concatenating symbols one at a time. Up to now, an inductive argument always added 1 to an integer-valued induction variable. Now the base case values will be atomic objects and the induction will be based on constructing larger objects from smaller objects for which a predicate is already known to be true.

So objects are defined by stipulating one or more *base case* objects and one or more *constructor cases*, which are rules for building objects from other objects. Such definitions are typically self-referential, so we must be careful to avoid infinite regress or circular reasoning. Larger things of a particular kind are defined in terms of smaller things of the same kind, and the smallest things are defined without any such self-reference.

Even the natural numbers fit this schema, since we can say that a number is either 0 (the base case), or 1 more than another natural number (the constructor case). For another example, the Thue sequence was defined by specifying a base sequence $T_0 = 0$, and then defining T_{n+1} in terms of T_n for every $n \geq 0$ by the constructor rule $T_{n+1} = T_n \overline{T_n}$.

✳

Any proof by structural induction begins with such an inductive definition. Let's consider the example of bit strings to drive home the way inductive definitions and proofs by structural induction go hand in hand.

We informally described bit strings as concatenations of 0s and 1s, using an intuitive idea of what it means to concatenate strings of bits. But to reason about bit strings more formally, we need more formal definitions of bit strings and the operations we might perform on them. A string of bits—or a string made up of any kind of individual symbols—can be defined inductively, as a special kind of ordered pair. What makes these ordered pairs "special" is not the formation of the pairs, but the effects of the concatenation operator, which combines two strings in such a way that the place where they were joined is "invisible" in the result. For example, if we use "·" for the concatenation operator rather than just writing the two strings next to each other, we can write

$$00 \cdot 11 = 001 \cdot 1 = 0011. \tag{8.2}$$

For (8.2) to be true, concatenating two strings can't be quite as simple as just making them into an ordered pair, because by the definition of ordered pair (page 54),

$$\langle 00, 11 \rangle \neq \langle 001, 1 \rangle,$$

because two ordered pairs are equal only if their first components are equal and their second components are equal.

So let's start from the beginning with strings. An *alphabet* is any finite set. We refer to the members of an alphabet as *symbols*, but there is really nothing special about symbols except that we can tell them apart. We have already worked with the binary alphabet $\{0, 1\}$. Words in the English language are written using the Roman alphabet $\{a, b, \ldots, z\}$ (perhaps along with 26 additional symbols for the uppercase letters).

If Σ is an alphabet, then Σ^* is the set of strings over Σ, commonly known as the *Kleene star* of Σ. The set Σ^* is defined as follows:

Base case (S1). The empty string λ is a string in Σ^*.

Constructor case (S2). If s is a string in Σ^* and a is a symbol in Σ, then $\langle a, s \rangle$ is a member of Σ^*.

Exhaustion clause (S3). Nothing else is in Σ^* except as follows from the base and constructor cases.

The exhaustion clause typically goes without saying; the base and constructor cases are given and it is understood that the things that can be created using them are the only things of the type being defined.

According to this definition, the one-bit string 0 is really $\langle 0, \lambda \rangle$, and the three-bit string 110 is really[1]

$$110 = \langle 1, \langle 1, \langle 0, \lambda \rangle \rangle \rangle.$$

We say that the *length* of the empty string is 0, and if s is any string and a is any symbol, then the length of $\langle a, s \rangle$ is one more than the length of s. Or in symbols,

Base case. $|\lambda| = 0$.

Constructor case. For any $a \in \Sigma$ and $s \in \Sigma^*$, $|\langle a, s \rangle| = |s| + 1$.

The payoff of this roundabout way of defining strings is in the definition of concatenation (Figure 8.1). We define concatenation of two strings, say $s \cdot t$, by structural induction on s, the *first* of the two strings being concatenated.

Base case (SC1). If t is any string in Σ^*, then $\lambda \cdot t = t$.

Constructor case (SC2). For any strings $s, t \in \Sigma^*$, and any symbol $a \in \Sigma$,

$$\langle a, s \rangle \cdot t = \langle a, s \cdot t \rangle. \tag{8.3}$$

Let's check that (8.3) makes sense—concatenation for a string of a given length should be defined only in terms of concatenation for strings of smaller lengths. Suppose string s is of length n. Then the left side is the concatenation

[1] This is essentially how lists are stored in programming languages such as LISP: as "dotted pairs" in which the second element is the tail of the list, or NIL to indicate the end of the list.

$$\overbrace{\langle a, sssss \rangle}^{n+1} \cdot ttttt$$
$$= \langle a, \underbrace{sssss \cdot ttttt}_{n} \rangle$$

Figure 8.1. Definition of string concatenation, by induction on the length of the first string. A string of length $n + 1$ is an ordered pair consisting of a symbol a and a string of length n, shown here as *sssss*. On the assumption that we already know how to concatenate a string of length n to another string, here shown as *ttttt*, we can define the concatenation of *asssss* to *ttttt* as the ordered pair in which the first element is a and the second element is the concatenation of *sssss* to *ttttt*.

of a string of length $n + 1$ (namely $\langle a, s \rangle$) to string t. The right side is an ordered pair, in which the second component is the concatenation of a string of length n with string t. That is, we are defining the concatenation of a string of length $n + 1$ with another string by referring to the concatenation of a string of length n with that other string.

This inductive definition of concatenation will enable us to carry out inductive proofs. In the language of Chapter 3, the mathematical induction is based on an integer induction variable, namely the length of the first string of the two being concatenated. In our new framework, we will simply say that the structural induction rests on the base and constructor cases in the definition of strings, (S1) and (S2).

Let's work through an example. $11 \cdot 00$ should be 1100. Is it?

$$
\begin{aligned}
11 \cdot 00 &= \langle 1, \langle 1, \lambda \rangle \rangle \cdot 00 && \text{(by definition of } 11) \\
&= \langle 1, \langle 1, \lambda \rangle \cdot 00 \rangle && \text{(by (SC2), with } a = 1, s = \langle 1, \lambda \rangle) \\
&= \langle 1, \langle 1, \lambda \cdot 00 \rangle \rangle && \text{(by (SC2), with } a = 1, s = \lambda) \\
&= \langle 1, \langle 1, 00 \rangle \rangle && \text{(by (SC1))} \\
&= 1100 && \text{(by definition of } 1100).
\end{aligned}
$$

So this one example seems to work as expected. But now we are in a position to prove the general fact that when concatenating strings with the same symbols in the same order, the result is always the same. That is, string concatenation is *associative*:

$$
s \cdot (t \cdot u) = (s \cdot t) \cdot u
$$

for any strings s, t, and u, in the same way that

$$
x + (y + z) = (x + y) + z
$$

for any three integers x, y, and z. The associative law for string concatenation justifies dropping any mention of grouping when writing a string such as 1100; it does not matter whether this is interpreted as $1 \cdot 100$, $11 \cdot 00$, or $110 \cdot 0$, for example, since all of these strings are the same. In fact, we could write the same string as $1 \cdot 1 \cdot 0 \cdot 0$, since the result does not depend on the grouping.

Theorem 8.4. *String concatenation is associative. That is, for any strings s, t, and u,*

$$
(s \cdot t) \cdot u = s \cdot (t \cdot u).
$$

Proof. The proof is by induction on the construction of s. We could carry out an ordinary induction on the length of s, like those of Chapter 3, but a more direct argument proceeds from (SC1) and (SC2): the first of the strings being concatenated is either the empty string or an ordered pair of a symbol and a string.

Base case. $s = \lambda$. Then

$$(\lambda \cdot t) \cdot u = t \cdot u \qquad \text{(by (SC1))}$$
$$= \lambda \cdot (t \cdot u) \quad \text{(by (SC1))}.$$

Induction step. We need to prove that $(s' \cdot t) \cdot u = s' \cdot (t \cdot u)$, on the induction
hypothesis that $(s \cdot t) \cdot u = s \cdot (t \cdot u)$ if s was constructed using fewer steps
than s' (fewer applications of (SC2)). Since s' must have been constructed
using (SC2) at least once, $s' = \langle a, s \rangle$ for some symbol a and some string s
that satisfies the induction hypothesis. Then

$$(\langle a, s \rangle \cdot t) \cdot u = \langle a, s \cdot t \rangle \cdot u \qquad \text{(by (SC2))}$$
$$= \langle a, (s \cdot t) \cdot u \rangle \quad \text{(by (SC2))}$$
$$= \langle a, s \cdot (t \cdot u) \rangle \quad \text{(by the induction hypothesis)}$$
$$= \langle a, s \rangle \cdot (t \cdot u) \quad \text{(by (SC2))}. \qquad \blacksquare$$

✳

In general, we will not describe structural induction proofs so elaborately.
The general schema goes as follows.

Suppose a set S is inductively defined as follows:

Base case. Certain base elements b are members of S.

Constructor cases. Certain constructor operations c produce more elements
of S from elements already known to be in S. That is, c is a k-place constructor operation with the property that if $x_1, \ldots, x_k$ are members of S,
then $c(x_1, \ldots, x_k) \in S$.

Exhaustion clause. Nothing else is in S except as follows from (1) and (2).

In Theorem 8.4, the only base element is λ, and the constructor operations (one per symbol in the alphabet) form the ordered pairs $\langle a, s \rangle$ from a
symbol a and a string s.

Then to prove that some predicate P holds for all $x \in S$, it suffices to prove
the following:

Base case. $P(b)$ holds for each base element $b \in S$, and

Induction step. For each k-place constructor c, if $x_1, \ldots, x_k$ are members of S,
and $P(x_1), \ldots, P(x_k)$ all hold, then $P(c(x_1, \ldots, x_k))$ also holds.

To see that these conditions suffice to show that $P(x)$ holds for all $x \in S$,
we can do an ordinary proof by induction on the number of times constructor operations were applied in the construction of x. If that number is zero,
then x must be a base element, so $P(x)$ holds by the base case. Now fix $n \geq 0$
and assume that $P(x)$ holds for all $x \in S$ constructed using at most n applications of constructor functions. Consider an element $y \in S$ constructed
using $n + 1$ applications of constructor functions. Then $y = c(x_1, \ldots, x_k)$
for some constructor function c and some $x_1, \ldots, x_k \in S$, each of which
was constructed using at most n applications of constructor functions. By

the induction hypothesis, $P(x_1)$, ..., $P(x_k)$ all hold, and therefore by the induction step $P(c(x_1, \ldots, x_k)) \equiv P(y)$ also holds.

Note that this simplified schema does not include the statement of the induction hypothesis, since the hypothesis is usually clear from the induction step (and indeed we skipped the hypothesis in the proof of Theorem 8.4). From now on, when presenting proofs by induction, we will generally not state the induction hypothesis explicitly, and move directly from proving the base case to proving the induction step.

Let's try out this simplified schema on another inductively defined set of objects. Many mathematical and programming languages allow the use of nested parentheses for grouping; for example, in the expression

$$(((3 + 4) \times (5 - 6))/(7 + 8)) \times (9 + 10).$$

But wait—do those parentheses balance? And what does "balance" even mean? For starters, there should be the same number of left and right parentheses, but there is more to it. Each left parenthesis has to match up with a particular subsequent right parenthesis (but what does *that* mean?).

Let's start over, and to simplify matters, just talk about balanced strings of parentheses (BSPs), without the other symbols. Rather than trying to explain after the fact why a string of parentheses is balanced, let's describe the rules by which such strings can be produced.

The alphabet contains just the left and right parenthesis symbols, and we can define a base case and constructor cases:

Base case. The empty string λ is a BSP.

Constructor cases.

(C1) If x is any BSP then so is (x); that is, the result of putting a left parenthesis before x and a right parenthesis after.

(C2) If x and y are BSPs then so is xy, the result of concatenating x and y in that order.

For example, let's show why "$(()())$" is a BSP. To start, λ is a BSP by the base case. Applying (C1) with $x = \lambda$, "$()$" is a BSP. Applying (C2) with $x = y = ()$, "$()()$" is a BSP. And then applying (C1) with $x = ()()$, "$(()())$" is a BSP, as required.

Now let's prove some facts about BSPs.

Example 8.5. *Every BSP has equal numbers of left and right parentheses.*

Solution to example. We prove this by structural induction.

Base case. The empty string λ has zero left parentheses and zero right parentheses, and therefore equal numbers of each.

Induction step. Suppose a BSP z was constructed using constructor case (C1) or (C2).

If the last step in the construction of z was (C1), then $z = (x)$ for some BSP x that was constructed using fewer steps. Then x has equal numbers of left and right parentheses by the induction hypothesis. Say x has n of each; then z has $n + 1$ left parentheses and $n + 1$ right parentheses, and so has equal numbers of each.

If z was constructed using (C2) as the final step, then $z = xy$ for some BSPs x and y, each of which was constructed using fewer applications of the constructor cases. So by the induction hypothesis, x and y each have equal numbers of left and right parentheses. If x has n left and n right parentheses and y has m left and m right parentheses, then z has $n + m$ left and $n + m$ right parentheses—equal numbers.   ∎

Example 8.5 gives a necessary but not sufficient condition for a string of parentheses to be balanced. Some strings with equal numbers of left and right parentheses are not balanced—")(", for example. As a final example, combining structural and ordinary induction, we examine a programming trick[2] for determining whether a string of parentheses is balanced: Start counting at 0. Read through the string, adding 1 for every left parenthesis and subtracting 1 for every right parenthesis. The string is balanced if the count is 0 at the end and never went negative along the way. Let's prove that this rule is correct—both necessary and sufficient.

First, let's define this rule as a property of strings of parentheses:

> *A string of parentheses is said to* satisfy the counting rule *if starting from* 0 *at the left end, adding* 1 *for each left parenthesis and subtracting* 1 *for each right parenthesis, gives* 0 *at the end of the string, without ever going negative.*

Example 8.6. *A string of parentheses is balanced if and only if it satisfies the counting rule.*

Solution to example. First we show that if x is any balanced string of parentheses, then x satisfies the counting rule. This is a structural induction.

Base case. $x = \lambda$. Then the count begins at 0 and ends immediately, so this string satisfies the counting rule.

Induction step, case 1. $x = (y)$ where y is balanced, and therefore satisfies the counting rule. Then the count for x is $+1$ after the first symbol, $+1$ again after the end of y, stays positive in between, and ends at 0 after the final right parenthesis.

[2] This is an old trick. A version appeared in 1954 in the original manual for the FORTRAN programming language, one of the first languages in which compound algebraic expressions could be used. *Working from left to right, number each parenthesis, right or left, as follows: Number the first parenthesis "1", label each left parenthesis with an integer one larger than the number of the parenthesis immediately to the left of it. Label each right parenthesis with an integer one less than the number of the parenthesis immediately to the left of it. Having done this, the mate of any left parenthesis labeled "n" will be the first right parenthesis to the right of it labeled n-1.*

Induction step, case 2. $x = yz$ where y and z are balanced and hence satisfy the counting rule. Then the count goes from 0 at the beginning to 0 after y to 0 again after z, without ever going negative in the middle of y or z.

Now for the other direction: If x satisfies the counting rule, then x is balanced. This part of the proof is an ordinary strong induction, on the length of x.

Base case. $|x| = 0$. Then $x = \lambda$, and x is balanced.

Induction hypothesis. Fix $n \geq 0$, and assume that for any $m \leq n$ and any string y of length m, if y satisfies the counting rule, then y is balanced.

Let x be a string of length $n + 1$ that satisfies the counting rule. There are two possibilities (see Figure 8.2).

Induction step, case 1. The count is never 0 except at the beginning and the end of x. Then $x = (y)$ for some string y of length $|x| - 2$, since the count went from 0 to $+1$ after the first symbol and went from $+1$ to 0 after the last symbol. So y satisfies the counting rule and by the induction hypothesis is balanced. By (C1), $x = (y)$ is also balanced.

Induction step, case 2. The count reaches 0 at some point between the beginning and end of x. Then we can write $x = yz$, where y and z are nonempty and the count goes to 0 after y. (If there are multiple points where the count reaches 0, we can choose any of them.) Then y and z are each shorter than x and each satisfies the counting rule, so by the induction hypothesis each is balanced. Then $x = yz$ is balanced by (C2). ■

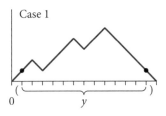

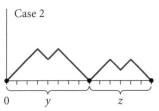

Figure 8.2. The two cases in the proof that any string that satisfies the counting rule is balanced. The count is 0 at the beginning and end of x and never goes negative. It either is never 0 in between (case 1) or it is 0 at some point in between (case 2).

Chapter Summary

- Structural induction generalizes proof by induction to nonnumerical objects, where larger objects can be expressed in terms of smaller ones.

- An *inductive definition* of a kind of object consists of *base case* objects, which are irreducible and defined without any self-reference, and *constructor cases*, which define objects in terms of smaller objects of the same kind.

- For example, the following sets can each be defined inductively:
  — Σ^*, the set of strings composed of *symbols* from the *alphabet* Σ.
  — String concatenation, the operation that joins two strings.
  — Balanced strings of parentheses.

- Structural induction is applied to sets that are defined inductively. If S is such a set, and $P(x)$ is a predicate to be proved about each element $x \in S$, then a proof by structural induction consists of:

Base cases. For each base case element $b \in S$, prove $P(b)$.

Induction step. For each constructor case $x = c(x_1, \dots, x_n)$: the induction hypothesis is $P(x_i)$ for $1 \leq i \leq n$; prove that the induction hypothesis implies $P(x)$.

Problems

8.1. If a is a symbol and s is a string, define $\#_a(s)$ to be the number of occurrences of a in s. For example, $\#_1(101100101) = 5$. Define $\#_a(s)$ inductively, and then prove by structural induction that for any strings s and t,

$$\#_a(s \cdot t) = \#_a(s) + \#_a(t).$$

8.2. (a) Give an inductive definition of the *reversal* x^R of a string x (for example, $11100^R = 00111$).

(b) Prove by induction that for any bit strings u and v, $(u \cdot v)^R = v^R \cdot u^R$.

(c) Prove that $(x^R)^R = x$ for any string x.

8.3. Let's call a string of 0s and 1s a *hedgehog* if it can be produced by the following rules.

1. λ is a hedgehog.

2. If x is a hedgehog, so is $0x1$.

3. If x is a hedgehog, so is $1x0$.

4. If x and y are hedgehogs, so is xy.

Prove that a binary string is a hedgehog if and only if it has equal numbers of 0s and 1s.

8.4. Prove that for any strings u and v, $|u \cdot v| = |u| + |v|$.

8.5. Prove that any balanced string of parentheses can be transformed into a string of left parentheses followed by an equal number of right parentheses by repeatedly replacing occurrences of ")(" by "( )".

8.6. Consider a set T defined as follows:

1. $0 \in T$

2. If $t, u \in T$, then so is $\langle t, u \rangle$.

Let the *weight* of an element $t \in T$ be the number of times rule 2 was used to construct it.

(a) List all elements of T of weight at most 3.

(b) Define the *height* $h(t)$ of an element $t \in T$ by the rules that $h(0) = 0$ and $h(\langle t, u \rangle) = 1 + \max(h(t), h(u))$. Prove that if t has weight n, then

$$\lceil \lg(n+1) \rceil \leq h(t) \leq n,$$

where $\lg x = \log_2 x$. Show that these bounds are tight: that is, for every n, show that there are weight-n elements of T achieving these lower and upper bounds.[3]

[3] A logarithmic function is the inverse of an exponential function, where the number in the subscript is the base of the exponent. That is, $\log_2 x = y$ means that $2^y = x$.

8.7. In this problem, you are going to define arithmetic using just structural induction and the basics of set theory, starting without any concept of the natural numbers. In essence, a natural number n will be defined as the set of all the n-element subsets of the set $U = \{\emptyset, \{\emptyset\}, \{\{\emptyset\}\}, \ldots\}$.

Say that a *singleton* is either $\{\emptyset\}$ or a set containing just one element that is itself a singleton, and let S be the set of all singletons. Define $\mathcal{N}$ to be the set including $\{\emptyset\}$, and also including, for any $n \in \mathcal{N}$,

$$\sigma(n) = \{A \cup B : A \in n, B \in S, \text{ and } A \cap B = \emptyset\}.$$

Then define, for $m, n \in \mathcal{N}$,

$$m + n = \{A \cup B : A \in m, B \in n, \text{ and } A \cap B = \emptyset\}.$$

(a) Let $\mathbf{0} = \{\emptyset\}$. Show that $\sigma(\mathbf{0}) = S$, and in general, for any $n \geq 0$,

$$\overbrace{\sigma(\sigma(\ldots \sigma(\mathbf{0}) \ldots))}^{n \text{ times}}$$

is the set of all n-element subsets of U.

The remaining parts of this problem are to be done without relying on part (a). That is, they require knowing only the definition of $+$ on members of $\mathcal{N}$, not the concept of a number.

(b) Show that if $m, n \in \mathcal{N}$, then $m + n \in \mathcal{N}$.

(c) Show that $m + n = n + m$ for any $m, n \in \mathcal{N}$.

(d) Show that $(m + n) + p = m + (n + p)$ for any $m, n, p \in \mathcal{N}$.

(e) How would you define subtraction on members of $\mathcal{N}$?

8.8. A *subsequence* of a string s is a sequence of the symbols from s in the order in which they appear in s, though not necessarily contiguously. If we write $SS(s)$ for the set of all subsequences of S, then, for example,

$$SS(101) = \{\lambda, 0, 1, 10, 01, 11, 101\}.$$

Define $SS(s)$ inductively.

Chapter 9

Propositional Logic

Precision is important when communicating with computers, but the English language is rife with ambiguities. Native speakers may barely be conscious of the way the same words can mean different things under different circumstances, because other native speakers know from the context how to disambiguate the meanings. But when those words are translated into data and rules for consumption by computers, the ambiguities can have real consequences.

Take a word as simple as "or." If there is a rule in the tax code that says

> *You can take this deduction if you are at least 18 years old or live in Arizona,*

there is no need to explain that 20-year-old Arizonans get the break. The "or" is *inclusive*; there is no need to add "or both." On the other hand, if your financial institution tells you that

> *You can take a lump sum payment or an annuity,*

there is not much doubt that taking both the lump sum and the annuity is not an option. That is the *exclusive* use of "or." And "or" can even imply a threat. For example,

> *Either you stop doing this or you will be in big trouble*

sounds menacing. It sounds less like an "or" than an implication—a statement that doing something has something else as a logical consequence:

> *If you don't stop doing this, then you will be in big trouble.*

Technically, such an implication is equivalent to the inclusive "or" (as we will see in Theorem 9.1). Contrary to what you might expect, the implication doesn't actually say that you *won't* be in big trouble even if you *do* stop doing what you are doing.

We'll give precise meaning to words like "or" and "if," and then look at the logical rules for understanding and manipulating propositions. You might recognize some of these logical rules from the proofs we did in Chapter 2, and we'll return to them again when we look at logic and computers in Chapter 11.

A *proposition* is just a statement that can be either true or false. Not every declarative English sentence is unambiguously either true or false. For example, "February has 28 days" is insufficiently specific: whether it is true or not depends on which year's February is meant. The statement

> *Hens are black*

is ambiguous in another way—perhaps it means "all hens are black" (certainly false), perhaps it means "some hens are black" (surely true), and perhaps you have to be there to know which hens are being discussed. We'll return in Chapter 12 to discuss the truth and falsity of such "quantified" statements (statements that talk about all or some members of a universe of discourse). The distinguishing feature of the *propositional logic* discussed in this and the next chapters is that propositions are unambiguously either true or false.

Finally,

> *Colorless green ideas sleep furiously*

can't be said to be either true or false, because it is meaningless (though it is grammatical, and may be evocative!).[1]

[1] This is Noam Chomsky's example from 1957 of a sentence that is syntactically correct but semantically meaningless. Since our subject is logic, not linguistics, we care a lot about syntax, but our view of semantics is limited to the way the truth or falsity of a compound proposition is determined by the truth or falsity of its components.

✳

We'll use symbols such as p, q, and r as *propositional variables*—that is, variables that represent propositions, in the same way that in ordinary algebra we use variables such as x and y to represent numbers. Propositional variables may also be called *atomic propositions*, as opposed to *compound propositions*, which are composed from other propositions using "and," "or," and the like. For example, let's say

> $p =$ "2000 was a leap year" (a true proposition) and
>
> $q =$ "16 is a prime number" (a false proposition).

We can negate any statement. Symbolically, we write $\neg p$ to mean "not p." This compound proposition is called the *negation* of p, as was anticipated (page 18) before we introduced logical notation. In our example where $p =$ "2000 was a leap year," $\neg p$ is "2000 was not a leap year," which is a false statement since p was true. Sometimes we write $\bar{p}$, putting a bar over the thing to be negated, instead of $\neg p$. So

> "$\neg p$" is the same as "$\overline{\text{2000 was a leap year}}$."

The negation of a statement is only as unambiguous as the original statement. For example, we might negate "hens are black" by saying "hens are not black," but does this mean that no hens are black, or that there is some hen somewhere that isn't black? In general, the negation of a proposition is exactly what we would have to observe in order to conclude that the proposition is false—in other words, for every proposition p, either p is true or $\neg p$ is true, but not both.[2] If r is the statement that *all* hens are black, then $\neg r$ is the statement that there is a hen that is not black.

We can also combine any two propositions to create a more complicated proposition, using relationships like "or" ($\vee$), "and" ($\wedge$), and "if ... then ..." ($\Rightarrow$). We define "or" ($\vee$) to mean "either one or both"—the "*inclusive or.*" When we want to say "one or the other but not both," we instead use *exclusive or*, symbolized by $\oplus$. For example, $p \vee q$ means "2000 was a leap year or 16 is a prime number," and $p \Rightarrow q$ means "if 2000 was a leap year then 16 is a prime number." Rather than saying "if p then q," we may say "p *implies* q." A statement of this form is called an *implication*; p is called the *premise* and q is called the *conclusion*. Finally, we use the symbol $\Leftrightarrow$ to mean "if and only if." The formula $p \Leftrightarrow q$ is shorthand for $(p \Rightarrow q) \wedge (q \Rightarrow p)$; that is, p implies q and q implies p. In sum:

Example	Name	Meaning
$\neg p, \overline{p}$	Negation	Not p
$p \vee q$	Inclusive Or	Either p or q or both
$p \wedge q$	And	Both p and q
$p \oplus q$	Exclusive Or	Either p or q, but not both
$p \Rightarrow q$	Implies	If p, then q
$p \Leftrightarrow q$	Equivalence	p if and only if q

As shown by the example in which p is "2000 is a leap year" and q is "16 is prime," formulas that are combined with logical operators need have no meaningful connection to each other.[3] They simply have to be unambiguously true or false individually. We analyze the truth or falsity of compound propositional formulas only insofar as that truth or falsity is determined by the truth or falsity of the component propositions.

When combining multiple propositions, the syntax needs to reflect precisely the way we mean to combine them. For example, does the statement $p \wedge q \Rightarrow r$ mean $p \wedge (q \Rightarrow r)$ (applying the $\wedge$ operator to p on the one hand and to $q \Rightarrow r$ on the other), or does it mean $(p \wedge q) \Rightarrow r$ (applying the $\Rightarrow$ operator to $p \wedge q$ on the one hand and to r on the other)? The two mean quite different things!

We can always add parentheses to make our meaning clear. But by convention, the operators have a *precedence* ordering, just as operators have in ordinary algebra: $\neg$ binds most tightly; then $\wedge$; then $\vee$, $\oplus$, and $\Rightarrow$ equally; then $\Leftrightarrow$. For example, in ordinary algebra

$$a \times b + c^2 \text{ means } (a \times b) + (c^2),$$

[2] That one and only one of p and $\neg p$ is true is known as the *law of the excluded middle*. It is not accepted in the constructive school of mathematics (see page 15).

[3] This is not a philosophy text, but it is worth noting that an extreme form of logical atomism, based on the notion that reality could be broken down into component facts, was the basis for Ludwig Wittgenstein's *Tractatus Logico-Philosophicus* (1922). Near the beginning of that work, he states "The world divides into facts. Any one can either be the case or not be the case, and everything else remain the same." Wittgenstein later repudiated this way of analyzing the world, but it lives on as a very useful way of representing information for processing by computers. Indeed, the method of truth tables was introduced in the *Tractatus*.

p	$\neg p$
T	F
F	T

Figure 9.1. Truth table for the negation operator.

p	q	$p \wedge q$
T	T	T
T	F	F
F	T	F
F	F	F

Figure 9.2. Truth table for the "and" operator.

p	q	$p \vee q$
T	T	T
T	F	T
F	T	T
F	F	F

Figure 9.3. Truth table for the "or" operator.

p	q	$p \oplus q$
T	T	F
T	F	T
F	T	T
F	F	F

Figure 9.4. Truth table for the "exclusive or" operator.

p	q	$p \Leftrightarrow q$
T	T	T
T	F	F
F	T	F
F	F	T

Figure 9.5. Truth table for the "if and only if" operator.

the result of multiplying a and b, and then adding to that product the square of c. Similarly,

$$p \wedge q \vee \neg r \text{ means } (p \wedge q) \vee (\neg r),$$

the result of "and"ing p and q, and then "or"ing that with the negation of r. The "and" operator $\wedge$ behaves like multiplication, the "or" operator $\vee$ behaves like addition, and $\neg$ behaves like squaring.

On the other hand, even logicians would likely be confused by $p \vee q \Rightarrow r$—better not to count on one interpretation or the other being "standard." If you mean that one or the other of p on the one hand, or $q \Rightarrow r$ on the other hand, is true, then write $p \vee (q \Rightarrow r)$; if you mean that if either of p or q is true, then r must also be true, then write $(p \vee q) \Rightarrow r$.

A *truth table* presents the truth values of a compound propositional formula in terms of the truth values of the components. In the margin, we show the truth tables for a number of simple propositions (Figures 9.1–9.6). Propositional logic is also referred to as *propositional calculus*, because it is a system for calculating the truth or falsity of compound propositions such as $p \oplus q$ from the truth or falsity of the components p and q.

Each row of a truth table corresponds to a single *truth assignment*; that is, to a single association of T and F values to each of the propositional variables. A truth assignment that makes the formula true is said to *satisfy* the formula, and a formula with at least one satisfying truth assignment is said to be *satisfiable*. An *unsatisfiable* formula, such as $p \wedge \neg p$, is one with no satisfying truth assignment—in this case, whichever value is given to p, one or the other of p and $\neg p$ will be false, so their logical "and" will be false. A *tautology* is a formula satisfied by *every* truth assignment—for example, $p \vee \neg p$. If a statement is unsatisfiable, its negation is a tautology, and vice versa, since all truth assignments make a statement false just in case all truth assignments make its negation true.

The truth table for $p \Rightarrow q$ may be puzzling (Figure 9.6). Why is it true that a false statement implies a true statement? Likewise, why is it also true that a false statement implies a false statement?

One way to look at it is to imagine an argument that begins "If p, then …," where we already know that p is false. An "if p then q" argument asks us to imagine that p is true and determine whether the consequence q follows. But if we already know that p is false, the argument is asking us to consider p to be *both true and false simultaneously,* violating the very foundation of the entire system of logic: every proposition is either true or false, but never both. Once a contradiction, such as "both p and $\neg p$ are true," is allowed, the whole logical system collapses: anything and its opposite can be inferred. So a false proposition implies any proposition, true or false.

Similarly, any proposition, true or false, implies a true proposition—since no premise is needed to infer the truth of something we already know is true. Any additional premise is irrelevant, and can't make a true statement false.

A truth table for a more complicated proposition can be built up one step at a time to see when the proposition is true or false. Let's consider the proposition $p \wedge q \Rightarrow r$ (that is, $(p \wedge q) \Rightarrow r$; Figure 9.7). First we create a column for each of our original propositions p, q, and r, and we create enough rows to analyze all possible combinations of values for these variables. In this example, we have three propositions, each of which has two possible values (true or false), so we create a table with 8 rows ($2 \times 2 \times 2$) and fill in each of the possibilities.

By convention, we fill in the columns for the propositional variables according to a regular pattern, so it is easy to see that we have covered all the possibilities: the first variable's column is halfway filled with T's and then halfway with F's; the next alternates from T to F twice as often, and the final one alternates T, F, …, T, F. In the next column, we put the proposition $p \wedge q$ and then fill in its value in each row, based on the values of p and q. Finally, we put the proposition $p \wedge q \Rightarrow r$ in the last column and then fill in its value in each row, based on the values of $p \wedge q$ and r. At each step, we only have to combine two of the previous values, so it is not hard to build up to complicated propositions that might be difficult to analyze all at once.

From Figure 9.7, we can see that $p \wedge q \Rightarrow r$ is true unless p and q are both true while r is false.

✳

As we saw on page 91, the "if and only if" operator $\Leftrightarrow$ is redundant: we can rewrite any statement that involves an "if and only if" by joining two "if…then…" statements with "and." We'll use the notation $\alpha \equiv \beta$ to mean that propositions α and β are *equivalent*; that is, that they have the same truth value for any possible assignment of truth values to the propositional variables. So, for example,

$$p \Leftrightarrow q \equiv (p \Rightarrow q) \wedge (q \Rightarrow p).$$

That is, $\equiv$ is English shorthand for a statement *about* two propositions; $\Leftrightarrow$ is an operator that can be used for constructing propositions. The standard terms for this distinction are that $\Leftrightarrow$ is a symbol in the *object language*, the system of signs and symbols we are studying, whereas $\equiv$ is part of the *metalanguage*—the language, including English, we are using to talk about propositions in the object language.

A simple way of seeing whether two formulas are equivalent is to write out a truth table for each one, and compare the columns headed by each: if the patterns of T and F in those columns are the same, the formulas are equivalent. For example, $\neg\neg p$ is equivalent to p (or in other words, $p \equiv \neg\neg p$): each negation flips the truth value, from true to false and false to true (Figure 9.8). This equivalence is known as the *double negation* law.

p	q	$p \Rightarrow q$
T	T	T
T	F	F
F	T	T
F	F	T

Figure 9.6. Truth table for the "implies" operator $\Rightarrow$. $p \Rightarrow q$ unless p is true and q is false.

p	q	r	$p \wedge q$	$p \wedge q \Rightarrow r$
T	T	T	T	T
T	T	F	T	F
T	F	T	F	T
T	F	F	F	T
F	T	T	F	T
F	T	F	F	T
F	F	T	F	T
F	F	F	F	T

Figure 9.7. Truth table for $p \wedge q \Rightarrow r$. Each possible combination of T and F for the truth values of p, q, and r is represented in one of the rows of the table, and the last two columns show the truth values of the components of the proposition as calculated from the truth values of its constituents. The right hand column represents the proposition itself, which is shown to be false under only one set of truth assignments, when p and q are both true and r is false.

p	$\neg p$	$\neg\neg p$
T	F	T
F	T	F

Figure 9.8. Truth table for the double negation law. The first and third columns are identical, so the propositions at the heads of those columns are equivalent.

It is not hard to see (Problem 9.7) that $\alpha \equiv \beta$ if and only if the proposition $\alpha \Leftrightarrow \beta$ is a tautology. Once we have established an equivalence for simple formulas involving just propositional variables, it follows that the same equivalence holds when more complex formulas are systematically substituted for the propositional variables. For example, from the double negation law it follows that $\alpha \equiv \neg\neg\alpha$ for any formula α. Taking α to be the formula $p \Rightarrow q$, for example, the double negation law says that

$$(p \Rightarrow q) \equiv \neg\neg(p \Rightarrow q).$$

In principle it always works to write out truth tables to determine satisfiability, equivalence, and so on, but it can be quite cumbersome, especially for formulas with many propositional variables. Each additional atomic proposition doubles the number of rows in the table—for a formula containing n propositional variables, we need a table with 2^n rows, which can be impractical even for moderately small values of n.

For example, $2^n = 1024$ when $n = 10$, so writing out a full truth table for a formula with 10 propositional variables is too hard to do by hand. When $n = 20$, 2^n is more than a million; and when $n = 300$, 2^n is more than the number of particles in the universe—so there is no way to write out the full truth table even using every quantum of energy and matter in every star and planet in every galaxy, and all the dark matter in between!

Rather than write out truth tables for every problem, we'll develop some rules for manipulating formulas symbolically so that we can transform an expression into one that is equivalent. Let's go back and take another look at some of our basic operators.

Theorem 9.1. $p \Rightarrow q \equiv \neg p \vee q.$

Proof. For a simple expression like this, we can prove equivalence by writing out a truth table. Comparing the columns for $p \Rightarrow q$ and $\neg p \vee q$, we see that they are the same.

p	q	$p \Rightarrow q$	$\neg p$	$\neg p \vee q$
T	T	T	F	T
T	F	F	F	F
F	T	T	T	T
F	F	T	T	T

■

Theorem 9.2. $p \oplus q \equiv (p \wedge \neg q) \vee (\neg p \wedge q).$

Proof. Again, we can write out a truth table:

p	q	$p \oplus q$	$\neg p$	$\neg q$	$p \wedge \neg q$	$\neg p \wedge q$	$(p \wedge \neg q) \vee (\neg p \wedge q)$
T	T	F	F	F	F	F	F
T	F	T	F	T	T	F	T
F	T	T	T	F	F	T	T
F	F	F	T	T	F	F	F

■

This truth table is a formal proof that these formulas are true under the same circumstances, but here it may be easier to grasp the intuition when stated informally: $\oplus$ means exactly one of the two propositions is true. Thus we need either p to be true and q to be false, or p to be false and q to be true.

Thus both $\Rightarrow$ and $\oplus$ are redundant; we can rewrite any formula that uses them using $\vee$, $\wedge$, and $\neg$ only. In the same way, $\Leftrightarrow$ is redundant, and it is conventional to think of our formulas as being built up from just the propositional variables and the three operators $\vee$, $\wedge$, and $\neg$, plus parentheses. Problems 9.1, 9.2, and 11.5 explore some other ways of writing formulas of propositional logic using fewer operators.

✳

So a completely formal description of what counts as a propositional formula follows from this inductive definition:

(PF1) Any propositional variable is a formula.

(PF2) If α and β are formulas, then so are

    (a) $(\alpha \vee \beta)$

    (b) $(\alpha \wedge \beta)$

    (c) $\neg \alpha$

For example, $\neg(p \vee q)$ is a formula, since p and q are formulas per (PF1), $(p \vee q)$ is a formula per (PF2a), and then $\neg(p \vee q)$ is a formula per (PF2c).

Formulas that strictly follow these rules may have more parentheses than are needed to be unambiguous. We have already indicated one way to use fewer parentheses—by exploiting the precedence of the operators. For example, $p \wedge q \vee r$ would, strictly speaking, be written as $((p \wedge q) \vee r)$, but the parentheses are understood without writing them since $\wedge$ has higher precedence than $\vee$. Another way of dropping parentheses is to use the *associative laws*, like those for set union and intersection (page 52) and string concatenation (page 82). For any formulas α, β, and γ,

(AL1) $(\alpha \vee \beta) \vee \gamma \equiv \alpha \vee (\beta \vee \gamma)$

(AL2) $(\alpha \wedge \beta) \wedge \gamma \equiv \alpha \wedge (\beta \wedge \gamma)$

We leave the proofs as exercises below (Problem 9.3). The associative laws make it possible to drop parentheses entirely in formulas such as $p \vee q \vee r$; it doesn't matter which way the subformulas are grouped, since both ways have the same truth tables.

The operators $\vee$ and $\wedge$ also obey *commutative laws* like those for set union and intersection: subformulas can be and-ed together in either order without changing the truth value of the formula, and the same holds for or-ing subformulas together. That is to say:

(CL1) $\alpha \vee \beta \equiv \beta \vee \alpha$

(CL2) $\alpha \wedge \beta \equiv \beta \wedge \alpha$

The proofs are trivial, but these rules are very useful when combined with the practice of dropping unnecessary parentheses. They imply that a formula with multiple occurrences of the same operator, such as $p \vee q \vee r \vee s$, can be reordered at will, for example, as $s \vee r \vee q \vee p$, without changing the meaning of the formula.

Finally, propositional logic has *distributive laws* that parallel the distributive laws for union and intersection of sets. The logical connectives $\vee$ and $\wedge$ each distribute over each other symmetrically. That is,

(DL1) $\alpha \vee (\beta \wedge \gamma) \equiv (\alpha \vee \beta) \wedge (\alpha \vee \gamma)$

(DL2) $\alpha \wedge (\beta \vee \gamma) \equiv (\alpha \wedge \beta) \vee (\alpha \wedge \gamma)$

As an ordinary-language example of (DL2), consider the question

> *Are you at least 21 years old, and either a US citizen or a US permanent resident?*

This question takes the same form as the left side of (DL2), and is equivalent to

> *Are you at least 21 years old and a US citizen, or at least 21 years old and a US permanent resident?*

which is the rewriting of the previous statement according to the rule (DL2). We used reasoning of exactly this kind when we proved Theorem 5.1 (page 53).

❋

Any formula α of propositional logic defines a *truth function* ϕ_α; that is, a function that takes bit arguments and produces a bit value. If α includes k propositional variables, say p_1, ..., p_k, then $\phi_\alpha : \{0,1\}^k \to \{0,1\}$ is a k-ary function; its value $\phi_\alpha(b_1, \ldots, b_k)$ is the truth value of α when p_i assumes the truth value b_i for each i. (By convention, we identify 0 with F and 1 with T.) For example, if α is $p_1 \Rightarrow p_2$, then $\phi_\alpha(0,1) = 1$ (third line of Figure 9.6, page 93). A truth function is commonly called a *Boolean function*.[4] *Boolean logic* is another name for propositional logic.

[4]Named for the British mathematician George Boole (1815–64), author of *The Laws of Thought*, which formalized these ideas.

The subject of Chapter 11 is how to implement truth functions in hardware, using circuit modules that compute basic truth functions such as those for $\land$, $\lor$, and $\lnot$.

Chapter Summary

- Propositional logic is used to give precise meaning to statements, which in ordinary language may be ambiguous.

- A *proposition* is a statement that is either true or false.

- A *propositional variable*, or *atomic proposition*, is a variable that represents a proposition, in the same way that a variable in ordinary algebra represents a number.

- For every proposition p, either p or its negation (written as $\lnot p$ or $\bar{p}$) is true, but not both.

- Two propositions can be combined to make a more complicated proposition, using the operators "inclusive or" ($\lor$), "and" ($\land$), "exclusive or" ($\oplus$), "implies" ($\Rightarrow$), and "equivalence" ($\Leftrightarrow$).

- By convention, the operators in a compound proposition have a precedence ordering: $\lnot$; then $\land$; then $\lor$, $\oplus$, and $\Rightarrow$; then $\Leftrightarrow$. Parentheses should be used if the order is ambiguous.

- A *truth table* shows the truth values of a compound propositional formula in terms of the truth values of its propositional variables.

- A *truth assignment* is an assignment of the values true and false to each of the propositional variables of a formula.

- A truth assignment that makes a formula true is said to *satisfy* the formula.

- A formula is *satisfiable* if at least one truth assignment satisfies it, and a *tautology* if every truth assignment satisfies it. It is *unsatisfiable* if no truth assignment satisfies it.

- The symbol $\equiv$ indicates that two propositions are equivalent; it is part of the *metalanguage*, not the *object language*, and so cannot be used within a propositional formula.

- Any propositional formula can be expressed using just the operators $\lnot$, $\lor$, and $\land$, along with its propositional variables.

- The operators $\lor$ and $\land$ obey associative, commutative, and distributive laws:

$$(\alpha \lor \beta) \lor \gamma \equiv \alpha \lor (\beta \lor \gamma)$$
$$(\alpha \land \beta) \land \gamma \equiv \alpha \land (\beta \land \gamma)$$
$$\alpha \lor \beta \equiv \beta \lor \alpha$$
$$\alpha \land \beta \equiv \beta \land \alpha$$

$$\alpha \vee (\beta \wedge \gamma) \equiv (\alpha \vee \beta) \wedge (\alpha \vee \gamma)$$
$$\alpha \wedge (\beta \vee \gamma) \equiv (\alpha \wedge \beta) \vee (\alpha \wedge \gamma).$$

Problems

9.1. The operators $\neg$ and $\vee$ are sufficient to define not just $\Rightarrow$, as we saw earlier in the chapter, but the rest of our operators as well. Using just $\neg$ and $\vee$ (and parentheses), write formulas involving p and q that are logically equivalent to

   (a) $p \wedge q$,

   (b) $p \oplus q$,

   (c) $p \Leftrightarrow q$.

9.2. Assuming that the operators $\neg$ and $\vee$ are sufficient to define the rest of the operators $\Rightarrow$, $\wedge$, $\oplus$, and $\Leftrightarrow$ (as shown in Problem 9.1), prove that the operators $\neg$ and $\wedge$ are also sufficient to define all of those same operators.

9.3. Prove the associative laws by comparing truth tables for the two expressions asserted in (AL1) and (AL2) to be equivalent.

9.4. (a) Give an ordinary-language example of (DL1).

    (b) Prove the distributive laws (DL1) and (DL2).

9.5. Using a truth table, determine whether each of the following compound propositions is satisfiable, a tautology, or unsatisfiable.

   (a) $p \Rightarrow (p \vee q)$

   (b) $\neg(p \Rightarrow (p \vee q))$

   (c) $p \Rightarrow (p \Rightarrow q)$

9.6. (a) Using the propositions $p =$ "I study," $q =$ "I will pass the course," and $r =$ "The professor accepts bribes," translate the following into statements of propositional logic:

      1. If I do not study, then I will only pass the course if the professor accepts bribes.

      2. If the professor accepts bribes, then I do not study.

      3. The professor does not accept bribes, but I study and will pass the course.

      4. If I study, the professor will accept bribes and I will pass the course.

      5. I will not pass the course but the professor accepts bribes.

   (b) Using the propositions $p =$ "The night hunting is successful," $q =$ "The moon is full," and $r =$ "The sky is cloudless," translate the following into statements of propositional logic:

      1. For successful night hunting, it is necessary that the moon is full and the sky is cloudless.

    2. The sky being cloudy is both necessary and sufficient for the night hunting to be successful.

    3. If the sky is cloudy, then the night hunting will not be successful unless the moon is full.

    4. The night hunting is successful, and that can only happen when the sky is cloudless.

9.7. Prove that $\alpha \equiv \beta$ if and only if the proposition $\alpha \Leftrightarrow \beta$ is a tautology.

9.8. Write a propositional formula that is true if and only if exactly one of p, q, and r is true.

9.9. Give real-world interpretations of p, q, and r such that $(p \wedge q) \Rightarrow r$ and $p \wedge (q \Rightarrow r)$ mean quite different things, and one is true while the other is false.

9.10. (a) Using the language of truth functions, explain $\alpha \equiv \beta$.
      (b) What is the relation of ϕ_α and ϕ_β if $\alpha \equiv \neg\beta$?

9.11. (a) Show how to write $p \Leftrightarrow q$ using $\oplus$ and the constant T.
      (b) Show that $\oplus$ and $\Leftrightarrow$ are associative.
      (c) We have shown (page 95) that any formula of propositional logic can be converted to an equivalent formula that uses only the operators $\{\wedge, \vee, \neg\}$. Another way to say the same thing is that these three operators suffice to express every possible truth function. Such a set of connectives is said to be *complete*. Prove that

$$\{\oplus, \Leftrightarrow, \neg, \mathrm{T}, \mathrm{F}\}$$

      is *not* a complete set by showing that certain truth functions cannot be expressed using just these operators.

Chapter 10

Normal Forms

Very different expressions can be equivalent, as we saw in Chapter 9. To take a simple example, $p \vee q$ and $q \vee p$ are different but equivalent formulas: for any of the four possible assignments of truth values to the two propositional variables p and q, the two formulas have the same truth value. As a more complicated example, consider these two formulas:

$$(p \vee q) \wedge (\neg p \vee q) \wedge (p \vee \neg q) \wedge (\neg p \vee \neg q) \qquad (10.1)$$

$$p \wedge \neg p. \qquad (10.2)$$

These formulas are equivalent since both are false under all possible assignments to the propositional variables. (Problem 10.2 asks you to prove this for (10.1).) It is sometimes convenient to use T and F as the propositions that are identically true and identically false, without any variables. Then we can say that (10.1) and (10.2) are both equivalent to F.

For a variety of reasons (including, for example, computer processing), it is useful to have standard ways of representing formulas. We'll look at two such *normal forms*, called *conjunctive normal form* (*CNF*) and *disjunctive normal form* (*DNF*).[1] In these forms,

- only the operators "and" ($\wedge$), "or" ($\vee$), and "not" ($\neg$) are used;
- any negations are attached to propositional variables, not to larger expressions; and
- the "or"s and "and"s are organized in regular patterns, as explained below.

Let's say that a *literal* is either a propositional variable (such as p) or the negation of a single variable (such as $\neg q$). Then a formula in CNF is an "and" of "or"s of literals. Equations (10.1) and (10.2) are in CNF, and so is

$$(p \vee \neg q \vee r) \wedge (\neg s \vee t).$$

A formula in DNF, on the other hand, is an "or" of "and"s of literals, such as

$$(p \wedge \neg q \wedge r) \vee (\neg s \wedge t).$$

[1] "Normal form" means that the formula is formatted according to a certain rule or "norm." A formula that is not in normal form is not "abnormal" in any pathological sense.

Sometimes these are referred to as *and-of-ors* form and *or-of-ands* form, respectively. Let's say that a formula such as $\alpha_1 \vee \alpha_2 \vee \ldots \vee \alpha_n$ is a *disjunction* of the α_i, which are called the *disjuncts*, and that $\alpha_1 \wedge \alpha_2 \wedge \ldots \wedge \alpha_n$ is a *conjunction* of the α_i, which are called the *conjuncts*. Then a CNF formula is a conjunction of disjunctions of literals, and a DNF formula is a disjunction of conjunctions of literals.

A formula such as $p \vee q$ is in both CNF and DNF, since we can regard it as either a conjunction of a single conjunct, which is a disjunction of two literals, or as a disjunction of two disjuncts, each of which is a conjunction of a single literal. For a similar reason, (10.2) is in both CNF and DNF.

In a CNF formula, the disjunctions of literals that are "and"-ed together are called the *clauses* of the formula. Similarly, in a DNF formula, the conjunctions of literals that are "or"-ed together are the clauses of the formula. For example, the CNF formula

$$(p \vee \neg q \vee r) \wedge (\neg q \vee t) \tag{10.3}$$

is composed of two clauses, $(p \vee \neg q \vee r)$ and $(\neg q \vee t)$.

Using this vocabulary, we can state a simple condition for the truth of a CNF formula. A CNF formula is true under a given truth assignment if and only if at least one literal from each clause is true, since those clauses are joined by "and." In (10.3), making q false makes the second literal in the first clause true and the first literal in the second clause true, and so makes the whole formula true, regardless of the truth values of p, r, and t.

Similarly, a formula in DNF is true if and only if there is at least one disjunct in which all the literals are true. For example, the DNF formula

$$(\neg p \wedge q) \vee (p \wedge r)$$

is true when p is false and q is true, regardless of the truth value of r, because these truth values make the first disjunct $(\neg p \wedge q)$ true.

In Chapter 9 we learned how to construct the truth table for a formula. It is also possible to construct a formula from a truth table—in fact, starting from a truth table we can easily construct a DNF formula with that truth table.

For each row of the truth table with a T in the final column, we create a conjunction of literals, negating any variable for which the value is stipulated as false. The disjunction ("or") of these conjunctions of literals has exactly the desired truth table. For example, let's take another look at the truth table for $p \Rightarrow q$ (shown again to the left).

p	q	$p \Rightarrow q$
T	T	T
T	F	F
F	T	T
F	F	T

Only the first, third, and fourth rows have T in the right hand column, so for the formula to be true, the truth values of the variables must be those represented in one of those rows. So $p \Rightarrow q$ is equivalent to the DNF formula

$$(p \wedge q) \vee (\neg p \wedge q) \vee (\neg p \wedge \neg q). \tag{10.4}$$

Constructing a DNF formula from a truth table is cumbersome—a truth table with n variables has 2^n rows, and the DNF formula constructed from it will have as many disjuncts as there are rows with T in the rightmost column. And the result may not be as simple as it could be—in this case we know that the formula $\neg p \vee q$ is equivalent to $p \Rightarrow q$ and hence to (10.4), and $\neg p \vee q$ is also in DNF.

❋

So starting with any formula, it is always possible to construct an equivalent formula in DNF: first write out the truth table, and then turn the truth table into a DNF formula with one disjunct for each row that has T in the right column of the truth table. But there is a more direct method based on structural induction, and it works equally well for turning formulas into CNF.

To begin, we'll assume that the formula has already been rewritten so that it does not use any connectives other than $\vee$, $\wedge$, and $\neg$ (see page 95), and there are never two $\neg$ signs in a row (since they can both be dropped by the double negation law). The table at right summarizes the replacement rules to rewrite a formula in this way.

Formula	Replacement
$\alpha \Rightarrow \beta$	$(\neg\alpha \vee \beta)$
$\alpha \oplus \beta$	$(\alpha \vee \beta) \wedge (\neg\alpha \vee \neg\beta)$
$\alpha \Leftrightarrow \beta$	$(\neg\alpha \vee \beta) \wedge (\alpha \vee \neg\beta)$
$\neg\neg\alpha$	α

Now given a formula containing only the $\vee$, $\wedge$, and $\neg$ operators, we convert it to CNF by structural induction on the construction of the formula. Along the way we will be using several laws already stated: the distributive, associative, and commutative laws (page 96). In particular, the commutative and associative laws make it possible to treat conjunctions and disjunctions of subformulas as sets of those subformulas: if several subformulas are and-ed or or-ed together, they can be rearranged in any order and duplicates can be dropped to produce an equivalent formula.

We need just one more logical law in our toolkit. *De Morgan's laws*[2] make it possible to drive in negation signs so they are attached only to propositional variables.

[2]Named for the British mathematician Augustus De Morgan (1806–71). De Morgan and his contemporary George Boole laid the groundwork for the modern field of logic.

Theorem 10.5. De Morgan's Laws.

$$\neg(p \wedge q) \equiv \neg p \vee \neg q \tag{10.6}$$

$$\neg(p \vee q) \equiv \neg p \wedge \neg q \tag{10.7}$$

Proof. We write out a truth table for each:

p	q	$p \wedge q$	$\neg(p \wedge q)$	$\neg p$	$\neg q$	$\neg p \vee \neg q$
T	T	T	F	F	F	F
T	F	F	T	F	T	T
F	T	F	T	T	F	T
F	F	F	T	T	T	T

p	q	$p \vee q$	$\neg(p \vee q)$	$\neg p$	$\neg q$	$\neg p \wedge \neg q$
T	T	T	F	F	F	F
T	F	T	F	F	T	F
F	T	T	F	T	F	F
F	F	F	T	T	T	T

∎

Of course, De Morgan's laws only formalize common sense. If it is not the case that both p and q are true, then either p is false or q is false. And if it is not the case that one or the other of p and q is true, then both p and q must be false. With these tools at hand, we can show how to put formulas in conjunctive or disjunctive normal form.

Theorem 10.8. *Every formula is equivalent to one in conjunctive normal form, and also to one in disjunctive normal form.*

Proof. We show the construction just for CNF; the procedure for DNF is symmetrical, with the roles of "$\vee$" and "$\wedge$" interchanged.

The construction has two stages. In the first stage, we drive in negation signs using De Morgan's laws so that they are bound only to propositional variables. In the second, we transform a formula into CNF on the assumption that all negation signs in the formula are attached to propositional variables.

Stage 1. We transform an arbitrary formula α into $T(\alpha)$, an equivalent formula in which all negation signs are attached to propositional variables. We proceed by structural induction on the formation of α, dividing the cases in which α is the negation of another formula into several subcases.

Base case. If α is p or $\neg p$, where p is a propositional variable, then $T(\alpha) = \alpha$.

Induction step, case 1. If $\alpha = \beta \vee \gamma$ or $\alpha = \beta \wedge \gamma$, then $T(\alpha) = T(\beta) \vee T(\gamma)$ or $T(\alpha) = T(\beta) \wedge T(\gamma)$, respectively.

Induction step, case 2. If $\alpha = \neg \beta$ for some formula β, then there are several subcases, depending on the structure of β.

　　Subcase a. If $\alpha = \neg\neg\gamma$, then $T(\alpha) = T(\gamma)$.

　　Subcase b. If $\alpha = \neg(\gamma \vee \delta)$, then $T(\alpha) = T(\neg\gamma) \wedge T(\neg\delta)$.

Subcase c. If $\alpha = \neg(\gamma \wedge \delta)$, then $T(\alpha) = T(\neg\gamma) \vee T(\neg\delta)$.

It follows by induction that $T(\alpha)$ is equivalent to α. The induction starts with the base cases in which α is a propositional variable or the negation of a propositional variable, and proceeds using De Morgan's laws in the induction step. In each case of the induction step, $T(\alpha)$ is defined in terms of the value of T on formulas with fewer connectives than α, so no infinite regress is possible. And finally, since the formula that is the value of $T(\alpha)$ includes a negation sign only in the base case, all negation signs in $T(\alpha)$, for any α, are attached to propositional variables.

Stage 2. Now we can assume that all negation signs in α are attached to propositional variables. CNF(α) will be the conjunctive normal form of α. We construct CNF(α) from the CNFs of the subformulas of α.

Base case. If α is a literal or a disjunction of literals, then CNF(α) = α.

Induction step, case 1. If $\alpha = \beta \wedge \gamma$, then CNF($\alpha$) = CNF($\beta$) $\wedge$ CNF(γ).

Induction step, case 2. Otherwise α is a disjunction $\beta \vee \gamma$, where at least one of β and γ is a conjunction (otherwise the base case would apply). Because of the commutative law for conjunctions, we can assume without loss of generality that γ is a conjunction, say $\gamma = \gamma_1 \wedge \gamma_2$. Then

$$\text{CNF}(\alpha) = \text{CNF}(\beta \vee \gamma_1) \wedge \text{CNF}(\beta \vee \gamma_2),$$

which is equivalent to α by induction and by the distributive law.

Any of the formulas β, γ_1, and γ_2 might themselves be conjunctions, but case 2 of the induction step defines CNF(α) in terms of the CNFs of formulas with fewer connectives, so again there can be no infinite regress.   ∎

Example 10.9. *Put*

$$p \vee (q \wedge (r \vee (s \wedge t)))$$

into conjunctive normal form.

Solution to example. The distributive law is applied twice. In each case, we show the occurrence of $\vee$ to which the distributive law is applied by circling it.

$p \varovee (q \wedge (r \vee (s \wedge t)))$

$\equiv (p \varovee q) \wedge (p \varovee (r \vee (s \wedge t)))$

$\equiv (p \vee q) \wedge (p \vee r \varovee (s \wedge t))$ ⠀⠀⠀⠀(dropping redundant parentheses)

$\equiv (p \vee q) \wedge (p \vee r \varovee s) \wedge (p \vee r \varovee t)$. ⠀⠀⠀⠀⠀⠀⠀⠀⠀∎

Example 10.10. *Write $(\neg p \wedge q) \Rightarrow \neg p$ in disjunctive normal form.*

Solution to example. Let's begin by getting rid of the $\Rightarrow$, following Theorem 9.1:

$$\neg(\neg p \wedge q) \vee \neg p.$$

Now we can apply De Morgan's laws (Theorem 10.5) to the first part of the expression, so we get

$$(\neg\neg p \vee \neg q) \vee \neg p.$$

We are left with just a number of literals joined by "or"s, so we don't need the parentheses anymore; and we can cancel out the double negation to get

$$p \vee \neg q \vee \neg p.$$

This is now a valid expression in DNF, since we have three disjuncts, each composed of just one literal, all joined by disjunction. Notice, however, that we can simplify this formula still further. We need to satisfy only one of these three literals, and p is allowed to take on either of the values true or false, so no matter what the value of p this expression is true—and we actually don't even need to know what q is. So we can get rid of $\neg q$ and end with the expression

$$p \vee \neg p,$$

and we can rewrite that simply as T if we are allowed to use T and F as propositions with no variables. ∎

As we mentioned on page 92, the expression $p \vee \neg p$ is a tautology—it holds true regardless of whether p is true or false. Any expression (such as Example 10.10) that can be simplified to become a tautology must be a tautology itself. Another way to check that a formula is a tautology is to write out a truth table for the expression and show that the final column is filled with T's—though as mentioned earlier, this method may be impractical if the formula has a large number of propositional variables.

Sometimes intuition suggests that an expression is tautological. In Example 10.10, it is evident that $\neg p \wedge q$ implies $\neg p$: if we know $\neg p$ is true plus the extraneous information that q is also true, then certainly we know that $\neg p$ is true.

Because so many phenomena can be described as logical formulas, determining whether a formula is a tautology is a fundamental problem of computer science. Recall that the negation of a tautology is unsatisfiable—so from a practical standpoint, any tautology-testing algorithm doubles as an unsatisfiability-testing algorithm and vice versa, simply by putting a negation sign in front of the formula. It turns out that this problem of determining what formulas are tautologies or unsatisfiable is trickier than it looks. In theory we can always find the answer, since we can just write out a truth table and check the final column. But truth tables are big and impractical: the truth table for a formula with 30 variables would have more than a billion rows! Is there a faster, smarter way to find out?

The answer to this question is unknown, in spite of decades of effort on the part of the computer science community. Satisfiability—or *SAT*, as it is known for short—belongs to a class of problems known as the *NP-complete* problems. NP-complete problems are known to be solvable algorithmically, but we don't know if it is possible to solve them without an exhaustive search that is exponentially costly. Many problems that on the surface look nothing like satisfiability are also NP-complete. A famous example is the problem of finding the shortest tour of a set of cities, given the distances separating each pair—commonly known as the *Traveling Salesperson Problem* or *TSP* (originally the *Traveling Salesman Problem*).

We do know, however, that if one of these NP-complete problems can be solved in a faster way, so can the rest of them. (That is the meaning of *complete* in this context—each problem has embedded within it the full complexity of all the others.) So if we found a way of answering the question of satisfiability in polynomial time—say, if the number of operations required were on the scale of n^3 or even n^{100} rather than 2^n—we would have proven that all NP-complete problems can be solved in polynomial time. The question of whether this is possible, often framed as "Is P equal to NP?," is one of the great unsolved problems of computer science—and indeed of mathematics.

Chapter Summary

- Different formulas are equivalent if they have the same truth value for every truth assignment.

- A *literal* is a propositional variable or its negation.

- A *disjunction* of several formulas is the formula that results from connecting them with ∨s.

- A *conjunction* of several formulas is the formula that results from connecting them with ∧s.

- Two standard ways of writing formulas are *conjunctive normal form* (CNF), also called *and-of-ors* form, or a conjunction of disjunctions; and *disjunctive normal form* (DNF), also called *or-of-ands* form, or a disjunction of conjunctions.

- A disjunction in a CNF formula, or a conjunction in a DNF formula, is called a *clause*.

- A CNF formula is true just in case at least one literal from each clause is true. A DNF formula is true just in case there is at least one clause in which each literal is true.

- *De Morgan's laws* are rules for driving negations from the outside of a formula inward to its subformulas:

$$\neg(p \wedge q) \equiv \neg p \vee \neg q$$

$$\neg(p \vee q) \equiv \neg p \wedge \neg q.$$

- Any formula can be written in DNF or CNF by converting all connectives to $\vee$, $\wedge$, and $\neg$; and applying the double negation law, De Morgan's laws, and the distributive laws.

- Determining whether a formula is a tautology is a fundamental problem of computer science. This problem belongs to a class of problems known as *NP-complete* problems, and it is not known whether efficient solutions to these problems exist.

Problems

10.1. Five of the following statements are equivalent to the negations of the other five. Pair each statement with its negation.

(a) $p \oplus q$

(b) $\neg p \wedge q$

(c) $p \Rightarrow (q \Rightarrow p)$

(d) $p \Rightarrow q$

(e) $p \wedge \neg q$

(f) $q \wedge (p \wedge \neg p)$

(g) $p \vee \neg q$

(h) $p \Leftrightarrow q$

(i) $p \wedge (q \vee \neg q)$

(j) $(p \Rightarrow q) \Rightarrow p$

10.2. Show that formula (10.1) is unsatisfiable.

10.3. For each of the following formulas, decide whether it is a tautology, satisfiable, or unsatisfiable. Justify your answers.

(a) $(p \vee q) \vee (q \Rightarrow p)$

(b) $(p \Rightarrow q) \Rightarrow p$

(c) $p \Rightarrow (q \Rightarrow p)$

(d) $(\neg p \wedge q) \wedge (q \Rightarrow p)$

(e) $(p \Rightarrow q) \Rightarrow (\neg p \Rightarrow \neg q)$

(f) $(\neg p \Rightarrow \neg q) \Leftrightarrow (q \Rightarrow p)$

10.4. Write the following formula:

$$(p \Rightarrow q) \wedge (\neg(q \vee \neg r) \vee (p \wedge \neg s))$$

(a) in conjunctive normal form.

(b) in disjunctive normal form.

10.5. (a) Show that for any formulas α, β, and γ,

$$(\alpha \wedge \beta) \vee \alpha \vee \gamma \equiv \alpha \vee \gamma.$$

(b) Give the corresponding rule for simplifying
$$(\alpha \vee \beta) \wedge \alpha \wedge \gamma.$$

(c) Find the simplest possible disjunctive and conjunctive normal forms of the formula $(p \wedge q) \Rightarrow (p \oplus q)$.

10.6. In each case, show that the formula is a tautology, or explain why it is not.

(a) $((p \wedge q) \Leftrightarrow p) \Rightarrow q$

(b) $(p \vee (p \wedge q)) \Rightarrow (p \wedge (p \vee q))$

(c) $(\neg p \Rightarrow \neg q) \Rightarrow (q \Rightarrow p)$

10.7. In this problem you will show that putting a formula into conjunctive normal form may increase its length exponentially. Consider the formula
$$p_1 \wedge q_1 \vee \ldots \vee p_n \wedge q_n,$$
where $n \geq 1$ and the p_i and q_i are propositional variables. This formula has length $4n - 1$, if we count each occurrence of a propositional variable or an operator as adding 1 to the length and ignore the implicit parentheses.

(a) Write out a conjunctive normal form of this formula in the case $n = 3$.

(b) How long is your conjunctive normal form of this formula, using the same conventions as above? For general n, how long is the conjunctive normal form as a function of n?

(c) Show that similarly, putting a formula into disjunctive normal form may increase its length exponentially.

(d) Consider the following algorithm for determining whether a formula is satisfiable: Put the formula into disjunctive normal form by the method of this chapter, and then check to see if all of the disjuncts are contradictions (containing both a variable and its complement). If not, the formula is satisfiable. Why is this algorithm exponentially costly?

10.8. Write in both conjunctive and disjunctive normal forms: "At least two of p, q, and r are true."

10.9. This problem introduces *resolution theorem-proving*.

(a) Suppose that $(e_1 \vee \ldots \vee e_m)$ and $(f_1 \vee \ldots \vee f_n)$ are clauses of a formula α, which is in conjunctive normal form, where $m, n \geq 1$ and each e_i and f_j is a literal. Assume that all the e_i are distinct from each other, and all the f_j are distinct from each other, so the clauses are essentially sets of literals. Suppose that e_m is a propositional variable p and $f_1 = \neg p$. Show that α is equivalent to the result of adding a new conjunct to α consisting of all the literals in these two clauses except for p from the first and $\neg p$ from the second. That is,
$$\alpha \equiv \alpha \wedge (e_1 \vee \ldots \vee e_{m-1} \vee f_2 \vee \ldots \vee f_n),$$
or to be precise, the result of dropping any duplicate literals from this clause. The new clause is derived from the other two by *resolution*, and is said to be their *resolvent*. To make sense of this expression in the case $m = n = 1$, we must construe the *empty clause* containing no literals as identically false—which makes sense since it is derived from the two clauses $(p) \wedge (\neg p)$. The empty clause, in other words, is another name for the identically false proposition F.

It follows that a formula is unsatisfiable if the empty clause results from the process of forming resolvents, adding them to the formula, and repeating.

(b) Show that the converse of part (a) is true; that is, that if the process of forming resolvents and adding them to the formula ends without producing the empty clause, then the original formula (and all the equivalent formulas derived from it by resolution) has a satisfying truth assignment. *Hint:* Let C be the formula formed by adding resolvents until no new ones are produced (why does this process terminate?). Suppose that C is unsatisfiable, but the empty clause is not among the clauses derived. If the original formula includes k propositional variables, $p_1, \ldots, p_k$, then let C_i be the formula consisting of just those clauses of C in which the propositional variables are among $p_i, \ldots, p_k$. (So $C_1 = C$.) Prove by induction on i that for each $i = 1, \ldots, k+1$, no truth assignment to the variables $p_i, \ldots, p_k$ makes C_i true. The only clause that could be false in C_{k+1} is the empty clause, a contradiction.

Chapter 11

Logic and Computers

Computers are machines for logical calculations. We may think they are calculating on numbers, but those numbers are represented as bits. Arithmetic operations are logical operations underneath.

In fact, computers are *built* out of logic. *Everything* in a computer is represented as patterns of 0s and 1s, which can be regarded as just other names for the two truth values of propositional logic, "false" and "true." Formulas of propositional logic derive their truth values from the truth values of their propositional variables. In the same way, computers manipulate the bits 0 and 1 (page 32) to produce more bits. Computers manufacture bits, using bits as the raw material. For this reason, the design of the physical stuff that produces bits from bits is called a computer's "logic."

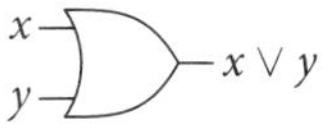

The conceptually simplest pieces of computer logic are *gates*. A gate produces a single bit of output from one or two bits of input (or sometimes more). Gates correspond to the operators of propositional logic, such as $\lor$, $\land$, $\oplus$, and $\lnot$. Pictures like those in Figures 11.1–11.4 should be interpreted as having direction: 1s and 0s enter on the "wires" on the left, and a 1 or a 0 comes out on the right.

These gates can be chained together in various ways to produce more complex results using only the simplest of components; such a computing device is called a *circuit*. But in fact, not all the illustrated gates are typically available to a circuit designer. For example, as shown on page 94, we can compute the "exclusive or" using just the "and," "or," and "not" gates, since

$$x \oplus y \equiv (x \land \lnot y) \lor (\lnot x \land y).$$

So a logic circuit to compute $x \oplus y$ can be read directly from that formula, as shown in Figure 11.5.

It is customary to show only one source for each input, and to allow branching "wires" when the same input is used in more than one place. When it is necessary to draw wires crossing even though they are not

Figure 11.1. "Or" gate.

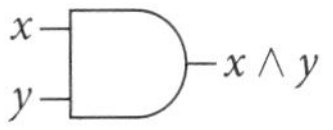

Figure 11.2. "And" gate.

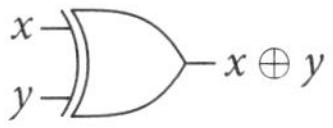

Figure 11.3. "Exclusive or" gate.

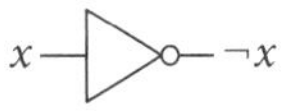

Figure 11.4. "Inverter," or "not" gate. The output is commonly written $\bar{x}$ instead of $\lnot x$.

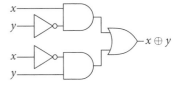

Figure 11.5. Constructing a circuit to compute "exclusive or" using only "and," "or," and "not" gates.

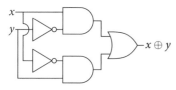

Figure 11.6. The same circuit, shown with one wire crossing over another.

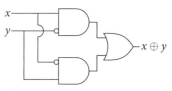

Figure 11.7. This is the same circuit, with negated inputs shown as "bubbles" (small circles) at inputs to the "and" gates.

Figure 11.8. "Nand" gate. Following the convention illustrated in Figure 11.7, the logical negation of the output of a gate is shown by adding a "bubble" at the output.

connected, the crossing is shown as a semicircle as in Figure 11.6, as though one wire were hopping over the other.

Also, negated inputs are often shown as a "bubble" at the gate, rather than as a separate negation gate. So the circuit of Figure 11.7 is functionally the same as those of Figures 11.5 and 11.6.

The "and," "or," "not," and "exclusive or" gates illustrated here correspond directly to the various logical operators introduced in Chapter 9. One very useful gate corresponds to the "not-and," also known as *nand* operator:

$$p \mid q \equiv \neg(p \land q).$$

This may seem to be a peculiar operator, but it has a lovely property for building logic circuits: any of the conventional operators can be expressed using it! (The "nor" operator of Problem 11.5 has the same property.) For example:

$$\neg p \equiv p \mid p$$

$$p \land q \equiv \neg(p \mid q) \equiv (p \mid q) \mid (p \mid q).$$

The | operator is known as the *Sheffer stroke*, and is sometimes written as ↑. We leave it as an exercise (Problem 11.6) to express $p \lor q$ using just the | operator. The corresponding logic gate is shown in Figure 11.8. If we think in terms of mass production, nand gates are ideal—if you can produce them cheaply, you don't need to produce anything else. Whatever you are trying to accomplish, you "just" have to figure out how to connect them together.

※

Binary notation is the name for this way that numbers are represented for processing by computer, using just 0s and 1s. Binary is the base-2 number system. The decimal or base-10 system we customarily use has the ten digits 0, 1, 2, 3, 4, 5, 6, 7, 8, and 9; the binary system is similar, but has only the two binary digits 0 and 1.

The number zero is represented in binary as 0 or a string of 0s, for example 00000000. Counting consists of repeatedly adding 1, subject to the basic rules of binary arithmetic:

$$\begin{aligned}
0+0 &= 0 \\
0+1 &= 1 \\
1+0 &= 1 \\
1+1 &= 0, \text{ carry } 1.
\end{aligned} \qquad (11.1)$$

That is, adding 1 to 1 in binary has the same effect as adding 1 to 9 in decimal; the result is 0, but a 1 is carried into the next position to the left. For example, adding 1 to 10111 produces the result 11000, with three carries.

$$
\begin{array}{ccccc}
 & 1 & 1 & 1 & \\
1 & 0 & 1 & 1 & 1 \\
+ & & & & 1 \\
\hline
1 & 1 & 0 & 0 & 0
\end{array}
$$

In exactly the same way, we can add two arbitrary numbers together by manipulating the bits of their binary representations. The only rule we need in addition to (11.1) is how to add $1 + 1$ plus a carry-in of 1. The answer is that the sum is 1 and there is a carry of 1 into the column to the left. So let's work one more example.

$$
\begin{array}{cccccc}
 & 1 & 1 & 1 & 1 & \\
 & 1 & 0 & 1 & 0 & 1 \\
+ & 0 & 1 & 1 & 1 & 1 \\
\hline
1 & 0 & 0 & 1 & 0 & 0
\end{array}
\tag{11.2}
$$

To interpret a number like 100100 without counting up, use the fact that each column represents a power of two, starting with the rightmost column, which represents 2^0, and adding one to the exponent as we move right to left from column to column. To get the numerical value of a bit string, add up the powers of two corresponding to the columns containing the 1 bits. For example, 100100 represents $2^5 + 2^2$:

$$
\begin{array}{cccccc}
2^5 & 2^4 & 2^3 & 2^2 & 2^1 & 2^0 \\
1 & 0 & 0 & 1 & 0 & 0 \\
\hline
2^5 & 0 & 0 & 2^2 & 0 & 0
\end{array}
$$

Since $2^5 + 2^2 = 32 + 4 = 36$, that is the value of the binary numeral[1] 100100. And (11.2) represents the sum of $2^4 + 2^2 + 2^0$ and $2^3 + 2^2 + 2^1 + 2^0$; that is, $21 + 15$—which is indeed 36.

Notice that this interpretation is the same one we use in the decimal system, except that in binary we use powers of 2 rather than powers of 10. For example, the numeral 100100 interpreted in decimal represents $10^5 + 10^2$.

The bit in the rightmost position—which represents 2^0 in binary—is known as the *low-order* bit. Similarly, the leftmost bit position is the *high-order* bit. Higher-order bits contribute more to the value of a numeral than lower-order bits, in exactly the way that in the decimal numeral 379, the 3 adds 300 to the value but the 9 adds only 9.

Armed with an understanding of logic gates and of binary arithmetic, how would we design hardware to do computer arithmetic? The simplest operation is the addition of two bits, as shown in (11.1). These equations

[1] A *numeral* is a name for a number. So we refer here to the binary numeral 100100 to emphasize that we are referring to a string of six bits, which represents the number 36.

really create *two* output bits from two input bits, since adding two bits produces both a sum and a carry bit. In the first three cases the carry bit wasn't mentioned—because the carry bit is 1 only when both input bits are 1. The sum, on the other hand, is 1 only when one of the inputs is 1 and the other is 0—in other words, just in case the two input bits are different. So the sum output is the logical "exclusive or" of the input bits and the carry bit is the logical "and" of the input bits. So the device in Figure 11.9, called a *half adder*, transforms input bits x and y into their sum s and their carry bit c.

There are other ways to wire gates together to produce the same outputs from the same inputs, and which way works the best in practice may depend on details of the particular implementing technology. So as we move from adding bits to adding longer binary numerals, it makes sense to represent the half adder as a box with inputs x and y and outputs s and c, hiding the internal details (Figure 11.10). Such a construct, with known inputs and outputs but hidden internal details, is often called a *black box*. If we construct more complicated circuits out of these boxes and then later decide to revise the internals of a box, our circuits will still work the same as long as the functions of the boxes remain the same.

Imagine we want to design a circuit that can compute binary sums such as (11.2). We can do this using several instances of a standard subcircuit. The half adder, as its name suggests, is not quite up to the job: it adds two bits, but when adding numbers with multiple digits we sometimes need three inputs: the original two bits plus a carry bit. A *full adder* takes in three inputs—x, y, and the carry-in bit c_{IN}—and puts out two bits, the sum s and the carry-out c_{OUT}. Figure 11.11 shows the truth table for computing the two outputs from the three inputs.

One way to construct a full adder is out of two half adders, as shown in Figure 11.12. This circuit produces the sum bit s by adding x and y and then adding c_{IN} to the result; the c_{OUT} bit is 1 if there is a carry from either the first or the second of these additions.

If we represent the full adder schematically as another black box, as shown in Figure 11.13, we can chain two of them together to produce a proper two-bit adder (Figure 11.14), which adds two two-bit binary numerals $x_1 x_0$ and $y_1 y_0$ to produce a two-bit sum $z_1 z_0$, plus a carry-out bit—which is 1 if the sum, such as $11 + 01$, can't be represented in two bits.

In exactly the same way, any number n of full adders could be chained together to produce an adder for two n-bit numbers. This kind of adder is called a *ripple-carry adder*; it implements in hardware the addition algorithm you were taught in grade school, which we used in (11.2).

But having gotten this far, we are confronted by lots of interesting questions. Does Figure 11.12 represent the most efficient implementation of the truth table in Figure 11.11? What if the "exclusive or" gate used in Figure 11.9 is not available, and we have to make do with some other set of primitives, perhaps just "nand" gates—what is then the simplest full adder? In designing real hardware to be fabricated as an integrated circuit, the number of

Figure 11.9. A half adder produces sum and carry bits from two input bits.

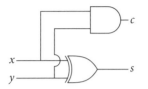

Figure 11.10. A half adder represented as a "black box."

x	y	c_{IN}	s	c_{OUT}
0	0	0	0	0
0	0	1	1	0
0	1	0	1	0
0	1	1	0	1
1	0	0	1	0
1	0	1	0	1
1	1	0	0	1
1	1	1	1	1

Figure 11.11. Truth table for the design of a full adder.

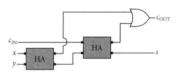

Figure 11.12. A full adder constructed from two half adders.

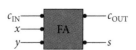

Figure 11.13. Abstract representation of a full adder.

gates is not even the most important metric—the wires actually cost more than the gates. Exploring that issue would take us too far afield, but we could certainly think about the *time* required for a circuit to compute its answer.

We can use *delay* as a simple measure of time. The delay of a circuit is the maximum number of gates between an input and an output. For example, the circuit of Figure 11.6 has delay 3.

A ripple-carry adder produces the output bits one at a time, from low order to high order (just like grade-school arithmetic). This means that a 64-bit adder would be eight times slower than an 8-bit adder—the delay doubles every time the number of input bits is doubled. That is unacceptable, and also unnecessary. By using extra hardware, the speed of such a "long" addition can be greatly increased; we explore this in Problem 11.10. Sometimes at least, it is possible to trade off hardware size against circuit speed.

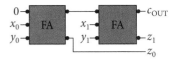

Figure 11.14. Two-bit adder. The addends are $x_1 x_0$ and $y_1 y_0$; x_0 and y_0 are the low-order bits. The carry-in to the first full adder is set to zero, the carry-out of the first adder becomes the carry-in to the second, and the carry-out of the second indicates overflow.

Chapter Summary

- In computers, all logic is represented as patterns of 0s and 1s.

- A *gate*, the simplest piece of computer logic, produces a bit of output by performing an operation (like "or," "and," "exclusive or," or "not") on one or more bits of input.

- A *circuit* is a logical device built out of gates.

- The "nand" operator (denoted by | or ↑) is convenient for building circuits, because all of the conventional operators can be expressed using only "nand" operators. The "nor" operator (denoted by ↓) has the same property.

- *Binary notation* is the name for the base-2 number system used by computers, consisting of just 0s and 1s (as opposed to decimal notation, our familiar base-10 system).

- In binary, each column represents a power of 2. The rightmost bit represents 2^0 and is called the *low-order* bit, while the leftmost bit represents the largest value 2^n and is called the *high-order* bit.

- The value of a bit string is the sum of the powers of 2 corresponding to the columns of the bit string that contain the digit 1.

- A *half adder* is a circuit that takes two input bits, the values to be summed; and returns two output bits, the sum and the carry bit.

- A *full adder* is a circuit that takes three input bits, the values to be summed and a carry-in bit; and returns two output bits, the sum and the carry-out bit. It can be built by combining two half adders.

- A *ripple-carry adder* is a circuit that adds two n-bit numbers. It can be built by combining n full adders.

- One way to measure a circuit's efficiency is its *delay*: the maximum number of gates between an input and an output.

Problems

[2]"Hexadecimal" comes from the Greek for "sixteen."

Hex	Binary	Decimal
0	0000	0
1	0001	1
2	0010	2
3	0011	3
4	0100	4
5	0101	5
6	0110	6
7	0111	7
8	1000	8
9	1001	9
A	1010	10
B	1011	11
C	1100	12
D	1101	13
E	1110	14
F	1111	15

Figure 11.15. The sixteen hexadecimal digits and their corresponding binary and decimal notations.

11.1. (a) Convert the binary numeral 111001 to decimal, showing your work.

(b) Convert the decimal numeral 87 to binary, showing your work.

11.2. *Hexadecimal notation* is a particular way of writing base-16 numerals, using sixteen hexadecimal[2] digits: 0 through 9 with the same meanings they have in decimal, and A through F to represent decimal 10 through 15 (Figure 11.15). It is easy to translate back and forth between hexadecimal and binary, simply by substituting hexadecimal digits for blocks of four bits or vice versa. For example, since 8 in hex is 1000 in binary and A in hex is 1010 in binary, the four-hex-digit numeral 081A would be written in binary as

$$0000\ 1000\ 0001\ 1010$$

where we have introduced small spaces between blocks of four bits for readability.

(a) Write the binary numeral 1110000110101111 in hexadecimal and in decimal.

(b) Write decimal 1303 in binary and in hexadecimal.

(c) Write hexadecimal ABCD in binary and in decimal.

11.3. We have described the use of sequences of n bits, for example $b_{n-1} \ldots b_0$, to represent nonnegative integers in the range from 0 to $2^n - 1$ inclusive, using the convention that this sequence represents the number

$$\sum_{i=0}^{n-1} b_i 2^i.$$

Alternatively, the 2^n sequences of n bits can be used to represent 2^{n-1} negative numbers, zero, and $2^{n-1} - 1$ positive integers in the range from -2^{n-1} to $2^{n-1} - 1$ inclusive by interpreting $b_{n-1} \ldots b_0$ to represent

$$-b_{n-1}2^{n-1} + \sum_{i=0}^{n-2} b_i 2^i.$$

This is known as *two's complement* notation. Let $n = 4$ for the purposes of this problem.

(a) What are the largest and smallest integers that can be represented?

(b) What is the two's complement numeral for -1?

(c) What number is represented by 1010?

(d) Show that a two's complement binary numeral represents a negative number if and only if the leftmost bit is 1.

(e) Convert these numbers to two's complement notation, work out their sum using the same rules for addition as developed in this chapter, and show that you get the right answer: $-5 + 4 = -1$.

11.4. A string of n bits can also be used to represent fractions between 0 and 1, including 0 but not including 1, by imagining the "binary point" (like a decimal point) to the left of the leftmost bit. By this convention, the bit string $b_1 \ldots b_n$

represents the number

$$\sum_{i=1}^{n} b_i \cdot 2^{-i}.$$

For example, if $n = 4$, the string 1000 would represent $\frac{1}{2}$.

  (a) Just as 0.9999 represents a number close to but less than 1 in decimal notation, so 1111 represents a number close to 1 in this binary notation. What is the exact value represented by 1111? In general, what is the value represented by a string of n 1s?

  (b) Work out the sum $\frac{1}{2} + \frac{3}{8} = \frac{7}{8}$ using this notation.

11.5. Let $p \downarrow q$ denote the "nor" operator: $p \downarrow q$ is true if and only if neither p nor q is true. The $\downarrow$ operator is known as *Peirce's arrow*.[3]

  (a) Write the truth table for $p \downarrow q$.

  (b) Using only the $\downarrow$ operator, write a formula equivalent to $\neg p$.

  (c) Show that $\vee$ and $\wedge$ can also be expressed using just the $\downarrow$ operator.

11.6. Write a formula involving only the | operator that is equivalent to $p \vee q$.

11.7. Write the logical formulas for the values of z_0, z_1, and c_{OUT} of the two bit adder of Figure 11.14, in terms of the inputs x_0, y_0, x_1, and y_1.

11.8. Consider the boolean variables a, b, c each representing a single bit. Write two propositional formulas: one for the result of the boolean subtraction $a - b$, and another for the "borrow bit" c, indicating whether a bit must be borrowed from an imaginary bit to the left of a (the case when $a = 0$ and $b = 1$) to ensure that the result of the subtraction is always positive. Start by creating a truth table for $a - b$ and the borrow bit c. The borrow bit c is always 1 initially, and is set to 0 only if borrowing was necessary.

11.9. A common way of displaying numerals digitally is to use a pattern of up to seven straight line segments (see Figure 11.16). By turning various segments on and off, it is possible to form recognizable representations of the digits 0 through 9. In the figure, the seven strokes have been labeled with the letters A through G. The numeral 8 has all seven strokes on; 1 has C and F on and the others off; and 2 has A, C, D, E, and G on and the others off.

  (a) Write out the strokes that should be turned on to represent each of the remaining digits.

  (b) Write a truth table with five columns, showing when the A segment should be turned on given the binary representation of a number in the range 0 through 9. The first four columns are the bits of the digit to be displayed, and the last column is 1 if and only if stroke A is on in the representation of that digit. For example, if the values in the first four columns are 0010, the binary representation of 2, the last column should be 1 since the A segment is on when 2 is displayed, as illustrated.

  (c) Write a DNF formula for A based on the truth table from part (b).

  (d) Draw a logic circuit that implements the formula from part (c). (A full design for this kind of display would need six other circuits, one for each of the other segments.)

[3] Named for the American mathematician, philosopher, and scientist Charles Sanders Peirce (1839–1914).

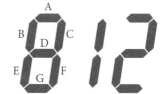

Figure 11.16. Digital numerals, composed of line segments A through G that can be turned on and off.

11.10. Your task is to design a four-bit adder; that is, a circuit to compute the four-bit sum $z_3z_2z_1z_0$ of two four-bit inputs $x_3x_2x_1x_0$ and $y_3y_2y_1y_0$. The circuit also produces a carry-out bit c.

  (a) Draw the two-bit adder of Figure 11.14 as a box with five input bits and three output bits (in the figure, the carry-in bit is set to 0). What is its delay, on the assumption that the half adder and full adder are implemented as shown in Figures 11.9 and 11.12 and each gate has delay 1? Show how to chain two of these boxes together to create a four-bit adder. What is the delay of your four-bit adder?

  (b) The solution of part (a) has the disadvantage that the delay keeps doubling every time the number of input bits is doubled. An alternative solution uses more hardware to reduce the delay. It uses three two-bit adders, instead of two, to compute simultaneously

    (A) the sum of the low-order bits (x_1x_0 and y_1y_0);

    (B) the sum of the high-order bits (x_3x_2 and y_3y_2) on the assumption that the carry-in is 0;

    (C) the sum of the high-order bits on the assumption that the carry-in is 1.

    It then uses a few more gates to output either the output of (B) or the output of (C), depending on the value of (A). Diagram this solution, and compute its delay.

  (c) If the methods of parts (a) and (b) were generalized to construct 2^k-bit adders, what would be the delays of the circuits?

11.11. This problem further explores the Thue sequence (page 33).

  (a) Prove that t_n, the nth bit in the Thue sequence, is the exclusive or of the bits of the binary notation for n. (The exclusive or of a single bit is the bit itself.)

  (b) Show that for every $n \geq 0$, $t_{2n} = t_n$, and t_{2n+1} is the complement of t_n.

11.12. A truth function $f : \{0,1\}^k \to \{0,1\}$ is said to be *monotone* if

$$f(x_1,\ldots,x_k) \leq f(y_1,\ldots,y_k) \text{ whenever } x_i \leq y_i \text{ for each } i = 1,\ldots,k.$$

That is, changing any argument from 0 to 1 cannot change the value from 1 to 0. Show that a truth function is monotone if and only if it can be expressed by a propositional formula using only $\vee$ and $\wedge$ (and without $\neg$). Equivalently, it can be computed by a circuit with only "and" and "or" gates.

Chapter 12

Quantificational Logic

Quantificational logic is the logic of expressions such as "for any," "for all," "there is some," and "there is exactly one," which we have been using quite a bit as ordinary English. With the aid of the vocabulary of sets and functions, we can discuss these expressions more formally, and be precise about the meaning of more complicated statements involving multiple uses of these terms. More importantly, by formalizing this logic we render it suitable for manipulation by computers—that is, for automated reasoning.

Quantificational logic extends the "propositional logic" of Chapter 9, and is so called because it is designed for quantifying how many of this or that kind of thing there may be. It also goes by other names: *first-order logic*, which suggests that propositional logic is zero-order logic, and there might be something called second-order logic too. (There is; it allows statements like "for every set *S*, ...") Quantificational logic is also referred to as *predicate logic* or the *predicate calculus*.

On page 90, we discussed the statement

> *Hens are black.* (12.1)

We observed that whether this is true or false depends on what (12.1) means. If it means that *all* hens are black, then the negation is something like "at least one hen is not black," since a single counterexample would establish the falsity of the statement that they are *all* black. On the other hand, if (12.1) means that most hens are black, but there are exceptions, then the negation would have to be "no hens are black," or words to that effect. Quantificational logic provides the tools for making such statements precisely, and for making logical inferences about them.

Recall that on page 12, we used the term *predicate* to describe a template for a proposition, with a placeholder for a variable that can assume different values, where the truth or falsity of the proposition depends on the value of the variable. We are now ready to generalize and formalize this idea. In quantificational logic, we use *predicate symbols* as names for properties of or relationships between things, so that we can make multiple statements using the same template, plugging in different things in similar contexts.

For example, if we write $H(x)$ for the predicate that x is a hen, then "Ginger is a hen" could be written as $H(\text{Ginger})$. Here "H" is a predicate symbol, and "Ginger" is being used as a *constant*, the name for a fixed thing, just as "1" and "2" are constants in the statement "$1 + 1 = 2$."

A predicate is essentially a boolean-valued function—that is, a function that maps its arguments to the values true and false. When applied to specific values, a predicate has a truth value, like an atomic proposition of propositional logic. $H(\text{Ginger})$ is true just in case Ginger is a hen; otherwise $H(\text{Ginger})$ is false. But of course we can't say whether $H(x)$ is true or false until we know the value of x.

Predicates, like functions, can have more than one argument, and when they do, the various argument positions refer to things in a consistent way. For example, we could write $S(x, y, z)$ for a three-place predicate meaning "$x + y = z$." Then, provided the variables refer to numbers, $S(1, 2, 3)$ is a true statement (since $1 + 2 = 3$) and $S(3, 2, 1)$ is a false statement (since it is not true that $3 + 2 = 1$).

The symbols $\forall$ and $\exists$ stand for "for all" and "there exists." So "Everything is a hen" could be rendered as $\forall x H(x)$ and "at least one hen exists" as $\exists x H(x)$. We can combine such statements using the ordinary logical connectives. For example, if we use $B(x)$ to mean "x is black," then B represents a 1-place relation. Then "all hens are black" might be rendered as

$$\forall x (H(x) \Rightarrow B(x)), \tag{12.2}$$

which literally says, "for any x, if x is a hen then x is black." On the other hand, "at least one black hen exists" could be captured by the statement

$$\exists x (H(x) \wedge B(x)), \tag{12.3}$$

that is, there is at least one thing x such that x is a hen and x is black.

✳

Let's return to the three-place predicate $S(x, y, z) \equiv x + y = z$. (Note our use of the metalinguistic expression $\equiv$ as we did with propositional logic. To write $\alpha \equiv \beta$ is simply to say that α and β are equivalent—in this case, because we have defined $S(x, y, z)$ to mean that $x + y = z$.)

We can quantify just some of the variables and have constants in the other places. For example, $\exists x S(x, 3, 5)$ means that there is a number x such that $x + 3 = 5$. This is a true statement since $x = 2$ works. What about the similar statement $\exists x S(x, 5, 3)$? Its truth value depends on something we haven't stipulated—the *universe*—that is, the set of allowed values of the variables, like the domain of a function. If the universe includes negative numbers, the statement is true, since $x = -2$ works; but if the universe includes only positive numbers, the statement is false. For the sake of argument, let's say that the values must be integers—positive, negative, or zero.

We can pile up quantifiers within a single statement. For example, $\forall x \exists y S(x, y, 0)$ is true, since whatever x may be, plugging in $-x$ as the value of y makes the statement true: $S(x, -x, 0)$ says that $x + (-x) = 0$. However, beware! The order of quantifiers can dramatically change the meaning of a statement. If the universe is the set of integers, then $\forall x \exists y S(x, y, 0)$ is true, as just noted. But $\exists x \forall y S(x, y, 0)$ is false, since it claims that there is some special number x that has the property that no matter what value of y is chosen, adding x to y produces a sum of 0. On the other hand, $\exists y \forall x S(x, y, x)$ is true, since taking y to have the value 0 makes $S(x, y, x)$ true for every number x.

To take just one more example, $\exists x S(x, y, z)$ is neither true nor false, because it has two *free variables*, y and z—that is, unquantified variables. Until some values or quantifiers are associated with those variables, the truth of the statement can't be assessed. A quantified variable, such as x in this formula, is said to be *bound*, while an unquantified variable, such as y or z in this formula, is said to be *free*.

Again, note the importance of observing the precise placement of quantifiers and parentheses in a quantified statement. For example, the formula (12.3) says something quite different from

$$\exists x H(x) \wedge \exists x B(x), \tag{12.4}$$

which states that a hen exists, and also that a black thing exists—without any implication that any hen is black. The fact that "x" is used in both halves of the formula is accidental; (12.4) means the same thing as

$$\exists x H(x) \wedge \exists y B(y), \tag{12.5}$$

which is actually a better way to write this formula, as it avoids the suggestion that the two uses of x in (12.4) refer to the same thing.[1]

More complex formulas express more complex ideas. For example, let $L(x, y)$ stand for "x loves y." Then we could express "Everybody loves Oprah" by the formula

$$\forall x L(x, \text{Oprah}),$$

where "Oprah" is a constant. But how would we write "Everybody loves somebody"? Being forced to write this as a logical formula presses us to be clear what it means. Does it mean that there is a particular person (such as Oprah) who is loved by every human being? Or does it mean that each human being has someone particular to him or her who is beloved? Almost certainly the latter, which amounts to $\forall x \exists y L(x, y)$. The former, if that is really what was intended, would be $\exists y \forall x L(x, y)$, and that is something quite different!

✢

[1] It is even legal, though unwise, to write $(\exists x H(x)) \wedge B(x)$. This statement says that a hen exists, and that x is black. The occurrence of x in $(\exists x H(x))$ has nothing to do with the occurrence of x as a free variable in the rest of the formula. So strictly speaking, *variables* are not free or bound; their *occurrences* are. The way to avoid such technicalities is to avoid reusing the same variable in a formula, except when the intended meaning of the formula requires it.

It is time to be clear and specific about the rules for constructing formulas of quantificational logic. First, the vocabulary of quantificational logic includes:

1. predicate symbols such as P, each of which has a fixed *arity*, which is an integer ≥ 0 (*unary* or 1-place, *binary* or 2-place, *ternary* or 3-place, etc.);

2. variables, such as x, y, z;

3. constants, such as "0" and "Sue", which name members of the universe;

4. function symbols, such as "+" or generic function names such as "f," which name functions (each with a fixed arity) from and to the universe;

5. terms that can be constructed from variables, constants, and function symbols, such as "$x + 0$" and "$f(y, z)$," which refer to members of the universe and can replace constants or variables as arguments of predicates;

6. the predicate "="; and

7. parentheses, quantifiers, and logical operators: "(," ")," "∀," "∃," "∧," "∨," and "¬."

This vocabulary allows for the use of special constant and function symbols with meanings that are particular to a domain. For example, the constant "0" and the function symbol "+" are needed in order to formalize statements about ordinary arithmetic. (In general we use notations such as $f(x, y)$ to denote a function f being applied to the variables x and y, but in the special case of arithmetic, we will use standard infix notation such as "$x + y$," rather than "$+(x, y)$.") The language of quantificational logic thus extended can be used to express familiar-looking statements such as

$$\forall x(x + 0 = x). \tag{12.6}$$

Now we can give an inductive definition of *formulas* and their *free variables*:

Base case. If P is a k-place predicate symbol ($k \geq 0$) and $t_1, \ldots, t_k$ are terms, then $P(t_1, \ldots, t_k)$ is a formula. The free variables of this formula are the variables occurring in $t_1, \ldots, t_k$. (For example, $P(x, f(y, x))$, where P is a binary predicate symbol and f is a binary function symbol, is a formula with free variables x and y.)

Constructor case 1. If F and G are formulas then so are $(F \vee G)$, $(F \wedge G)$, and $\neg F$. The free variables of $(F \vee G)$ and $(F \wedge G)$ are the free variables of F and the free variables of G, and the free variables of $\neg F$ are the free variables of F.

Constructor case 2. If F is a formula and x is a variable, then $\forall x F$ and $\exists x F$ are formulas. Their free variables are all the free variables of F except x, which is *bound*.

We refer to F and G as *subformulas* of $(F \vee G)$ in constructor case 1, and similarly F is a subformula in constructor case 2.

A formula with no free variables, such as (12.3), is said to be *closed*. Intuitively, such a formula has no dangling references. For example, to decide whether or not (12.3) is true, we need to know something about the color of hens, and that will settle the question. But we don't need any further information about what "x" refers to; x is a *quantified* variable and could be systematically replaced by any other variable. So (12.3) says the same thing as

$$\exists y (H(y) \wedge B(y)). \tag{12.7}$$

On the other hand, a formula with a free variable, such as $(H(y) \wedge B(y))$, invites the question, "But what is y?"

As in the case of propositional logic, parentheses are formally required in certain places so the grouping is maintained. For example, the outer parentheses in $(H(y) \wedge B(y))$ from (12.7) specify that when this formula is quantified, the quantifier applies to the whole formula.

In fact, quantificational logic is an extension of propositional logic. A 0-place predicate symbol takes no arguments and is exactly what we have called a propositional variable. If all the predicate symbols are 0-place, then there is no use for variables or quantifiers, and all the formulas are just formulas of propositional logic.

We can use propositional variables together with predicates that themselves take variables. For example, the statement "If it is raining, then anything that is outside will get wet" can be represented as

$$r \Rightarrow \forall x (Outside(x) \Rightarrow Wet(x)), \tag{12.8}$$

where r is a propositional variable representing "it is raining," $Outside(x)$ and $Wet(x)$ are unary predicates, and x is a bound variable of the formula. (We are not attempting to capture the subtleties of tense in this paraphrase.) Note that both r and x are referred to as "variables," but they take on different roles: r represents a statement, which on its own can be evaluated as true or false; whereas x represents an object in the universe, which is supplied as an argument to a predicate.

As examples, let us use the language of quantificational logic to state some of the properties of functions defined in Chapter 6. Recall that a function f is *surjective* if every element of the codomain is the value of the function for some argument in the domain. In logical language, $f : A \to B$ is surjective if and only if

$$\forall b (b \in B \Rightarrow \exists a (a \in A \wedge f(a) = b)).$$

Because we so commonly want to restrict variables to be members of particular sets, a statement like this is abbreviated as

$$(\forall b \in B)(\exists a \in A)f(a) = b.$$

The equals sign comes in handy when we need to stipulate that two things are different. For example, to say that f is injective, we need to say that all values of f are distinct. We can say this two ways:

$$(\forall a_1 \in A)(\forall a_2 \in A)(f(a_1) = f(a_2) \Rightarrow a_1 = a_2)$$

says that if f takes the same value on two arguments, the two arguments are actually the same. The contrapositive may be clearer, however. It says that different arguments yield different values:

$$(\forall a_1 \in A)(\forall a_2 \in A)(a_1 \neq a_2 \Rightarrow f(a_1) \neq f(a_2)).$$

Of course, we are using "$a_1 \neq a_2$" as an abbreviation for "$\neg(a_1 = a_2)$." A further notational simplification abbreviates the last formula as

$$(\forall a_1, a_2 \in A)(a_1 \neq a_2 \Rightarrow f(a_1) \neq f(a_2)),$$

using just one occurrence of the universal quantifier to bind two different variables.

Another example uses $L(x, y)$ once again to mean that x loves y. To say that everybody loves someone other than him- or herself, we could write:

$$\forall x \exists y (x \neq y \wedge L(x, y)).$$

To write simply

$$\forall x \exists y L(x, y)$$

would admit the possibility of people loving themselves and no one else. Similarly, to say that there is someone who loves two different people (neither of whom is the someone in question), we could write

$$\exists x, y, z (x \neq y \wedge x \neq z \wedge y \neq z \wedge L(x, y) \wedge L(x, z)).$$

Finally, to say that everyone loves at most one person we could write

$$\forall x, y, z (L(x, y) \wedge L(x, z) \Rightarrow y = z). \tag{12.9}$$

This admits the possibility that someone (some x) might love no one. The premise $L(x, y) \wedge L(x, z)$ would then be false regardless of the values of the values of y and z, so the implication would be true (because a false proposition implies any proposition). To say instead that everyone has a unique beloved, we could write

$$\forall x \exists y (L(x, y) \wedge \forall z (L(x, z) \Rightarrow y = z)). \qquad (12.10)$$

An implication that is true because nothing in the universe satisfies the premise is said to be *vacuously* true. For example, in (12.9), if X is a constant representing someone who loves no one, then no values of y and z satisfy the premise $L(X, y) \wedge L(X, z)$ and so the conclusion $y = z$ is vacuously true. An example of a vacuously true proposition in ordinary language is "All flying pigs are green," which we might write as

$$\forall x (Pig(x) \wedge Flying(x) \Rightarrow Green(x)).$$

This is vacuously true since there are no flying pigs. What comes as the conclusion is irrelevant.

✳

We have yet to give any formal rules for interpreting such statements and determining whether or not they are true, or what exactly it means for two formulas to be equivalent or to "say the same thing." For that we need machinery that is more complex than the truth tables of Chapter 9.

In essence, an *interpretation* of a quantificational formula has to specify the following elements:

1. the *universe U*, the nonempty set from which the values of the variables are drawn;

2. for each, say k-ary, predicate symbol P, which k-tuples of members of U the predicate is true of; and

3. what elements of the universe correspond to any constant symbols, and what functions from the universe to itself correspond to any function symbols mentioned in the formula.

Let's take a simple example:

$$\forall x \exists y P(x, y). \qquad (12.11)$$

In some interpretations (12.11) is true and in some it is false. For example,

- if the universe is $\{0, 1\}$ and P is the less-than relation:
  - $P(0, 0)$ is false
  - $P(0, 1)$ is true
  - $P(1, 0)$ is false
  - $P(1, 1)$ is false

then (12.11) is false in this interpretation, because there is no value of y for which $P(x, y)$ is true when x is 1;

- on the other hand, if the universe is the same but P is the not-equal relation:
  - $P(0, 0)$ is false
  - $P(0, 1)$ is true
  - $P(1, 0)$ is true
  - $P(1, 1)$ is false

then (12.11) is true in this interpretation.

In general, the universe of an interpretation will be an infinite set and it will be impossible to list the value of the predicate for every combination of elements. But we can use the mathematical notion of a *relation*. To restate the definition in terms of relations, an *interpretation* of a closed logical formula consists of:

1. a nonempty set called the universe;

2. for each k-place predicate symbol, a k-ary relation on the universe;

3. for each k-place function symbol, a k-ary function from the universe to itself.

Note that we have restricted this definition to closed formulas to avoid the problem of dangling references to free variables.

For example, we can interpret (12.11) over the natural numbers, taking P to be the less-than relation; under this interpretation (12.11) is true. However, if the universe is the natural numbers and P is interpreted instead as the greater-than relation, then (12.11) is false, since when $x = 0$ there is no $y \in \mathbb{N}$ such that $x > y$.

For an example involving functions, consider the interpretation of (12.6) in which U is the set of natural numbers, the constant 0 is interpreted as zero, and the binary function symbol $+$ represents addition. Then (12.6) is true under this interpretation. The formula

$$\forall x \exists y (x + y = 0) \tag{12.12}$$

is false; since, for example, when $x = 1$, there is no natural number y such that $x + y = 0$. But if U is the set of all integers, the comparable interpretation of (12.12) is true.

Two formulas are *equivalent* if they are true under the same interpretations. For example, $\forall x \exists y P(x, y)$ and $\forall y \exists x P(y, x)$ are equivalent; changing the variable names in a systematic way may be confusing to the reader, but doesn't change the underlying logic. Following the notation we used for propositional logic, we write $F \equiv G$ to denote the fact that formulas F and G are equivalent.

A *model* of a formula is an interpretation in which it is true. A model of a quantificational formula is analogous to a satisfying truth assignment of a propositional formula, so a *satisfiable* formula of quantificational logic is one that has a model. A *valid* formula, also known as a *theorem*, is a formula that is true under every interpretation (except for ones that don't assign meanings to all the predicate and function symbols). Valid formulas are the quantificational analogs of tautologies in propositional logic.

An example of a valid formula is $\forall x(P(x) \wedge Q(x)) \Rightarrow \forall y P(y)$.[2] An example of an unsatisfiable formula is $\forall x P(x) \wedge \exists y \neg P(y)$.

✳

Everything we learned about logical connectives in Chapter 9 carries over to quantificational logic. The logical connectives bind subformulas in the same way; so, for example,

$$\forall x(P(x) \Rightarrow \neg Q(x) \wedge R(x)) \text{ means } \forall x(P(x) \Rightarrow ((\neg Q(x)) \wedge R(x))).$$

But quantifiers are not themselves subformulas and cannot be manipulated in this way. For example, as noted earlier,

$$\exists x(H(x) \wedge B(x)) \qquad (12.13)$$
$$\text{is not the same as } \exists x H(x) \wedge \exists x B(x).$$

And quantifiers themselves bind tightly, like ¬. So (12.13) is something very different from

$$\exists x H(x) \wedge B(x),$$
$$\text{which would mean } (\exists x H(x)) \wedge B(x),$$

a formula in which the "x" in $B(x)$ is free.

The distributive laws hold; so, for example,

$$\forall x(P(x) \wedge (Q(x) \vee R(x))) \qquad (12.14)$$

is equivalent to

$$\forall x((P(x) \wedge Q(x)) \vee (P(x) \wedge R(x))). \qquad (12.15)$$

In general, we can regard whole subformulas as propositional variables, and rewrite a formula purely using equivalences of propositional logic, to get an equivalent quantificational formula.

Quantificational Equivalence Rule 1. *Propositional substitutions.* Suppose F and G are quantificational formulas, and F' and G' are propositional formulas that result from F and G, respectively, by replacing each subformula by a corresponding propositional variable at all of its occurrences in both F

[2] Because the universe cannot be empty, the formula $\forall x P(x) \Rightarrow \exists y P(y)$ is also valid. If we did allow for the empty universe, it would be merely satisfiable. Excluding the possibility of the empty universe not only makes this intuitively obvious formula true in all interpretations, it avoids the metaphysical question of the need for a logic in a world in which there are no logicians and nothing for them to reason about!

and G. Suppose $F' \equiv G'$ as formulas of propositional logic. Then replacing F by G in any formula results in an equivalent formula.

For example, $\forall x \neg\neg P(x) \equiv \forall x P(x)$, since $p \equiv \neg\neg p$, and therefore $\neg\neg P(x)$ can be replaced by $P(x)$. Similarly, (12.14) is equivalent to (12.15) because replacing $P(x)$ by p, $Q(x)$ by q, and $R(x)$ by r turns $(P(x) \wedge (Q(x) \vee R(x))$ into $(p \wedge (q \vee r))$ and $((P(x) \wedge Q(x)) \vee (P(x) \wedge R(x))$ into the equivalent propositional formula $(p \wedge q) \vee (p \wedge r)$.

There are also important laws of logical equivalence that pertain directly to the use of quantifiers. We have already suggested one, namely that variables can be systematically renamed; here is a general statement.

Quantificational Equivalence Rule 2. *Change of variables.* Let F be a formula containing a subformula $\Box x G$, where $\Box$ is one of the quantifiers $\forall$ or $\exists$. Assume G has no bound occurrences of x and no occurrences of y, and let G' be the result of replacing x by y everywhere in G. Then replacing $\Box x G$ by $\Box y G'$ within the formula results in an equivalent formula.

It is by this principle that (12.3) is equivalent to (12.7). Remove the quantifier $\forall x$ from (12.3), substitute y for x, and restore the quantifier as $\forall y$. The resulting formula is equivalent to the original.

Next we have the rules about negations and quantifiers. The negation of "all hens are black" is "some hen is not black." More generally, the negation of an existential ($\exists$) statement is the universal ($\forall$) of the negation of that statement, and the negation of a universal statement is the existential of the negation of that statement. That is quite a mouthful, but the principle is quite natural. We can think of a universal quantifier as a very large "and": "$\forall x P(x)$" means P is true of this and that and this other thing and ... of every element of the universe. Similarly, an existential quantifier is akin to a very large "or": "$\exists x P(x)$" means P is true either of this or of that or of ... some member of the universe. Of course the universe generally is infinite so the quantifiers can't literally be replaced by conjunctions and disjunctions, but if we think of them as big ands and ors, this principle is just a version of De Morgan's laws.

Quantificational Equivalence Rule 3. *Quantifier negation.*

$$\neg \forall x F \equiv \exists x \neg F, \text{ and} \tag{12.16}$$
$$\neg \exists x F \equiv \forall x \neg F. \tag{12.17}$$

This principle is used when we wish to "slide" quantifiers in or out across negation signs.

A final rule describes when quantifiers can be moved across logical connectives. Basically the answer is that the scope of a quantifier can be enlarged, as long as doing so does not "capture" an occurrence of a variable in a way that would change the meaning of the formula.

Quantificational Equivalence Rule 4. *Scope change.* Suppose the variable x does not appear in G. Let $\Box$ denote either the existential quantifier $\exists$ or the universal quantifier ∀, and let ◇ denote ∨ or ∧. Then

$$(\Box x F \diamond G) \equiv \Box x (F \diamond G) \tag{12.18}$$

$$(G \diamond \Box x F) \equiv \Box x (G \diamond F). \tag{12.19}$$

We can apply this to our previous example, "If it is raining, then anything that is outside will get wet." The variable x does not appear in the formula r, so we can "pull out" the quantifier for x:

$$r \Rightarrow \forall x (Outside(x) \Rightarrow Wet(x))$$

$$\equiv \forall x (r \Rightarrow (Outside(x) \Rightarrow Wet(x))).$$

(Recall from Theorem 9.1 that $F \Rightarrow G \equiv \neg F \vee G$, so pulling out the quantifier over $\Rightarrow$ works just like pulling it out over $\vee$.) In plain English, this transformed statement is less natural than the original, but it says something like, "Any object, in case of rain, will get wet if it is outside."

Anther consequence of scope change is less obvious: it means that

$$(\forall x P(x) \vee \exists y Q(y)) \equiv \forall x \exists y (P(x) \vee Q(y)) \tag{12.20}$$

$$\equiv \exists y \forall x (P(x) \vee Q(y)) \tag{12.21}$$

since the quantifiers can be pulled out in either order. Note, however, that if the formula with which we started was

$$(\forall x P(x) \vee \exists x Q(x)), \tag{12.22}$$

neither of the quantifiers could be moved, because the quantified variable x appears in both subformulas. However, the variable renaming rule can be applied first to turn (12.22) into the form shown in (12.20).

Through repeated use of the quantificational equivalence rules, all quantifiers can be pulled out to the beginning of the formula. Such a formula is said to be in *prenex normal form*. For example, a prenex normal form of (12.10) is

$$\forall x \exists y \forall z (L(x, y) \wedge (L(x, z) \Rightarrow y = z)).$$

Example 12.23. *Translate into predicate logic, and put into prenex form: "If there are any ants, then one of them is the queen."*

Solution to example. Let's write $A(x)$ for "x is an ant," and $Q(x)$ for "x is a queen." Implicit in the English-language statement are two facts that need to be captured in the predicate logic version. First is that there is only one queen (that is the implication of the definite article "the" in the phrase "the queen"), and second is that the queen is an ant (that is the implication of the phrase "of them"). So the sentence can be formalized as

$$\exists x A(x) \Rightarrow \exists x (A(x) \wedge Q(x) \wedge \forall z (Q(z) \Rightarrow z = x))$$
$$\equiv \exists x A(x) \Rightarrow \exists y (A(y) \wedge Q(y) \wedge \forall z (Q(z) \Rightarrow z = y)),$$

where we have renamed one variable so no two quantifiers use the same variable. This statement, in much more formal English, can be read as, "If there exists an x such that x is an ant, then there exists a y such that y is an ant and y is a queen and any z that is a queen is equal to y."

Now, replacing the first "$\Rightarrow$" by its definition in terms of $\neg$ and $\vee$ and pulling out the quantifiers one at a time,

$$\exists x A(x) \Rightarrow \exists y (A(y) \wedge Q(y) \wedge \forall z (Q(z) \Rightarrow z = y))$$
$$\equiv \neg \exists x A(x) \vee \exists y (A(y) \wedge Q(y) \wedge \forall z (Q(z) \Rightarrow z = y))$$
$$\equiv \forall x \neg A(x) \vee \exists y (A(y) \wedge Q(y) \wedge \forall z (Q(z) \Rightarrow z = y))$$
$$\equiv \forall x \exists y \forall z (\neg A(x) \vee (A(y) \wedge Q(y) \wedge (Q(z) \Rightarrow z = y))).$$

If A and Q are replaced by their English meanings, it may be hard to read the last version as equivalent to the first. But because the quantifiers have been isolated and the rest of the formula can now be manipulated using the laws of propositional logic, it may be more convenient for processing by computer. ∎

Chapter Summary

- A *predicate* describes a property of, or relationship between, things; for example, the predicate $P(x)$ is a property of x.

- The *universe* of a predicate is the set of possible values of its variables.

- The symbol $\forall$ denotes "for all," as in the statement $\forall x P(x)$: "for all x, the property $P(x)$ is true."

- The symbol $\exists$ denotes "there exists," as in the statement $\exists x P(x)$: "there exists an x for which the property $P(x)$ is true."

- A *formula* can be defined inductively: a predicate is a formula; and we can create a formula by combining two formulas with $\vee$ or $\wedge$, or by negating a formula ($\neg$), or by adding a quantifier ($\forall$ or $\exists$) for a variable in a formula.

- The truth value of a formula can be determined only if all of its variables are quantified, or *bound*.

- The order of quantifiers in a formula is important to its meaning.

- An *interpretation* of a quantificational formula specifies the universe from which the variable values are drawn; and the meaning of each predicate symbol, constant, and function symbol.

■ A *model* of a formula is an interpretation in which it is true, analogous to a satisfying truth assignment of a propositional formula. A formula is *satisfiable* if it has a model, and *valid* if it is true under every interpretation.

■ Quantificational formulas obey the same associative, commutative, and distributive laws as propositional formulas.

■ A quantificational formula can be rewritten as an *equivalent* formula, using *propositional substitutions*, *change of variables*, *quantifier negation* ($\neg(\forall xF) \equiv \exists x\neg F$ and $\neg(\exists xF) \equiv \forall x\neg F$), and *scope change*.

■ Any quantificational formula can be written in *prenex normal form*, with its quantifiers pulled out to the beginning of the formula, which is useful for automated processing.

Problems

12.1. Write as quantificational formulas:
   (a) There is somebody who loves everyone.
   (b) There is somebody who loves no one.
   (c) There is no one person who loves everyone.

12.2. Write the following statements using quantificational logic. Use $S(x)$ to denote that x is a student and $H(x)$ to denote that x is happy. The universe is the set of all people.
   (a) Every student is happy.
   (b) Not every student is happy.
   (c) No student is happy.
   (d) There are exactly two unhappy people, at least one of whom is a student.

12.3. Consider the formula
$$\exists x \exists y \exists z (P(x,y) \wedge P(z,y) \wedge P(x,z) \wedge \neg P(z,x)).$$
Under each of these interpretations, is this formula true? In each case, R is the relation corresponding to P.
   (a) $U = \mathbb{N}$, $R = \{\langle x,y \rangle : x < y\}$.
   (b) $U = \mathbb{N}$, $R = \{\langle x, x+1 \rangle : x \geq 0\}$.
   (c) $U = $ the set of all bit strings, $R = \{\langle x,y \rangle : x$ is lexicographically earlier than $y\}$.
   (d) $U = $ the set of all bit strings, $R = \{\langle x,y \rangle : y = x0$ or $y = x1\}$.
   (e) $U = P(\mathbb{N})$, $R = \{\langle A, B \rangle : A \subseteq B\}$.

12.4. Write quantificational formulas that state the following.
   (a) The unary function f is a bijection between the universe and itself.
   (b) The binary function g does not depend on the second argument.

12.5. Using the binary predicate $\in$ for set membership and the binary predicate $\subseteq$ for the subset relation, write formulas stating these basic properties of set membership.

(a) Any two sets have a union; that is, a set containing all and only the members of the two sets.

(b) Every set has a complement.

(c) Any member of a subset of a set is a member of that set.

(d) There is a set which has no members and is a subset of every set.

(e) The power set of any set exists.

12.6. Describe models for these formulas.

(a) $\forall x \exists y \exists z P(x, y, z) \land \forall u \forall v (P(u, u, v) \Leftrightarrow \neg P(v, v, u))$

(b) $\forall x \exists y \forall z ((P(x, y) \Leftrightarrow P(y, z)) \land (P(x, y) \Rightarrow \neg P(y, x)))$

12.7. Prove that this formula is satisfiable but has no finite model.

$$\forall x \forall y (P(x, y) \Leftrightarrow \neg P(y, x))$$
$$\land \forall x \exists y P(x, y)$$
$$\land \forall x \forall y \forall z (P(x, y) \land P(y, z) \Rightarrow P(x, z))$$

12.8. (a) In the formula of Problem 12.7, it was clearer to reuse variables than to use different variables in each quantified conjunct. Rewrite this formula using different variables in each part, and then put the formula in prenex normal form.

(b) Write in prenex normal form:

$$\neg \forall x (P(x) \Rightarrow \exists y Q(x, y)).$$

12.9. The *cardinality* of a model is the size of its universe. Show that there is a formula that has a model of cardinality 3, but no model of any smaller size. Do not use the equals sign (that is, do not simply say that there exist 3 elements, no two of which are equal).

12.10. (a) Suppose $P(x, y)$ is true if and only if y is a parent of x. Write a quantificational formula such that $A(x, y)$ is true if and only if y is an ancestor of x.

(b) Using $A(x, y)$, write a formula such that $R(x, y)$ is true if and only if x and y are blood relatives; that is, one is an ancestor of the other or they have a common ancestor.

Chapter 13

Directed Graphs

Directed graphs represent binary relations. Directed graphs can be visualized as diagrams made up of points and arrows between the points. We refer to the points as "vertices" and to the arrows as "arcs," and draw an arc from vertex v to vertex w to represent that the ordered pair $\langle v, w \rangle$ is in the relation. We have already used this graphical convention on page 60, where Figure 6.2 represents part of the binary relation (6.2). A simpler example is shown in Figure 13.1, which represents the binary relation

$$\{\langle a, b\rangle, \langle b, c\rangle, \langle a, c\rangle, \langle c, d\rangle, \langle c, c\rangle, \langle d, b\rangle \, \langle b, d\rangle\}. \qquad (13.1)$$

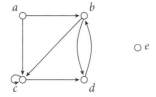

Figure 13.1. A directed graph with five vertices and seven arcs.

Vertices may appear in a graph even if they are not present in any arc of the relation—Figure 13.1 shows a vertex e which is at neither the head nor the tail of any arc. A vertex may have an arc to itself—the arc from c to c is present in this example, but no other vertex has such a *self-loop*. Between any pair of distinct vertices, there can be an arc in either direction, neither, or both—between b and d in the example there are two distinct arcs, one in each direction. But there cannot be more than one arc in the same direction between two vertices, since the collection of arcs is a set, and therefore cannot contain the same element more than once. All this leads us to define directed graphs formally as follows:

A *directed graph*, or *digraph*, is an ordered pair $\langle V, A \rangle$, where V is a nonempty set and A is a subset of $V \times V$. The members of V are called *vertices* and the members of A are called *arcs*.[1] We'll generally write arcs as $v \to w$ rather than the clumsier notation $\langle v, w \rangle$. Both the vertex set and the arc set can be infinite, but for the most part we will be discussing finite digraphs.

Transportation and computer networks have natural representations as digraphs. For example, the vertices might represent cities, with an arc from one city to another if there is a nonstop flight from the one to the other. So that we can discuss movements within such networks, we'll begin by defining a few basic properties of digraphs.

A walk in a digraph is simply a way of proceeding through a sequence of vertices by following arcs. For example, there are several walks from b

[1] The term *node* is sometimes used instead of "vertex," and *edge* is sometimes used instead of "arc." The plural of "vertex" is "vertices."

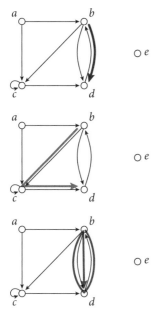

Figure 13.2. Three walks from b to d. The first and second are paths, but the third contains the cycle b, d, b and therefore is a walk but not a path.

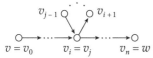

Figure 13.3. Shortening a walk by omitting a cycle.

to d in the example: one goes straight from b to d along the arc between them that points in the correct direction; the sequence b, c, d is another; and the sequence b, d, b, d is a third (Figure 13.2). But there is no walk from b to a, or indeed from any of the other vertices to a. Formally, a *walk* in a digraph $\langle V, A \rangle$ is a sequence of vertices $v_0, \ldots, v_n \in V$ for some $n \geq 0$, such that $v_i \to v_{i+1} \in A$ for each $i < n$. The *length* of this walk is n, the number of arcs (one less than the number of vertices).

A *path* is a walk that doesn't repeat any vertex. A walk in which the first and last vertices are the same is called a *circuit*, and is called a *cycle* if that is the only repeated vertex. So $b \to c \to d \to b$ is a cycle of length 3, and $b \to d \to b$ is a cycle of length 2, and $c \to c$ is a cycle of length 1. There is a path of length 0 and a cycle of length 0 from any vertex to itself; such a path or cycle is said to be *trivial*, and any path or cycle of length greater than 0 is said to be *nontrivial*. A digraph without any nontrivial cycles is said to be *acyclic*.

If there is a walk from one vertex to another, then there is a path from one to the other, because any nontrivial cycle along the way can be left out. To see this, suppose (Figure 13.3) there is a walk from v to w, say

$$v = v_0 \to \ldots \to v_n = w,$$

which includes a cycle, say $v_i \to v_{i+1} \to \ldots \to v_j$, where $i < j$ and $v_i = v_j$. Then

$$v = v_0 \to \ldots \to v_i \to v_{j+1} \to \ldots \to v_n = w$$

is a shorter walk from v to w. The resulting walk may yet contain a cycle, but since removing cycles shortens the walk, eventually this process will yield a walk with no cycles—that is, a path. For example, we noted above that $b \to d \to b \to d$ was a walk from b to d. This walk includes the cycle $d \to b \to d$, and removing it from the walk leaves the path consisting of one arc, $b \to d$.

Vertex w is *reachable* from vertex v if there is a path from v to w—or, equivalently, if there is a walk from v to w. The *distance* from vertex v to vertex w in a digraph G, denoted by $d_G(v, w)$, is the length of the shortest path from v to w, or ∞ if there is no such path. For example, the distance from a to d in Figure 13.1 is 2. We'll say that w is *nontrivially reachable* from v if w is reachable from v by a walk of length greater than 0. This is not quite the same thing as saying that w is reachable from v and $w \neq v$, since it is possible for there to be a nontrivial cycle from v to itself. In that case, v is nontrivially reachable from itself by following that cycle.

Because any walk from v to w includes among its arcs a path from v to w, we have the following simple result.

Lemma 13.2. *The distance from one vertex of a graph to another is at most the length of any walk from the first to the second.*

✳

A digraph in which every vertex is reachable from every other vertex is said to be *strongly connected*. To take a real-world example, the digraph representing nonstop flights between cities is strongly connected, because there is always some way to get from any point of departure to any destination! On the other hand, the digraph of Figure 13.1 is not strongly connected. Vertex *e*, for example, is not reachable from any of the other vertices. Even if we set vertex *e* aside and consider the part of the digraph consisting of just the four vertices *a*, *b*, *c*, and *d*, the digraph is not strongly connected since vertex *a* is not reachable from *b*, *c*, or *d*.

Let's be explicit about that idea of a "part of" a digraph. If $G = \langle V, A \rangle$ is a digraph, and $V' \subseteq V$ and $A' \subseteq A$, and every arc in A' is between vertices that are in V', then $\langle V', A' \rangle$ is said to be a *subgraph* of *G*. So $\langle V, \emptyset \rangle$ is a subgraph of *G*, and so is $\langle \{v, w\}, \{v \to w\} \rangle$ where $v \to w$ is any arc in *A*. If $V' \subseteq V$ is any subset of the vertices of *G*, the *subgraph induced* by V' includes all the arcs in *A* with both endpoints in V'; that is,

$$\langle V', \{v \to w \in A : v \text{ and } w \text{ are both in } V'\} \rangle.$$

So a more succinct statement about the digraph $\langle V, A \rangle$ shown in Figure 13.1 is that the subgraph induced by the vertex set $V - \{e\}$ is not strongly connected. By contrast, the subgraph induced by $\{b, c, d\}$ is strongly connected.

An acyclic digraph is generally called a *directed acyclic graph* or *DAG*. DAGs are a useful representation when there are are some constraints on the order of things but not a fully prescribed sequencing. For example, Figure 4.12 on page 43 showed the positions and moves of a game as a DAG, though we did not use that terminology at the time. As another example, consider how you might represent the rules for which college courses must be taken before which others. Suppose CS2 and CS3 are prerequisites for CS10, and that CS2 and CS3 can be taken in either order, but neither can be taken without taking CS1 first. The DAG of Figure 13.4 shows these relations in a natural way. Of course it would be a problem if the diagram for an actual course catalog contained a cycle!

The *out-degree* of a vertex *v* is the number of arcs leaving it; that is, $|\{w : v \to w \text{ is in } A\}|$. Similarly, the *in-degree* of a vertex is the number of arcs entering it.

Theorem 13.3. *A finite DAG has at least one vertex of out-degree 0 and at least one vertex of in-degree 0.*

Proof. Let $G = \langle V, A \rangle$ be a finite DAG. We'll prove by contradiction that *G* has a vertex of out-degree 0. Suppose to the contrary that *G* has no vertex of out-degree 0. Pick any vertex v_0. Since v_0 has positive out-degree, there is an arc $v_0 \to v_1$ for some vertex v_1. By the same token, v_1 has positive out-degree,

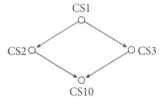

Figure 13.4. A DAG representing the prerequisites for four courses. CS1 has none; CS2 and CS3 each require that CS1 be taken first; and CS10 can't be taken until both CS2 and CS3 have been taken.

and there is an arc $v_1 \rightarrow v_2$ for some vertex v_2. There are, in fact, walks of arbitrary length starting from v_0, since every vertex has positive out-degree and therefore an arc to some other vertex. But V is finite, so some vertex will be repeated, creating a cycle. That contradicts the assumption that the digraph was acyclic.

A very similar argument proves that G has a vertex of in-degree 0, since if G had no vertex of in-degree 0, we could start from any vertex v_0 and follow arcs backward until a vertex was repeated, forming a cycle. ∎

In a DAG, a vertex of in-degree 0 is called a *source* and a vertex of out-degree 0 is called a *sink*. The DAG of Figure 13.4 has one source (CS1) and one sink (CS10), but DAGs may have more than one of either or both. For example, the subgraph of Figure 13.4 induced by omitting CS10 has one source and two sinks (Figure 13.5).

Let's pull these ideas together by talking about one special kind of digraph. A *tournament graph* is a digraph in which every pair of distinct vertices are connected by an arc in one direction or the other, but not both. As the name suggests, tournament graphs are a natural representation of a round-robin athletic tournament in which every team plays every other exactly once (and each contest results in one team winning—no draws). Figure 13.6 shows a tournament graph with four vertices and six arcs; the number of arcs in a tournament graph with n vertices will always be exactly $\frac{n(n-1)}{2}$ since there is exactly one arc between each pair of distinct vertices.

If the tournament depicted in Figure 13.6 was actually meant to determine a champion among the four teams, the result would be inconclusive. Each team played three others, and no team won all its contests. H, Y, and D stand in an awkwardly cyclic relationship to each other—H beat Y, Y beat D, but D beat H. The same holds for H, P, and D. There is no obvious way just on the basis of the teams' victories and losses to decide which should be considered the best.

On the other hand, the tournament depicted in Figure 13.7 has a clear winner: H beat every other team. Moreover, this tournament has clear second, third, and fourth place finishers. Y lost only to H, so should be second; P lost to H and to Y, so should be third; and D lost to all of H, Y, and P, and should be fourth.

Let's say that a *linear order*, denoted by $\preceq$, is a binary relation on a finite set S with the property that the elements of S can be listed $S = \{s_0, \ldots, s_n\}$ in such a way that $s_i \preceq s_j$ if and only if $i \leq j$. (We'll generalize this concept to infinite sets in the next chapter.) For example, take the set S to be all words in the English language; then we can define the linear order $s_i \preceq s_j$ to mean that either s_i appears before s_j alphabetically, or they are equal. (This is one possible linear order on S, but there are many others.)

Figure 13.5. A DAG with one source and two sinks.

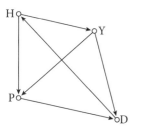

Figure 13.6. A tournament in which H beats Y and P, Y beats D and P, P beats D, and D beats H.

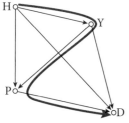

Figure 13.7. A tournament in which H beats Y, P, and D and is the clear winner of the tournament. In fact there are clear second, third, and fourth place finishers, as shown by the red path from H to D.

A *strict linear order* of a finite set, denoted by $\prec$, is an ordering such that $s_i \prec s_j$ if and only if $i < j$. That is, a strict linear order $\prec$ is like a linear order $\preceq$, except that it is always false that $x \prec x$ while it is always true that $x \preceq x$. Again taking S to be the set of all words in the English language, the analogous strict linear order $s_i \prec s_j$ is the one that means s_i appears before s_j alphabetically.

A set is strictly linearly ordered by a binary relation if there is one and only one way to list it so that earlier elements in the list always stand in the $\prec$ relation to later elements in the list. If $x \prec y$ among distinct teams is defined to mean that x beats y, and $\prec$ is a strict linear order, then there are well defined first-, second-, ..., and last-place teams.

Theorem 13.4. *A tournament digraph represents a strict linear order if and only if it is a DAG.*

Proof. Let $G = \langle V, A \rangle$ be a tournament digraph. If G represents the strict linear order $v_0 \prec v_1 \prec \ldots \prec v_n$, then every arc goes from a vertex v_i to a vertex v_j where $i < j$. Then the digraph is acyclic, since any cycle would have to include at least one arc $v_j \to v_i$ where $j > i$.

Suppose on the other hand that G is a DAG; we show that it represents a strict linear order by induction on the number of vertices. If G has but one vertex, it is trivially a strict linear order. If G has more than one vertex, at least one of the vertices, say v_0, has in-degree 0 by Theorem 13.3. Since G is a tournament digraph, G must contain each of the arcs $v_0 \to v_i$ for $i \neq 0$. Consider the subgraph induced by $V - \{v_0\}$. This induced subgraph is also a DAG, since all its arcs are arcs of G, which contains no cycles. And it is a tournament digraph, since G is a tournament digraph, and the subgraph includes all the arcs of G except those with v_0 as an endpoint. By the induction hypothesis, the induced subgraph represents a strict linear order, say $v_1 \prec v_2 \prec \ldots \prec v_n$. From our original analysis of G, $v_0 \prec v_i$ for $i \neq 0$, so putting v_0 first yields the strict linear order for all of V, namely $v_0 \prec v_1 \prec \ldots \prec v_n$. ■

Applying this method to linearly order the vertices of Figure 13.7, the first element is H since H has in-degree 0. Removing H from the digraph leaves the digraph induced by $\{Y, P, D\}$, which is a tournament digraph in which Y has in-degree 0. Removing Y yields the arc $P \to D$, in which P has in-degree 0, and finally removing P leaves the isolated vertex D. So the linear order of the four vertices is

$$H \prec Y \prec P \prec D.$$

※

Digraphs and DAGs can be infinite. For example, the natural numbers form an infinite DAG by interpreting $m \to n$ as the predecessor relation:

pred(*m*, *n*) if and only if $n = m + 1$. The DAG for *pred* looks like

$$0 \rightarrow 1 \rightarrow 2 \rightarrow 3 \rightarrow \dots \tag{13.5}$$

This digraph is a DAG, because if $n = m + 1$ so that $m \rightarrow n$, then there can be no number $p \geq n$ such that $p \rightarrow m$. This DAG has a source (0) but no sink (since there is no largest natural number). Similarly, if the arrows in this digraph were reversed, it would be a DAG with a sink (0) but no source.

Chapter Summary

- Directed graphs, or *digraphs*, are composed of *vertices* and *arcs*. Formally, a digraph is an ordered pair $\langle V, A \rangle$ where V is the set of vertices and $A \subseteq V \times V$ is the set of arcs.

- A digraph may include *self-loops* from a vertex to itself, as well as arcs in both directions between two vertices.

- A digraph can be interpreted as a binary relation, where an arc from a to b indicates that $\langle a, b \rangle$ is in the relation.

- A *walk* in a digraph is a sequence of vertices where each is connected to the next by an arc, and a walk's *length* is the number of those arcs. A *path* is a walk with no repeated vertices. A *circuit* is a walk that starts and ends at the same vertex, and a circuit is a *cycle* if it contains no other duplicate vertices. Each vertex forms its own *trivial* cycle; that is, a cycle with length 0.

- A digraph is *acyclic* if it contains no *nontrivial* cycles. Such a graph is called a *directed acyclic graph*, or *DAG*.

- A vertex w is *reachable* from another vertex v if there is a path from v to w, and the *distance* between v and w is the length of the shortest such path.

- A digraph is *strongly connected* if every vertex is reachable from every other vertex.

- The *in-degree* of a vertex is the number of arcs entering it, and its *out-degree* is the number of arcs leaving it.

- A finite DAG has at least one *source*, a vertex with in-degree 0, and at least one *sink*, a vertex with out-degree 0. This is not true of infinite DAGs; for example, the digraph representing the predecessor relation on the natural numbers has no sink.

- A *linear order* on a finite set is a binary relation $\preceq$ with the property that the elements can be listed in such a way that $x \preceq y$ if and only if x is listed before y, or is equal to y. A *strict linear order* $\prec$ is similar, except that $x \prec x$ is always false while $x \preceq x$ is always true.

Problems

13.1. Find another walk of length 3 from b to d in the digraph of Figure 13.2 (other than $b \to d \to b \to d$).

13.2. Prove that for any digraph G and any vertices u, v, w of G,

$$d_G(u, w) \le d_G(u, v) + d_G(v, w).$$

(This is called the *triangle inequality* for directed graphs.)

13.3. Consider the graph G of Figure 13.8.
  (a) What is the in-degree and out-degree of each vertex?
  (b) List the cycles of G. How many cycles pass through each vertex? (Do not count two cycles as different if they simply trace through the same arcs starting at a different vertex.)
  (c) Find the distances between each pair of vertices (25 values).
  (d) What is the length of the *longest* path in G? List all paths of that length.

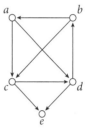

Figure 13.8. A directed graph.

13.4. Let G be a digraph with $V = \{a, b, c, d, e\}$. For which of the following sets of arcs does $G = \langle V, A \rangle$ contain a nontrivial cycle?
  (a) $A = \{\langle a, b \rangle, \langle c, a \rangle, \langle c, b \rangle, \langle d, b \rangle\}$
  (b) $A = \{\langle a, c \rangle, \langle b, c \rangle, \langle b, d \rangle, \langle c, d \rangle, \langle d, a \rangle\}$
  (c) $A = \{\langle a, c \rangle, \langle b, d \rangle, \langle c, b \rangle, \langle d, c \rangle\}$
  (d) $A = \{\langle a, b \rangle, \langle a, d \rangle, \langle b, d \rangle, \langle c, b \rangle\}$

13.5. Consider the set of prerequisites for CS classes shown in Figure 13.9.
  (a) Draw the directed graph representing these prerequisites, where the vertices represent classes and where an arc from x to y means that x is a prerequisite for y.
  (b) Is this graph a DAG? Why does it not make sense for a graph like this representing prerequisites to contain cycles?
  (c) If you must take all of a course's prerequisite courses before taking the course, what is the smallest number of semesters it would take to complete all these courses? Assume all courses are offered in every semester, and that you are able to handle an unlimited workload in a semester.

Course	Prerequisites
CS182	CS51, CS121
CS121	CS20
CS124	CS50, CS51, CS121, Stat110
CS51	CS50
CS61	CS50
CS20	None
CS50	None
Stat110	None

Figure 13.9. Courses and prerequisites.

13.6. Prove that in any digraph, the sum of the in-degrees of all vertices is equal to the sum of the out-degrees of all vertices.

13.7. (a) What is the minimum number of arcs in any strongly connected digraph with n vertices? What does that digraph look like? Prove your answer.
  (b) What is the maximum distance between any two vertices in the digraph of part (a)?
  (c) Show that with seven arcs more than the minimum from part (a), there is a digraph for which the maximum distance between two vertices is a little more than $\frac{n}{2}$.

(d) Show that for any n, there is a digraph with n vertices and $2n - 2$ arcs such that the maximum distance from any vertex to any other is 2.

13.8. In Lemma 13.2, we argued that if there is a walk between two vertices then there is a path between them, because cycles could be removed repeatedly until no cycles remain. The argument was really a proof by induction or the Well-Ordering Principle. Write out such a proof.

13.9. (a) Prove that any digraph with n vertices and more than $\frac{n(n-1)}{2}$ arcs contains a nontrivial cycle.

(b) What is the minimum value of m that makes the following statement true? Any digraph with n vertices and more than m arcs contains a cycle of length at least 2.

13.10. Let $\langle V, A \rangle$ be a finite DAG. The proof of Theorem 13.3 shows that there is an upper bound b on the length of any walk in G. What is that upper bound b in terms of V and A? What does the DAG look like in which the maximum value of that bound is met?

13.11. Consider the digraph $D = \langle \mathbb{N} - \{0, 1\}, A \rangle$, where $u \rightarrow v \in A$ if and only if $u < v$ and $u \mid v$.

(a) Draw the subgraph induced by the vertex set $\{2, \ldots, 12\}$.

(b) Which vertices of the infinite digraph D have in-degree 0? 1?

(c) What are the minimum and maximum out-degrees of any vertex in D?

(d) Prove that D is acyclic.

13.12. Consider the digraph $\langle V, A \rangle$, where V is the set of all bit strings and $u \rightarrow v \in A$ if and only if $v = u0$ or $v = u1$.

(a) What are the in-degrees and out-degrees of the vertices?

(b) Under what circumstances is vertex v reachable from vertex u?

Chapter 14

Digraphs and Relations

Let's take another look at the predecessor relation, (13.5) on page 138: m is the predecessor of n, symbolically $pred(m, n)$, if and only if $n = m + 1$. This relation is not a linear order (page 136), because, for example, it is not the case that $pred(0, 2)$—in other words, there is no arc from 0 to 2 in (13.5)—even though $0 < 2$. In the full digraph of the less-than relation, every vertex would have infinite out-degree (since for any number m, there are infinitely many n greater than m) and finite in-degree (because n is greater than exactly n natural numbers, $0, \ldots, n - 1$) (Figure 14.1).

And yet these two relations, predecessor and less-than, are closely connected: we might say that the less-than relation is the result of "extending" the predecessor relation. We need some new terminology to make this idea precise.

A binary relation $R \subseteq S \times S$ is *transitive* if for any elements x, y, $z \in S$, if $R(x, y)$ and $R(y, z)$ both hold, then $R(x, z)$ also holds (Figure 14.2). The less-than relation on natural numbers is transitive, since if $x < y$ and $y < z$ then $x < z$.

All linear orders are transitive, but many relations that are not linear orders are also transitive. For example, consider the relations *child* and *descendant* between people, where $child(x, y)$ is true if and only if x is a child of y, and $descendant(x, y)$ is true if and only if x is a descendant of y. So $descendant(x, y)$ holds if $child(x, y)$ holds. The relation $descendant(x, y)$ also holds if x is a grandchild of y; that is, there is a z such that $child(x, z)$ and $child(z, y)$. In general, *descendant* is transitive, because a descendant of a descendant of a person is a descendant of that person.

In any directed graph G, reachability is a transitive relation, since if there is a walk from u to v and a walk from v to w, then there is a walk from u to w, simply by connecting the end of the first walk to the beginning of the second.

The *transitive closure* of a binary relation $R \subseteq S \times S$ is that relation, denoted by R^+, such that for any $x, y \in S$,

$R^+(x, y)$ if and only if there exist

$\qquad x_0, \ldots, x_n \in S$ for some $n > 0$

such that $x = x_0$,

Figure 14.1. The digraph representation of the less-than relation on natural numbers.

Figure 14.2. The transitivity property: For any x, y, and z, if $x \rightarrow y$ and $y \rightarrow z$, then $x \rightarrow z$.

$$y = x_n,$$
and $R(x_i, x_{i+1})$ for $i = 0, \ldots, n - 1$.

Intuitively, the transitive closure is the "minimal" extension of R that is transitive—the additional ordered pairs that R^+ includes (beyond those in R itself) are only those that are necessary to make it transitive. If we represent R as a digraph, then R^+ is the nontrivial reachability relation on R. That is, $x \to y$ in R^+ if and only if there is a nontrivial path from x to y in R, or in symbols, if and only if

$$x = x_0 \to x_1 \to \ldots \to x_n = y \text{ in } R$$

for some $n > 0$ and some $x_0, \ldots, x_n \in S$. For example, the less-than relation on natural numbers is the transitive closure $pred^+$ of the predecessor relation. The descendant relation is the transitive closure of the *child* relation: $descendant = child^+$.

✳

A relation $R \subseteq S \times S$ is said to be *reflexive* if $R(x, x)$ holds for every $x \in S$. For example, the equality relation is reflexive, since it is always the case that $x = x$. So is the less-than-or-equal-to relation $\leq$ on the natural numbers, since $x \leq x$ for every x.

The less-than relation is not reflexive, because for instance $0 \not< 0$. Note that it takes but a single counterexample to show that a relation is not reflexive, since it is reflexive only if *every* member of S holds the relation to itself. The less-than relation has the stronger property that *no* natural number stands in the relation to itself. A relation R is said to be *irreflexive* if it is never the case that $R(x, x)$ for any $x \in S$.

We can form the *reflexive closure* of a relation R by adding to R any ordered pairs $\langle x, x \rangle$ that are not already in R:

the reflexive closure of $R = R \cup \{\langle x, x \rangle : x \in S\}$.

In the digraph representation, the reflexive closure of R includes loops from every vertex to itself.

The reflexive closure of the transitive closure of R is called the *reflexive, transitive closure* of R and is denoted by R^*. For example, the less-than-or-equal-to relation is the reflexive, transitive closure of the predecessor relation on the natural numbers, since $x \leq y$ if and only if there is a series of numbers $x_0, \ldots, x_n$ for some $n \geq 0$ such that

$$x = x_0$$
$$x_1 = x_0 + 1$$
$$x_2 = x_1 + 1$$

$$\vdots$$
$$x_n = x_{n-1} + 1$$
$$= y.$$

In the $n = 0$ case, this simply states that $x = x_0 = x_n = y$.

In terms of the digraph representations of R and R^*, there is an arc $x \to y$ in R^* if and only if there is a path (trivial or nontrivial) from x to y in R. The reflexive, transitive closure of a digraph is therefore its reachability relation.

The reflexive, transitive closure of a binary relation holds a special place in the theory of computation. To anticipate the next chapter, consider a huge digraph in which the vertices represent states of a computer (the exact values in all the memory cells, the memory address of the instruction that is about to be executed, and so on), and an arc connects one state to another if in a single instruction step the machine passes from the first state to the second. Then it is in principle easy to check whether two states are connected by an arc: just check what the executed instruction does. It may be far less obvious whether one state is *reachable* from another if the program is allowed to run long enough—that is, whether the first state is related to the second in the reflexive, transitive closure of the single-step relation!

A word on notation: The symbol "*" can be understood as denoting "0 or more of," since R^* is the relation representing that one vertex is reachable from another by a path of 0 or more arcs in R. This is consistent with our previous use of * in Σ^*, to denote a string of zero or more symbols from Σ. In the same way, "+" means "one or more of," and Σ^+ is the set of strings of nonzero length, sometimes called the *Kleene plus* of Σ.

A binary relation $R \subseteq S \times S$ is *symmetric* if $R(y, x)$ holds whenever $R(x, y)$ holds. In terms of the digraph representation, a symmetric relation has an arc $x \leftarrow y$ whenever it has an arc $x \to y$. We sometimes abbreviate such a pair of arcs by a single line with arrowheads at both ends: $x \leftrightarrow y$, but that is merely a notational convenience. If G is a digraph $\langle V, A \rangle$ with such a pair of arcs, then A contains both the arcs $\langle x, y \rangle$ and $\langle y, x \rangle$.

Think of a digraph representing the roadways near you. The vertices are the intersections and the arcs represent the direction in which traffic flows along the streets and roads. If you are in a small town, the digraph is probably symmetric, because all the roads are two-way. In a metropolitan area, there may be some one-way streets, and in that case the digraph will be not be symmetric. For most airlines, the digraph representing nonstop air flights is symmetric—if you can fly from X to Y, then there is a return flight directly from Y to X. But some airlines fly triangle routes in some areas, where there are flights from city X to Y, Y to Z, and Z to X, but to get from Y to X you have to make a stop in Z (Figure 14.3).

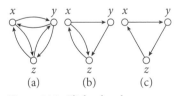

Figure 14.3. Flight plans between three cities. (a) is symmetric; (b) and (c) are not symmetric; and (c) is asymmetric and antisymmetric.

A relation is *asymmetric* if it is never the case that both $x \rightarrow y$ and $y \rightarrow x$ for any elements x and y. An asymmetric relation is irreflexive, since any self-loop $x \rightarrow x$ would violate the asymmetry condition. A relation is *antisymmetric* if $x \rightarrow y$ and $y \rightarrow x$ both hold only if $x = y$. So a reflexive relation can be antisymmetric, but not asymmetric—the less-than-or-equal-to relation on the integers is an example of such a relation.

The three basic properties—transitivity, reflexivity, and symmetry—are independent of each other, in the sense that there exist relations that satisfy any one of these properties but not the other two, and also relations that satisfy any two of these properties but not the third. (See Figures 14.4 and 14.5, for example.)

It would be tempting, but incorrect, to reason that any relation that is both symmetric and transitive must also be reflexive, since if there are arcs $x \rightarrow y \rightarrow x$ then there must be an arc $x \rightarrow x$ by transitivity. But that logic assumes that every vertex has positive out-degree (or equivalently, since the digraph is symmetric, positive in-degree). The trivial digraph, with one vertex and no arcs, is symmetric and transitive trivially. That is, it is trivially true that there is an arc $x \leftarrow y$ whenever there is an arc $x \rightarrow y$, and that there is an arc $x \rightarrow z$ whenever there are arcs $x \rightarrow y$ and $y \rightarrow z$, since the digraph has no arcs at all. But this symmetric, transitive digraph is not reflexive, since there is no loop from the unique vertex to itself.

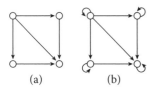

(a)   (b)

Figure 14.4. Two transitive, antisymmetric relations. The relation of (a) is asymmetric; the relation of (b) is reflexive and not asymmetric.

Figure 14.5. A relation that is symmetric without being either reflexive or transitive.

✳

A binary relation that is symmetric, reflexive, *and* transitive is called an *equivalence relation*.

Example 14.1. *Let's assume that every person has one and only one hometown. Then the binary relation between people with the same hometown is an equivalence relation: it is symmetric, since if your hometown is the same as mine then mine is the same as yours; it is reflexive, since I live in the same place as myself; and it is transitive, since if X and Y share a hometown and Y and Z share a hometown then X and Z also have the same hometown.*

Something deeper is lurking under this discussion of equivalence relations. Any equivalence relation on a set S splits S into subsets that are disjoint from each other and together make up all of S. Such a collection of subsets of S is called a *partition* of a set S, and the sets in the partition are called its *blocks*. That is, $\mathcal{T} \subseteq \mathcal{P}(S)$ is a partition of S if and only if

- each block is nonempty; that is, if $X \in \mathcal{T}$, then $X \neq \emptyset$;
- the blocks are disjoint; that is, if $X_1, X_2 \in \mathcal{T}$ and $X_1 \neq X_2$, then $X_1 \cap X_2 = \emptyset$; and
- the blocks exhaust S; that is, $S = \bigcup_{X \in \mathcal{T}} X$.

For example, the populations of the various possible hometowns comprise a partition of the set of all people; each town corresponds to a block that includes all the residents of that town. It is not an accident that there is a close connection between the equivalence relation "has the same hometown as" and the partition in which all the people with the same hometown are collected together. Let's pull these concepts together.

Theorem 14.2. *Let ↔ be any equivalence relation on a set S. Then the equivalence classes of ↔ are a partition of S. Conversely, any partition of S, say 𝒯, yields an equivalence relation ↔ on S in which all elements in the same block of 𝒯 are equivalent to each other under ↔.*

Proof. If ↔ is an equivalence relation on S, then

$$\mathcal{T} = \{\{y : y \leftrightarrow x\} : x \in S\}$$

is a partition of S, since every element of S is in one and only one block in $\mathcal{T}$. Conversely, if $\mathcal{T}$ is a partition of S, then the relation ↔ such that $x \leftrightarrow y$ if and only if x and y belong to the same block of $\mathcal{T}$ is reflexive, symmetric, and transitive, by reasoning like that of Example 14.1. The relation ↔ is therefore an equivalence relation.   ∎

✳

On page 135, we called a digraph "strongly connected" if any two vertices are reachable from each other. We now use this notion as the basis for partitioning arbitrary digraphs.

As we observed on page 141, reachability is a transitive relation. It is not necessarily symmetric—sometimes y is reachable from x but x is not reachable from y. But in any digraph, *mutual reachability* of two vertices—the relation that holds between two vertices if each is reachable from the other—is symmetric as well as transitive. And because every vertex is trivially reachable from itself, mutual reachability is transitive, symmetric, and reflexive—in other words, an equivalence relation.

It is therefore possible to partition the vertices of any digraph according to the mutual reachability relation. In the digraph of Figure 13.1 on page 133, the partition according to mutual reachability is $\{\{a\}, \{b, c, d\}, \{e\}\}$—vertices b, c, and d are all reachable from each other, but neither a nor e is reachable from any vertex but itself (see Figure 14.6). This graph might represent, for example, a neighborhood of houses and one-way streets between them (though a poorly planned neighborhood, since not every house is reachable from the others).

The subgraphs induced by the blocks of the mutual reachability relation are called the *strongly connected components* of the digraph, or simply *strong components*. Each strongly connected component is a strongly connected digraph, and no two vertices in different strong components are mutually reachable. In fact, the strong components of a digraph are connected to each other as a DAG (see Figure 14.7).

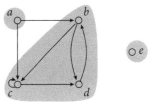

Figure 14.6. The strongly connected components of the digraph of Figure 13.1.

Figure 14.7. The DAG of the strong components of the digraph of Figure 14.6. The strong component $\{b, c, d\}$ has been reduced to a single vertex bcd. There is an arc $a \rightarrow bcd$ because in the graph of Figure 14.6, there is an arc $a \rightarrow b$ (and also an arc $a \rightarrow c$). Arcs lying completely within a strong component are irrelevant.

Theorem 14.3. *Let $G = \langle V, A \rangle$ be any digraph, and let $\mathcal{T}$ be the partition of V into the vertex sets of the strong components of G. Construct a new digraph $G' = \langle \mathcal{T}, A' \rangle$, where for any $X, Y \in \mathcal{T}$, $X \to Y \in A'$ if and only if $X \neq Y$ and $x \to y \in A$ for some $x \in X$ and some $y \in Y$. Then G' is a DAG.*

Proof. G' is acyclic because if there were a cycle in G', there would be a cycle $x_0 \to x_1 \to \ldots \to x_k \to x_0$ in G for some $x_0, \ldots, x_k$ belonging to different strong components of G. But then all these vertices would be mutually reachable in G and would not belong to distinct strong components of G.   ∎

✳

A *partial order* is a reflexive, antisymmetric, transitive binary relation. For example, $\leq$ is a partial order on integers, and the subset relation $\subseteq$ is a partial order on $\mathcal{P}(\mathbb{N})$. A *linear order*, which we defined in the previous chapter for finite sets, is a partial order such that any two distinct elements are related one way or the other. That is, if we use $\preceq$ to denote a partial order, then for any elements x and y, either $x \preceq y$ or $y \preceq x$. So any linear order is a partial order. For example, $\leq$ on the integers is both. But not every partial order is a linear order. For example, $\subseteq$ is not a linear order on $\mathcal{P}(\mathbb{N})$ since neither $\{0\} \subseteq \{1\}$ nor $\{1\} \subseteq \{0\}$ is true.

A *strict partial order* is an irreflexive, antisymmetric, transitive binary relation—in other words, the result of removing the self-loops $\langle x, x \rangle$ from a partial order. So $<$ is a strict partial order on the natural numbers, and "proper subset" $\subsetneq$ is a strict partial order on $\mathcal{P}(\mathbb{N})$. In parallel with the relationship between linear orders and partial orders, a *strict linear order* is a strict partial order such that any two distinct elements are related, one way or the other.

The graph of any strict partial order is a DAG, and any partial order is the reflexive, transitive closure of a DAG (Problem 14.8). Figure 14.8 is the reflexive, transitive closure of the graph from Figure 13.4, representing prerequisites among computer science courses. The red arrows alone show a DAG as might be stated in the course catalog: CS2 has prerequisite CS1, CS3 has prerequisite CS1, and CS10 has prerequisites CS2 and CS3. The red and green arrows together form a strict partial order, which specifies not just the immediate prerequisites but also the prerequisites of any prerequisites— in this case, that all three of CS1, CS2, and CS3 must be taken before CS10. Note that this relation is not a strict linear order, since neither CS2 nor CS3 must be taken before the other. All the arrows together (red, green, and blue) comprise the reflexive, transitive closure of the DAG represented by the red arrows.

In a partial order, a *minimal* element is one that is not greater than any other element—since not all elements are comparable to one another, there may be multiple such elements. In Figure 14.8, there happens to be only one minimal element, since CS1 is the only class with no prerequisites. But, for

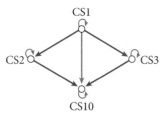

Figure 14.8. A partial order on computer science courses.

instance, among the subset of courses {CS2, CS3, CS10} there are two minimal elements, CS2 and CS3, since neither must be taken before the other. This contrasts with the usual meaning of *minimum* as applied to, say, the $<$ relationship (or any other linear order), in which there can be no more than one minimum value in a set of elements. For example, in the set of numbers {2, 3, 10}, ordered by $<$, 2 is the minimum element.

In this sense, the red arrows of Figure 14.8 are in fact a minimal DAG for which the digraph including all arcs in Figure 14.8 is its reflexive, transitive closure, if we consider DAGs as sets of arrows ordered by the strict partial order $\subsetneq$. That is, there is no subset of the red arrows that would have the same reflexive, transitive closure.

Chapter Summary

- A binary relation R on a set S is *transitive* if for any $x, y, z \in S$ such that $R(x, y)$ and $R(y, z)$ hold, $R(x, z)$ holds as well.

- The *transitive closure* of a relation R, denoted R^+, is the extension of R to the minimal transitive relation that includes R.

- A binary relation R on a set S is *reflexive* if $R(x, x)$ holds for every $x \in S$. R is *irreflexive* if $R(x, x)$ never holds, for any $x \in S$.

- The *reflexive closure* of a relation R is the extension of R to the minimal reflexive relation that includes R.

- The reflexive closure of the transitive closure of a relation R is the *reflexive, transitive closure* of R, denoted R^*. R^* is the reachability relation of the digraph representing R.

- A binary relation R on a set S is *symmetric* if $R(y, x)$ holds whenever $R(x, y)$ holds, for every $x, y \in S$. R is *asymmetric* if it is never the case that $R(x, y)$ and $R(y, x)$ both hold, for any $x, y \in S$. R is *antisymmetric* if $R(x, y)$ and $R(y, x)$ both hold only if $x = y$. Over the integers, $<$ is asymmetric but $\leq$ is not. Both $<$ and $\leq$ are antisymmetric.

- Transitivity, reflexivity, and symmetry are independent of each other.

- A binary relation that is symmetric, reflexive, and transitive is an *equivalence relation*.

- An equivalence relation on a set S corresponds to a *partition* of S, which is a set of *blocks* that are nonempty and disjoint, and together cover the entire set. These blocks are also called *equivalence classes*.

- In a digraph, *mutual reachability* of two vertices is an equivalence relation. The subgraphs induced by the blocks of the corresponding partition are the *strongly connected components* of the graph.

- A *partial order* is a reflexive, antisymmetric, transitive binary relation. A *strict partial order* is an irreflexive, antisymmetric, transitive binary relation.

- A (strict) linear order is a (strict) partial order such that any two distinct elements are related one way or the other.

- A *minimal* element of a partial order is one that is not greater than any other element; there may be multiple minimal elements since not all elements are comparable to one another.

Problems

14.1. True or false? Explain and/or correct. "In any digraph, if there is a path from u to v and a path from v to w, then connecting the end of the first path to the beginning of the second creates a path from u to w."

14.2. Is the reflexive closure of the transitive closure of a binary relation always the same as the transitive closure of the reflexive closure of that relation? Prove it or give a counterexample.

14.3. Let $f : S \rightarrow T$ be any surjection from set S to set T.
 (a) Prove that $\{f^{-1}(y) : y \in T\}$ is a partition of S, where $f^{-1}(y) = \{x \in S : f(x) = y\}$.
 (b) Why would the statement in part (a) be false if we said "function" from S to T rather than "surjection"?
 (c) What are the blocks of $\{f^{-1}(y) : y \in T\}$ if f is a bijection?
 (d) Interpret Example 14.1 as an instance of this phenomenon.
 (e) The floor function $f(x) = \lfloor x \rfloor$ is a surjection from the nonnegative real numbers onto $\mathbb{N}$. What are the blocks of $f^{-1}[\mathbb{N}]$?

14.4. Draw relations, like that of Figure 14.5, that are
 (a) reflexive without being either symmetric or transitive;
 (b) reflexive and transitive without being symmetric;
 (c) reflexive and symmetric without being transitive;
 (d) transitive and symmetric without being reflexive.

14.5. Which of the properties reflexive, symmetric, and transitive hold of the digraph with two vertices and a single arc between them?

14.6. Say that a string x *overlaps* a string y if there exist strings p, q, r such that $x = pq$ and $y = qr$, with $q \neq \lambda$. For example, *abcde* overlaps *cdefg*, but does not overlap *bcd* or *cdab*.
 (a) Draw the overlap relation for the four strings of length 2 over the alphabet $\{a, b\}$.
 (b) Is overlap reflexive? Why or why not?
 (c) Is overlap symmetric? Why or why not?
 (d) Is overlap transitive? Why or why not?

14.7. For each of the following relations, determine which of the properties reflexive, symmetric, and transitive hold, and why.
 (a) The subset relation.
 (b) The proper subset relation.

(c) The relation a set has to its power set.

(d) The relation "shares a class with," where two people share a class if there is a class that they are both enrolled in this semester.

(e) The relation R on $\mathbb{Z}$, where $R(a, b)$ if and only if b is a multiple of a (that is, if and only if there is an $n \in \mathbb{Z}$ such that $b = na$).

(f) The relation R on ordered pairs of integers, where $R(\langle a, b \rangle, \langle c, d \rangle)$ if and only if $ad = bc$.

14.8. Prove that a binary relation is a partial order if and only if it is R^* for some DAG R.

14.9. Figure 14.9 shows a set of vertices, the subsets of $\{0, 1, 2\}$. Copy the figure and add arrows representing the $\subseteq$ relation, as follows: With one color, add arrows forming a minimal DAG such that $\subseteq$ is its reflexive, transitive closure. With a second color, add the arrows needed to form the transitive closure $\subsetneq$. With a third color, add the arrows needed to form the reflexive closure $\subseteq$.

$$\begin{array}{llll} & \{0\} & \{0, 1\} & \\ \emptyset & \{1\} & \{0, 2\} & \{0, 1, 2\} \\ & \{2\} & \{1, 2\} & \end{array}$$

Figure 14.9. The subsets of $\{0, 1, 2\}$.

14.10. What are the strong components of the digraph of Figure 13.8 (page 139)? Draw the DAG of those components.

14.11. Show that the divisibility relation $(p \mid q)$ is a partial order on $\mathbb{N} - \{0\}$. What is the minimal DAG such that the divisibility relation is its reflexive, transitive closure?

14.12. What are the minimal elements of

$$S = \{\{1\}, \{2, 3\}, \{1, 4\}, \{3, 4\}, \{3, 5\}, \{1, 2, 4\}, \{2, 4, 5\}, \{1, 2, 3, 4, 5\}\},$$

ordered by $\subsetneq$?

Chapter 15

States and Invariants

The states of a digital computer are *discrete*. Computers operate by jumping directly from one state to another; they can't occupy any in-between state. An integer-valued variable, for example, or the *register* (memory cell) where it is stored, might contain the value 1 or 2, but not 1.5 or $\sqrt{2}$. Even so-called "floating point" values can have only one of a discrete set of values. By contrast, old-fashioned slide rules and radio dials are *analog* devices— they represent continuous quantities, and in principle could have any value in between two others they can represent.[1]

So an analog device can be in any one of infinitely many different states, in the same way that there are infinitely many real numbers between 0 and 1, while digital devices are made up of components that can have only two states. We generally call the states 0 and 1; there is no value of 0.5 or $\frac{2}{3}$.[2]

Digital computers are powerful because they are made up of a great many such binary components. With larger and larger numbers of components, the number of states of the entire digital device increases exponentially. If we think of the external storage and input devices and other computers to which a computer is connected as all being part of the system, a digital system could be of practically unlimited size, but its states are nonetheless discrete.

In this chapter we will develop a useful mathematical model of discrete-state systems such as digital computers. The model will work regardless of the internal structure of the device it represents, or indeed whether the system being represented is finite or infinite. The underlying structure of any discrete-state system will be a digraph. The vertices represent states and an arc from one state to another represents a transition—a "jump"—from the first state to the second. Typically, a transition happens in one time step or one instruction execution. To take a very simple example, consider Figure 15.1, which represents a system that can store a single but arbitrarily large natural number, and add 1 to it. So the states correspond to the numbers 0, 1, ..., and the arcs show the transitions by adding 1: from 0 to 1, from 1 to 2, and so on.

To represent a system with more than one kind of state transition, we need to label the arcs. For example, Figure 15.2 shows the same system with two operations, adding 1 and subtracting 1. Note that there is no transition

[1] The position of an analog device is "analogous" to the value of the physical quantity it represents.

[2] At least there shouldn't be. Ultimately transistors and other digital components are in fact analog devices, and it is a fact of electrical engineering that they can occupy intermediate states due to manufacturing or environmental problems. How to make analog components behave digitally is an important area of computer science and engineering.

Figure 15.1. A system in which a single natural number is stored and the only operation on it is to add 1.

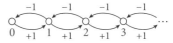

Figure 15.2. A system in which a single natural number is stored and the operations are adding and subtracting 1.

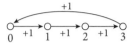

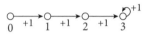

Figure 15.3. The state representation of a two-bit counter, in which adding 1 to 3 results in the value 0.

Figure 15.4. A system capable of storing a number in the range 0, …, 3 in which adding 1 when in state 3 leaves the system in state 3.

out of the 0 state by subtracting 1. The diagram suggests that it is simply impossible to subtract 1 if the system is in the 0 state.

Figure 15.3 shows a system with exactly four possible states, named 0, 1, 2, and 3. The transitions represent adding 1 as in Figure 15.1, except that adding 1 when in state 3 results in a transition to state 0. This is more or less the way a two-bit register would work, if the carry-out of the leftmost bit position is ignored and discarded.

Diagrams like these are useful for representing the intended *behavior* of a digital system, independent of its internal structure. Figure 15.3 represents just the fact that the device has four states and is therefore capable of "remembering" a number between 0 and 3 inclusive—it does not describe the implementation, which might be two bits or might not. Figure 15.4 shows the behavior of a similar system in which adding 1 to 3 leaves the system in state 3 instead of transitioning to state 0.

✳

It's time to be more precise, in particular about what a "labeled arc" is.

A *state space* is a pair $\langle V, A \rangle$, where V and A are sets. The members of V are called *states*. The members of A, which are called *labeled arcs*, are triples of the form $\langle x, a, y \rangle$, where x and y are states and a is a member of some finite set Σ, called the *labels*.

If all the labels were dropped, and each labeled arc $\langle x, a, y \rangle$ was replaced by the pair $\langle x, y \rangle$, then the result would be an ordinary digraph, what we call the digraph *underlying* the state space. Because a state space can have multiple labeled arcs with different labels connecting the same two states, the underlying digraph may have fewer arcs than the state space had labeled arcs. We write a labeled arc $\langle x, a, y \rangle$ as $x \overset{a}{\to} y$. So the state space of Figure 15.2 is $\langle V, A \rangle$, where

$$V = \mathbb{N},$$

$$\Sigma = \{+1, -1\}, \text{ and}$$

$$A = \{n \overset{+1}{\longrightarrow} n + 1 : n \in \mathbb{N}\}$$

$$\cup \{n + 1 \overset{-1}{\longrightarrow} n : n \in \mathbb{N}\}.$$

A state space is a useful representation of a digital system that moves from state to state in response to some series of external events. For example, let's imagine a computing device with two possible inputs, *a* and *b*. At every time step, the device gets an input—these could be printed symbols that the device reads one at a time, or buttons being pushed by an operator, or some other source of binary events. The state space illustrated in Figure 15.5 keeps track of how many of each input symbol have been encountered. So the states will represent two numerical values, unlike the single numerical value of Figure 15.1: $\langle m, n \rangle$ represents the state in which m of the inputs seen so far are *a* and n of them are *b*. Formally,

$$V = \mathbb{N} \times \mathbb{N},$$

$$\Sigma = \{a, b\}, \text{ and}$$

$$A = \{\langle m, n \rangle \xrightarrow{a} \langle m+1, n \rangle : m, n \in \mathbb{N}\}$$

$$\cup \{\langle m, n \rangle \xrightarrow{b} \langle m, n+1 \rangle : m, n \in \mathbb{N}\}. \tag{15.1}$$

Figure 15.5 is a static representation of all the possible states of our simple counting device and how they relate to each other, at least for single-step transitions. We'll now move to a more global statement about how those states relate to the operation of the device over arbitrarily large numbers of steps.

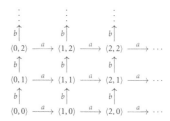

Figure 15.5. A state space for counting two symbols.

Example 15.2. *Suppose the machine starts in state $\langle 0, 0 \rangle$. Then after reading any input composed of m occurrences of a and n occurrences of b, the machine will be in state $\langle m, n \rangle$.*

Solution to example. This seems almost obvious, but it's important to realize that it is true by mathematical induction. The induction variable is the length of the input string read so far; let's call that length ℓ. The length ℓ is a nonnegative integer.

Base case: $\ell = 0$. Then the input string read thus far is λ; that is, the machine is just starting. We stipulated that the machine starts in state $\langle 0, 0 \rangle$. The empty string λ has 0 occurrences of a and 0 occurrences of b, so the state indeed represents the pair consisting of the numbers of a and b symbols seen so far.

Induction hypothesis. Let's assume that for some fixed but arbitrary $\ell \geq 0$, after any input string of length ℓ, the machine will be in state $\langle m, n \rangle$, where m is the number of occurrences of a seen so far and n is the number of occurrences of b (with $m + n = \ell$).

Induction step. Now consider any string w of length $\ell + 1$, where $\ell \geq 0$. Since $|w| > 0$, w can be written as $w = u\sigma$, where u is a string of length ℓ and σ is either a or b. By the induction hypothesis, after the input string u the system is in state $\langle m, n \rangle$, where m and n are respectively the numbers of occurrences of a and b in u. By (15.1), if $\sigma = a$, then the device transitions to state $\langle m+1, n \rangle$, and if $\sigma = b$, then the device transitions to state $\langle m, n+1 \rangle$. In either case, the new state is the ordered pair consisting of the numbers of occurrences of a and b in $w = u\sigma$. ∎

Let's present in more abstract and general terms what is really going on in the inductive proof of Example 15.2. We ascribed a general property to states, namely that the system reaches state $\langle m, n \rangle$ if and only if the input read thus far comprises m occurrences of a and n of b. We then proved that this property was true of every state by induction. First we showed that it was true of the start state $\langle 0, 0 \rangle$. And then we proved that if it was true of every state after a computation of length ℓ, then it was true of every state after any computation of length $\ell + 1$. By a simple mathematical induction, it is true of all states in all computations that start as stipulated.

This is an example of the *Invariant Principle*. Informally, something that is true at the beginning of a computation and is preserved under all state transitions is true at any point during a computation.[3] Formally,

Theorem 15.3. Invariant Principle. *Let $\langle V, A \rangle$ be a state space, and let $P(v)$ be a predicate on states. Suppose $P(v_0)$ is true, where v_0 is a designated start state. And suppose that P is preserved by state transitions; that is, suppose that for any states x and y and any transition $x \overset{a}{\to} y$, if $P(x)$ is true, then $P(y)$ is true. Then $P(v)$ holds for any state v that is reachable from v_0 in the digraph underlying $\langle V, A \rangle$.*

The predicate P is called an *invariant*. We omit the proof of the Invariant Principle, which closely tracks that of Example 15.2. But note that the reachability condition is not to be taken lightly. There will be cases in which a state space and a start state are designated, and the predicate is true of the start state and preserved under state transitions, and yet the predicate is not true of every state because certain states are unreachable from the start state. Consider, for example, Figure 15.6, which is like Figure 15.1 but includes states corresponding to the negative integers. If state 0 is the designated start state, then the predicate "state n is the state reached after n state transitions" is true if $n \geq 0$, but is not true of state -1, which cannot be reached from state 0.

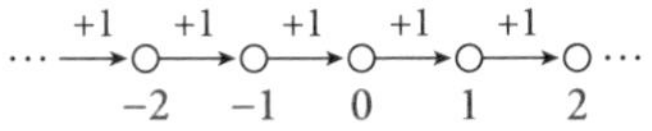

Now let's look at a "real" algorithm and its corresponding state space. The algorithm is *Euclid's algorithm* for computing the greatest common divisor (GCD, for short) of two positive integers m and n; that is, the largest integer that divides both. Now there is an obvious way to find that number (which always exists, since it is 1 if no larger number divides them both): just start with either of the numbers, check if it divides both, and if not keep subtracting 1 and doing the same check. In what was arguably the first result in what we now call number theory, Euclid realized that there was a much more efficient way of solving this problem.[4] Here it is:

Euclid's algorithm for computing the GCD of m and n:

1. $\langle p, q \rangle \leftarrow \langle m, n \rangle$ (simultaneously assign m to p and n to q)

2. While $q \neq 0$

    (a) $\langle p, q \rangle \leftarrow \langle q, p \bmod q \rangle$

3. Return p

[3] Formulated in early papers by Robert Floyd and C.A.R. Hoare. However, Floyd himself attributes the principle to Saul Gorn, and Gorn, in turn, ascribes its earliest instantiation to a 1947 programming manual by Herbert Goldstine and John von Neumann. Robert Floyd, "Assigning Meanings to Programs," in *Mathematical Aspects of Computer Science*, vol. 19 of Proceedings of Symposia in Applied Mathematics (Providence, RI: American Mathematical Society, 1967), 19–32; C.A.R. Hoare, "An Axiomatic Basis for Computer Programming," *Communications of the ACM* 12, no. 10 (October 1969): 576–83.

Figure 15.6. This state space has a state for every integer, but those representing negative numbers are not reachable from state 0.

[4] Euclid, who lived around the turn of the third century BCE, probably got this algorithm from older sources, perhaps the school of Pythagoras. He describes it, in Book VII, Proposition 2 of the *Elements*, using repeated subtraction rather than division.

We're using an informal programming notation to explain the algorithm. The statement inside the "while" uses the binary operation mod, where p mod q is the remainder when p is divided by q. So 17 mod 3 is 2, for example, since when 17 is divided by 3, the quotient is 5 and the remainder is 2: $17 = 5 \cdot 3 + 2$. Since the remainder p mod q is one of the numbers 0, 1, …, $q - 1$, it is always the case that $(p \bmod q) < q$, and

$$p = \left\lfloor \frac{p}{q} \right\rfloor \cdot q + (p \bmod q).$$

In our algorithmic notation, the left arrow "←" assigns the value on the right to the variable on the left. In the first step, we are assigning the value of the ordered pair $\langle m, n \rangle$ to the ordered pair $\langle p, q \rangle$; that simply means assigning $p \leftarrow m$ and $q \leftarrow n$ simultaneously. In the same way, $\langle p, q \rangle \leftarrow \langle q, p \bmod q \rangle$ means that the values on the right are both calculated and then both simultaneously assigned to the variables on the left. That is, this statement says that the "new" value of p is the "old" value of q, and the "new" value of q is the "old" value of p mod the "old" value of q.

To see the algorithm in action, let's try the values $m = 20$ and $n = 14$. The successive values of p and q are

p	q	
20	14	
14	6	(since 20 mod 14 = 6)
6	2	(since 14 mod 6 = 2)
2	0	(since 6 mod 2 = 0),

so the greatest common divisor of 20 and 14 is 2.

Note that if we had instead started out with $m = 14$ and $n = 20$, then initially p would be 14 and q would be 20, but on the first iteration of the "while" statement the values would have been reversed, since 14 mod 20 = 14 ($14 = 0 \cdot 20 + 14$).

Theorem 15.4. *Euclid's algorithm computes the greatest common divisor.*

Proof. We look at the state space, define an appropriate invariant, and apply the Invariant Principle. The states are pairs of natural numbers; that is, $V = \mathbb{N} \times \mathbb{N}$, representing the value of $\langle p, q \rangle$. The transitions are between the values of these variables on successive executions of the "while" statement: $\langle p, q \rangle \to \langle q, p \bmod q \rangle$. For example, the transitions for the example just shown are

$$\langle 20, 14 \rangle \to \langle 14, 6 \rangle$$
$$\to \langle 6, 2 \rangle$$
$$\to \langle 2, 0 \rangle.$$

The start state is $\langle m, n \rangle$. And the invariant is that *in any reachable state $\langle p, q \rangle$, the greatest common divisor of p and q is equal to the greatest common divisor of m and n.*

The invariant holds initially, when $p = m$ and $q = n$. Now suppose that the greatest common divisor of p and q is the greatest common divisor of m and n. What about the greatest common divisor of q and $(p \bmod q)$?

Let $r = p \bmod q$, so that $p = kq + r$ for some $k \in \mathbb{N}$. We show that the set of divisors of p and q is exactly the same as the set of divisors of q and r, so the largest members of those sets are equal.

Suppose d is any divisor of p and q. Then d is also a divisor of r, since $r = p - kq$. Likewise, if d is any divisor of both q and r, then d is also a divisor of p, since $p = kq + r$. So d is a divisor of both p and q if and only if d is a divisor of both q and $(p \bmod q)$. Therefore the *greatest* common divisor of p and q is also the greatest common divisor of q and $(p \bmod q)$, so the invariant is preserved under state transitions.

When and if the algorithm reaches a state with second component 0, say $\langle r, 0 \rangle$, r must be the greatest common divisor of m and n. This is because the GCD of the components of any reachable state is the GCD of m and n, and the GCD of r and 0 is r. So by the Invariant Principle, Euclid's algorithm computes the GCD of m and n—if it terminates. And the algorithm does terminate, since the value of the second component decreases at each step: invariably $(p \bmod q) < q$ at each step, and $(p \bmod q)$ is never negative, so eventually $(p \bmod q) = 0$. ∎

Chapter Summary

- Digital devices like computers have a finite number of *discrete* states, whereas analog devices have an infinite number of states and move continuously between them.

- The behavior of a digital device can be represented by a *state space*, which consists of a set of *states* and a set of *labeled arcs* that connect those states.

- A labeled arc from x to y, labeled with the symbol a, is denoted $x \xrightarrow{a} y$. Labeled arcs represent the possible *transitions* between states.

- The digraph *underlying* a state space is the digraph such that its vertices are the states of the state space, containing an arc from state x to y if there is any labeled arc from x to y in the state space.

- The labels of the labeled arcs represent input to the device, which drive the device from one state to another.

- A predicate is an *invariant* if it is true of the start state, and is preserved under all state transitions.

- The *Invariant Principle* says that an invariant is true of all reachable states.

Problems

15.1. Show the state transitions of Euclid's algorithm during its computation on the numbers 3549 and 462.

15.2. Show the entire state space for the game of Example 4.1 (page 39), starting with seven coins. Show all transitions, and distinguish winning from losing states for the first player.

15.3. Consider a two-bit register with an overflow bit. The register can store the nonnegative integers up to 3, and both adding 1 and subtracting 1 are available operations. Adding 1 to 3 gives the value 0 but turns on the overflow bit, and subtracting 1 from 0 gives the value 3 but turns on the overflow bit. Once the overflow bit is turned on, it remains on through any further operations. Draw the complete state and transition diagram for this system.

15.4. The 15-puzzle is a 4×4 grid with tiles numbered 1 to 15, and one square empty. The tiles can be moved up, down, left, or right, up to the edges of the grid. Another way to look at it is that the empty square can be moved up, down, left, or right.

The 16 grid positions also have numbers from 1 to 16, starting at the top left, across the top row, then from left to right in the second row, and so on. At the start of the game, tile i is in position i for $1 \le i \le 15$, and position 16 is empty (Figure 15.7(a)). We aim to show that no matter how the tiles are moved around, they can't be repositioned so that the empty spot is in the upper left corner and all the tiles are still in order (that is, tile i is in position $i + 1$ for $1 \le i \le 15$, as in Figure 15.7(b)).

Given any configuration of the tiles on the grid, let d_i, for $1 \le i \le 15$, be the number of tiles which are in grid positions later than the position of tile i and which have tile numbers less than i. For example, in the initial position, d_i would be 0 for every i, but if tile 15 were in the top left corner, then d_{15} would be 14. Now let $D = r + \sum_{i=1}^{15} d_i$, where r is the row number of the empty space. (The top row is row 1 and the bottom row is row 4.)

(a) Show that if the empty space moves left or right, D does not change.

(b) Show that if the empty space moves up or down, D changes by an even number.

(c) Show that the state space of the 15-puzzle satisfies the invariant that D is even.

(d) Show that the configuration in which the empty space is in the top left corner but the tiles are in order cannot be reached from the initial configuration.

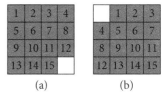

Figure 15.7. (a) Initial configuration of the 15-puzzle. (b) Is this configuration possible?

15.5. Problem 1.4 (page 9) showed that if one pigeon is placed on each square of a 9×9 grid and all pigeons move up, down, left, or right simultaneously, some grid square winds up with at least two pigeons on it. Now consider instead an $n \times n$ grid for any n and imagine that such moves are repeated indefinitely.

(a) Find and prove an invariant from which you can conclude that it is not possible for all n^2 pigeons to wind up on a single square.

(b) Is it possible for all pigeons to wind up on a set of squares that are all aligned on one diagonal line?

15.6. (a) Prove that for any $a, b, c \in \mathbb{N}$ such that $c > 0$,

$$(a + b) \bmod c = ((a \bmod c) + (b \bmod c)) \bmod c.$$

(b) *Throwing out nines.* Consider the following algorithm for determining whether a decimal numeral represents a number divisible by 9 without having to do arithmetic on any number larger than 16, by going through the digits one at a time from left to right.

1. $r \leftarrow 0$

2. While any digits remain

   (a) $d \leftarrow$ the next digit

   (b) $r \leftarrow r + d \bmod 9$

3. If $r = 0$ then the number was divisible by 9, else it was not.

This algorithm has a state space of size 9, corresponding to the values of r. There are 10 possible input digits. Write the transitions as a 9×10 table showing the next state for each possible current state and input symbol.

(c) Write an invariant describing what is true of a string that has been read at the time the automaton enters state r, $0 \leq r < 9$.

(d) Prove that the algorithm maintains the invariant, and conclude that the algorithm is correct.

15.7. Tic-tac-toe is played on a 3×3 grid, which starts out empty. X moves first and chooses a square, then O moves and chooses an unoccupied square. The players alternate until there are three occurrences of X in a row or three of O, in which case that player wins; or the board is full, in which case the game is a draw. See Figure 15.8.

Describe the game of tic-tac-toe formally as a state space with transitions. State and prove an invariant from which you can infer that the state in Figure 15.9 is unreachable.

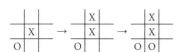

Figure 15.8. The third and fourth moves in a game of tic-tac-toe.

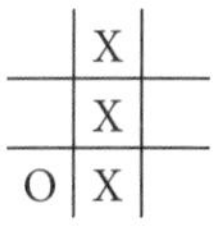

Figure 15.9. An unreachable state.

15.8. Bulgarian solitaire is a game played by one player. The game starts with 2 piles of 3 coins each. Then the player repeats the following step:

• Remove one coin from each existing pile and form a new pile.

The order of the piles doesn't matter, so the state can be described as a sequence of positive integers in non-increasing order adding up to 6. For example, the first two moves are $(3, 3) \to (2, 2, 2)$ and $(2, 2, 2) \to (3, 1, 1, 1)$. On the next move, the last three piles disappear, creating piles of 4 and 2 coins.

(a) Trace the sequence of moves until it repeats.

(b) Draw as a directed graph the complete state space with six coins and initial piles of various sizes.

(c) Repeat part (b) with 7 coins.

(d) Show that if the stacks are of heights $n, n - 1, \ldots, 1$ for any n, the next configuration is the same.

(e) Prove that for any $n \geq 1$, if $k = \sum_{i=1}^{n} i$, then starting the game with a single pile of k coins eventually yields a configuration in which there are piles of n, $n-1$, ..., and 1 coin, which then repeats itself. (In fact, starting from *any* distribution of $\sum_{i=1}^{n} i$ coins into piles eventually results in that same configuration $n, n-1, \ldots, 1$, but proving that is not straightforward!)

Chapter 16

Undirected Graphs

An undirected graph is a collection of vertices, together with edges connecting them. Figure 16.1[1] shows one of the first uses of undirected graphs in a scientific paper, by J. J. Sylvester in 1878, to suggest the structure of "an isolated element of carbon." This graph has eight vertices and twelve edges; in modern depictions, like the other ones in this chapter, circles or bullets are used to highlight the vertices.

Undirected graphs are different from digraphs in two ways. First, edges have no direction—both ends are the same. (That is why we refer to them as "edges" rather than "arcs.") And second, an edge cannot connect a vertex to itself—there are no self-loops. In an undirected graph there cannot be two edges between the same pair of vertices, in much the same way that in a digraph there cannot be two arcs in the same direction between two vertices. Figure 16.2 shows an undirected graph with nine vertices—note that they do not all need to be "connected together," a concept we will make precise shortly.

Figure 16.3 shows features that *cannot* appear in any undirected graph. The concept of a graph can be generalized to include these features—multiple edges, loops, and even a mixture of directed arcs and undirected edges—but for us, an undirected graph has none of these "extras."

Formally, an *undirected graph*, or simply a *graph*, is a pair $\langle V, E \rangle$ where V is a set of *vertices* and E is a set of *edges*. An edge is simply a two-element subset of V, though we will ordinarily write an edge as $x{-}y$ rather than $\{x, y\}$. We will refer to x and y as the *endpoints* of the edge $x{-}y$, which *joins* x and y. We also say that x and y are *incident* on the edge $x{-}y$.

The graph of Figure 16.2 has nine vertices and eight edges:

$$V = \{1, 2, 3, 4, 5, 6, 7, 8, 9\},$$
$$E = \{1{-}4, 2{-}4, 4{-}5, 2{-}5, 5{-}3, 5{-}6, 3{-}6, 7{-}8\}.$$

Because E is a set, we include only one of $x{-}y$ and $y{-}x$, since they refer to the same edge $\{x, y\}$.

[1] J. J. Sylvester, "On an Application of the New Atomic Theory to the Graphical Representation of the Invariants and Covariants of Binary Quantics, with Three Appendices," *American Journal of Mathematics* 1, no. 1 (January 1, 1878): 64–104, doi:10.2307/2369436.

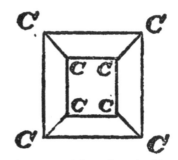

Figure 16.1. Sylvester's graph of the structure of a carbon atom.

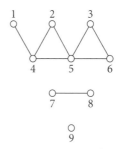

Figure 16.2. An undirected graph with three connected components. Or, three connected undirected graphs.

Figure 16.3. Undirected graphs cannot have multiple edges between the same vertices, or an edge from a vertex to itself, or any directed arcs.

Walks and paths in graphs mean almost the same thing as they mean for directed graphs (see page 134), but there are some subtleties, so it is better to restate the definitions from scratch for the undirected case.

A *walk* in a graph is a series of one or more vertices in which successive vertices are joined by an edge. The *length* of the walk is the number of edges. A *path* is a walk that does not repeat any edge. A *circuit* is a path that ends at its beginning. A *cycle* is a circuit that repeats no vertex, except that its first and last vertices are the same. So in Figure 16.2,

$$1-4-2-5-4-1 \text{ is a walk,}$$
$$1-4-5-3 \text{ is a path,}$$
$$2-4-5-3-6-5-2 \text{ is a circuit,}$$
$$\text{and } 4-2-5-4 \text{ is a cycle.}$$

On the other hand, $7-8-7$ is a walk but is not a path, and is therefore neither a circuit nor a cycle, because the edge $7-8$ is the same as the edge $8-7$. There is no path from 9 to any other vertex. Unlike the convention for digraphs, we don't consider a single vertex to be a trivial cycle (though we do still consider it to be a trivial path).

Theorem 16.1. *Any circuit includes a cycle.*

Proof. Suppose $C = v_0 - v_1 - \ldots - v_n = v_0$ is a circuit of length n. We find a cycle within this circuit by finding a subsequence of these vertices comprising a path of length at least 3, with no duplicate vertices except that the first and last vertices are the same. Let $0 \leq i < j \leq n$ be such that $v_i = v_j$ and $j - i$ is minimal—there are no k and ℓ, $0 \leq k < \ell \leq n$, such that $v_k = v_\ell$ and $\ell - k < j - i$. We can find such i and j by the Well-Ordering Principle, since the set of values of $j - i$ such that $i < j$ and $v_i = v_j$ has a smallest element. Moreover, $j - i \geq 3$ since $j - i = 1$ would require there being an edge from a vertex to itself, and $j - i = 2$ would mean that $v_i - v_{i+1}$ and $v_{i+1} - v_{i+2}$ would be the same edge. Then $v_i - v_{i+1} - \ldots - v_j$ is a cycle: it is a path of length at least 3 that ends at its beginning, and if there were a pair of duplicates among the vertices $v_{i+1}, \ldots, v_j$, the difference of their indices would be less than $j - i$, contradicting the minimality of $j - i$. ∎

Two vertices are *connected* if there is a path between them. Connectedness is an equivalence relation on vertices (note in particular that every vertex is connected to itself by a path of length zero). The equivalence classes are called the *connected components* of the graph. So the graph of Figure 16.2 has three connected components, with 6, 2, and 1 vertices. A graph is *connected* if it has only one connected component, so this graph is not connected—it is *disconnected*.

Let's compare this notion of connectedness to that of the strongly connected components of a digraph. Any undirected graph has a corresponding directed graph in which the edge $x-y$ is replaced by the two directed edges $x \leftrightarrow y$. Two vertices are connected in an undirected graph if they are connected in the corresponding digraph, and the connected components of the graph are the same as the strongly connected components of its corresponding digraph.

Conversely, we can turn a digraph into a corresponding undirected graph: any directed arc $x \rightarrow y$ is replaced by an edge $x-y$, duplicates are merged, and self-loops are removed. That graph is called the graph *underlying* the digraph (Figure 16.4). The strongly connected components of the digraph do *not* correspond to the components of the underlying graph. The components of the underlying graph are referred to as the *weakly connected components* of the digraph. That is, two vertices of a digraph are weakly connected if it is possible to get from one to the other while following directed arcs *in either direction*. In Figure 16.4, the digraph has two strong components but only a single *weak component*, since all three vertices can be reached from each other if the direction of the arcs is ignored.

The *degree* of a vertex is the number of edges incident on it. The graph of Figure 16.2 has one vertex of degree 4, one of degree 3, three of degree 2, three of degree 1, and one of degree 0. The degrees of all the vertices total 16, which is twice the number of edges. That makes sense, since each edge contributes 1 to the degrees of each of two vertices. For that reason, the sum of the degrees of the vertices of *any* graph must be even—to be precise, equal to twice the number of edges.

If $G = \langle V, E \rangle$ is a graph, $v \in V$, and e is a set of two vertices in V, then $G - v$ is the result of removing v and any edges incident on v, $G - e$ is the result of removing e (assuming it was present in E) but leaving its endpoints, and $G + e$ is the result of adding e to E. That is,

$$G - v = \langle V - \{v\}, E - \{\{v, y\} : y \in V\} \rangle$$
$$G - e = \langle V, E - \{e\} \rangle$$
$$G + e = \langle V, E \cup \{e\} \rangle.$$

With just these few concepts, we can state and prove the oldest theorem in graph theory.

An *Eulerian circuit*[2] is a circuit that includes every edge exactly once.

Theorem 16.2. Euler's Theorem. *A connected graph has an Eulerian circuit if and only if every vertex has even degree.*

Proof. Suppose that a connected graph has an Eulerian circuit. At every vertex, the number of edges by which the circuit enters the vertex is equal to the

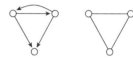

Figure 16.4. A digraph and its underlying undirected graph. The digraph has two strongly connected components (the top two vertices are reachable from each other, and the bottom vertex is a separate strong component). It has only one weak component, which is the one connected component of the underlying graph.

[2] Named after *Leonhard Euler*, an extremely prolific eighteenth-century mathematician who made important contributions to many areas of mathematics, including graph theory. Theorem 16.2 was inspired by Euler's study of the seven bridges of Königsberg, a city in what was then Prussia.

number of edges by which the circuit leaves the vertex, since a circuit returns to its start vertex; and this accounts for every edge in the graph, since each edge appears exactly once in an Eulerian circuit. So the total number of edges incident on each vertex is even.

Now suppose that G is a connected graph and each vertex has even degree. We prove that G has an Eulerian circuit by induction on the number of edges of G. No graph with fewer than 3 edges can have an Eulerian circuit, and if G has 3 edges, it is a triangle and plainly has an Eulerian circuit. So suppose G has more than 3 edges, and assume as the induction hypothesis that there is an Eulerian circuit in any connected graph with fewer edges than G in which every vertex has even degree.

We'll start by proving that we can always construct a circuit, by picking any vertex and then wandering around the graph, following edges but not reusing any edge that has already been traversed, until we return to the original vertex. First, it is possible to leave any vertex other than the original vertex via an unused edge: its count of unused incident edges was even before it was reached, became odd when the walk reached the vertex (so in particular, it was not 0), and became even again upon departure from the vertex. Second, we will eventually reach the original vertex again: it is possible to leave any other vertex that is reached, so if we are forced to end the walk, it must be because we have returned to the original vertex. So a circuit exists; however, it may not include every edge. Let's call this circuit $C = c_0 - c_1 - \ldots - c_\ell$ (Figure 16.5).

Removing all the edges of C breaks the graph into a set S of connected subgraphs. Each of these graphs has fewer edges than G. Every vertex in each graph in S has even degree, since each vertex lost an even number of incident edges when C was removed from G (Figure 16.6). By induction, each $H \in S$ has an Eulerian circuit C_H. It remains only to glue C and the C_H together to create an Eulerian circuit of G.

Each $H \in S$ shares a vertex with C since G was a connected graph. Let's say that v_H is a vertex that H shares with C. (Some of the v_H could be the same for different H.) Construct an Eulerian circuit of G by starting at c_0 and following the edges of C until $c_\ell = c_0$ is reached, taking a detour at each vertex c_i to follow the Eulerian circuit of any graph $H \in S$ such that $v_H = c_i$. ∎

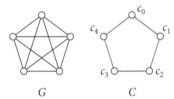

Figure 16.5. Finding an Eulerian circuit. G is on the left. Every vertex has degree 4 so this graph should have an Eulerian circuit. On the right, one circuit found by visiting the vertices in the order c_0, c_1, c_2, c_3, c_4. This is the circuit C of the proof.

Figure 16.6. The result, H, of removing the edges of C from G. Each vertex of H has degree 2 and therefore (by induction) H has an Eulerian circuit.

In the example of Figures 16.5 and 16.6, suppose that circuit of H visits the vertices in the order c_0, c_2, c_4, c_1, c_3, c_0. If the chosen vertex v_H that is common to G and H is c_0, then the Eulerian circuit for G is

$$c_0 - c_2 - c_4 - c_1 - c_3 - c_0 - c_1 - c_2 - c_3 - c_4 - c_0,$$

where we have colored the circuit of H in black and the circuit C in red to correspond to the colors of the edges in the figures.

Figure 16.7 shows a graph that is nearly identical to that of Figure 16.5, except that it is missing one of the edges. This graph has vertices of odd degree, so it does not have an Eulerian circuit.

Figure 16.7. A graph with no Eulerian circuit.

✳

Two graphs are *isomorphic* if there is a bijection between their vertices that preserves their edges. That is to say, an isomorphism between two graphs $G = \langle V, E \rangle$ and $G' = \langle V', E' \rangle$ is a bijection $f : V \to V'$ such that $x-y \in E$ if and only if $f(x)-f(y) \in E'$. Informally, isomorphic graphs are the same "up to renaming vertices." For example, the two graphs in Figure 16.8 don't look very much alike, but are essentially the same, under the isomorphism

$$1 \leftrightarrow A$$
$$2 \leftrightarrow D$$
$$3 \leftrightarrow C$$
$$4 \leftrightarrow B$$
$$5 \leftrightarrow E.$$

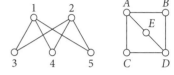

Figure 16.8. Isomorphic graphs.

In many cases, we care only about properties of a graph that don't depend on what the particular vertices are called or how it is drawn, and in that case we say that we care about the graph only "up to isomorphism." For example, a graph is said to be *planar* if it is possible to draw it in the plane (on a piece of paper) in such a way that no two edges cross. Figure 16.8 shows a graph that at first doesn't look planar, but is—planarity is a property of the graph itself, not of the particular way it is drawn. That is, if a graph is planar, then any graph isomorphic to it is planar too.

If two graphs are isomorphic, they must have the same number of vertices and the same number of edges. They must also have the same number of vertices of each degree. For example, in Figure 16.8, both graphs have three vertices of degree 2 and two vertices of degree 3. So it is sometimes possible to prove quickly that two graphs are *not* isomorphic, even if they have the same number of vertices and edges, by counting the degrees of all the vertices and finding a discrepancy. For example, the two graphs of Figure 16.9 have the same number of vertices and edges, but can't be isomorphic since the one on the left has a vertex of degree 4 but the one on the right does not.

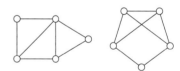

Figure 16.9. Nonisomorphic graphs with the same numbers of vertices and edges.

Unfortunately, having the same number of vertices of each degree is not a sufficient condition for two graphs to be isomorphic (see Figure 16.10). If the simple degree-counting test fails to demonstrate that graphs are not isomorphic, there may be no obviously better way to check whether they are isomorphic than the brute-force approach: try every possible way of matching the vertices of one graph with the vertices of the other, and for each, check whether the edges also match up. Finding the simplest test for two graphs to be isomorphic is a classic problem of the field, still unresolved although significant progress has been made in recent years.

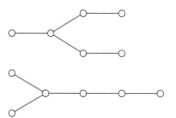

Figure 16.10. Nonisomorphic graphs with the same numbers of vertices and edges, and the same number of vertices of each degree.

✳

The *distance* between two vertices is the length of the shortest path between them. In Figure 16.2, vertices 2 and 3 are distance 2 apart, vertices 1 and 6 are at distance 3, and vertices 6 and 7 are infinitely far apart since there is no path connecting them. The *diameter* of a graph is the greatest distance between any two vertices. Both graphs in Figure 16.9 have diameter 2, since in each graph any pair of vertices is connected by a path of length at most 2.

As an example, consider graphs of social networks, where the vertices represent people and an edge connects each pair of people who are friends. (Implicit in this representation is the symmetry of the friendship relation: if *A* is a friend of *B*, then *B* is a friend of *A*, so it is appropriate to represent a social network using an undirected graph rather than a digraph.) In such a social network, your friends are at distance 1 from you, and your friends-of-friends are at distance 2. The maximum-degree vertex is the person with the most friends, and the diameter of the graph is the smallest number *d* such that *d* "friend of" steps suffice to connect any person to any other. A graph with many vertices may have small diameter, but only if it has many edges—in which case at least some people must have lots of friends.

The graph with *n* vertices and all possible edges is called the *complete graph* on *n* vertices, and is denoted by K_n. The first few complete graphs are shown in Figure 16.11; we have already seen K_5 as the graph *G* of Figure 16.5. K_n has $\frac{n \cdot (n-1)}{2}$ edges (10 in the $n = 5$ case), and (when $n > 1$) has diameter 1 since every vertex is connected to every other. Every vertex has degree $n - 1$; in social network terms, each person knows all of the $n - 1$ other people.

A *tree* is a connected acyclic graph. A *forest* is an acyclic graph—in other words, a graph in which the connected components are trees (see Figure 16.12). Trees have many applications, because the minimal graph connecting a set of vertices is a tree. Let's go through a few basic facts about trees.

Theorem 16.3. *A tree with n vertices has n − 1 edges.*

Proof. The unique tree with $n = 1$ vertex has $n - 1 = 0$ edges—it is the trivial graph.

Now consider a tree *T* with $n + 1 \geq 2$ vertices, on the assumption that any tree with *n* vertices has $n - 1$ edges.

First, we claim that *T* must have at least one vertex of degree 1. We'll prove this claim by contradiction, so suppose it does not. Since *T* is connected, it has no vertex of degree 0—otherwise that vertex would be its own connected component, and *T*, which has at least two vertices, would not be connected. And if every vertex had degree 2 or greater, then starting from any vertex v_0, we could construct a sequence of vertices $v_0 - v_1 - v_2 - \ldots$, such that successive edges are distinct (that is, $v_i \neq v_{i+2}$ for each *i*). But since *T* is finite, some vertex must be repeated eventually (Figure 16.13): there exist *i* and *j*, where $0 \leq i < j \leq n$, such that $v_i = v_j$ (and in fact $j - i > 2$). Then $v_i - v_{i+1} - \ldots - v_j$ is a cycle, contradicting the assumption that *T* is a tree.

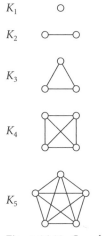

Figure 16.11. Complete graphs on *n* vertices, for $n = 1, \ldots, 5$.

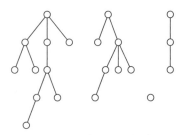

Figure 16.12. A forest. Each of its four connected components is a tree.

Figure 16.13. If every vertex in a finite graph has degree at least 2, the graph contains a cycle.

So T has a vertex of degree 1. Let v be such a vertex. Then $T - v$ is a tree: it is connected, since v had degree 1, and was therefore joined by a single edge to some other vertex; and $T - v$ is acyclic, since T was acyclic, and we could not have created a cycle by removing an edge. So $T - v$ is a tree with n vertices; by the induction hypothesis, it has $n - 1$ edges. But T has exactly one more edge than $T - v$, namely the edge that was incident on v, so T has n edges. ∎

Theorem 16.4. *Removing an edge from a tree without removing any vertices disconnects it. Adding an edge to a tree without adding any vertices creates a cycle.*

Proof. If T is a tree with n vertices and $n - 1$ edges, then removing an edge creates a graph with n vertices and $n - 2$ edges, which by Theorem 16.3 is no longer a tree. Since no cycle could have been created by removing an edge, the graph is acyclic, and must be disconnected since it is no longer a tree (Figure 16.14). Adding an edge to T creates a graph with n vertices and n edges. It cannot have become disconnected by adding an edge. Since by Theorem 16.3 it is no longer a tree, it must have a cycle. ∎

Theorem 16.5. *Between any two vertices in a tree, there is one and only one path.*

Proof. There is at least one path between two distinct vertices in a tree because a tree is connected by definition.

Suppose there were more than one path between vertices v and w of a tree T, say

$$v = v_0 - v_1 - \ldots - v_m = w$$

$$v = u_0 - u_1 - \ldots - u_n = w.$$

We will find segments of these two paths that can be concatenated (part of the first path followed by the reversal of part of the second path) to create a cycle.

Let i be the smallest number such that $v_i = u_i$ but $v_{i+1} \neq u_{i+1}$ (some such i must exist since otherwise the two paths would be the same). Since the two paths converge at the end, we can find the earliest points in each where they come together. That is, there exist $j > i$ and $k > i$ such that $v_j = u_k$ but the paths $v_i, v_{i+1}, \ldots, v_j$ and $u_i, u_{i+1}, \ldots, u_k$ share only the vertices at the beginning and end. Then

$$v_i - v_{i+1} - \ldots - v_j = u_k - u_{k-1} - \ldots - u_i = v_i$$

is a cycle, contradicting the assumption that T was a tree. ∎

❋

We can label the edges of graphs, just as we labeled the arcs of directed graphs on page 152. If the label of edge $e = u - v$ is a, we write a next to the

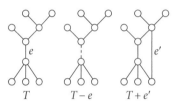

Figure 16.14. Removing an edge from a tree disconnects it, and adding an edge to a tree creates a cycle.

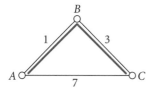

Figure 16.15. A graph with three spanning trees, $A-B-C$ (weight 4), $B-C-A$ (weight 10), and $C-A-B$ (weight 8). In this case, the minimum spanning tree is unique, and is shown in blue.

Figure 16.16. The graph G for which we are to find a minimum spanning tree. The edges will be processed in order of their weight, 1 through 6.

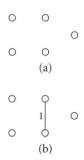

Figure 16.17. The algorithm starts with the empty graph (a), and the first edge added is the edge of lowest weight, which is 1 (b).

[3] Named for the American mathematician and computer scientist Joseph Kruskal (1928–2010), who published it in 1956.

line connecting u and v in a diagram. If the labels are drawn from a set D, then we can use a function $w : E \to D$ to indicate the label $w(e)$ of edge e. If D is a set of numbers such as $\mathbb{N}$ or $\mathbb{R}$, the numerical labels may denote capacities of a pipeline network, or distances in a road map, or some other kind of cost associated with connections between junction points. A numeric value associated with an edge is generally called its *weight*, and many important optimization problems can be described as finding a set of edges in a graph that minimizes the total weight of the selected edges.

We close this chapter with a simple but important example: finding a "minimum spanning tree" of a graph in which each edge has been assigned a weight. For example, if the weights are road distances between cities, a minimum spanning tree would correspond to the shortest network of roadways connecting all the cities—which tells us, for example, the minimum number of miles of road that need to be plowed after a snowstorm in order to allow traffic between all the cities. We need a few useful definitions first.

For any graph $G = \langle V, E \rangle$, a *subgraph* of G is any graph $H = \langle V', E' \rangle$ such that $V' \subseteq V$ and $E' \subseteq E$. We write $H \subseteq G$ when H is a subgraph of G. (Of course, for H to be a graph, every edge in E' must have both endpoints in V'.) A *spanning tree* of G is a subgraph of G that is a tree and includes every vertex of G (that is, $V' = V$). If the edges of G are labeled with numbers, then a minimum-weight spanning tree is a spanning tree $T \subseteq G$ such that the sum of the labels of the edges is minimal. That is, if we write $w(e)$ for the weight associated with edge e, define the weight of any tree $T = \langle V_T, E_T \rangle$ as the sum

$$w(T) = \sum_{e \in E_T} w(e).$$

Then T is a minimum-weight spanning tree of G if T is a spanning tree and there is no other spanning tree T' for G such that $w(T') < w(T)$. We also call T simply a *minimum spanning tree*. For example, of the three spanning trees of the graph of Figure 16.15, the tree of minimum weight is $A-B-C$, which has weight 4.

We always say "a" minimum spanning tree rather than "the" minimum spanning tree, because there may be several spanning trees with the same, minimum weight. For example, if all the edges of a graph have weight 1, then any spanning tree is a minimum spanning tree, and the weight of the tree is the number of edges; that is, one less than the number of vertices.

A labeled graph with more than a few edges may have a great many spanning trees, and at first it might seem necessary to examine them all in order to find the one of lowest weight. In fact, several straightforward algorithms are guaranteed to find a minimum spanning tree quite efficiently. We here explain Kruskal's algorithm,[3] and prove that it works. In our example, Figure 16.16 is a labeled graph in which the edges have labels 1, 2, 3, 4, 5, and 6. Kruskal's algorithm constructs the minimum spanning tree, shown in Figure 16.19; it has total weight 11.

Kruskal's algorithm is simplicity itself. Starting from an empty graph—that is, a graph with no edges—with the same vertices as G (Figure 16.17), add edges one at a time in order of increasing weight, skipping any edge that would form a cycle. (See Figures 16.18 and 16.19.) The algorithm grows a forest, starting with n trivial trees, and gradually merges trees together to form larger trees by adding edges connecting different trees. The algorithm ends after $n - 1$ edges have been added, when the forest has become a single tree. More precisely:

Kruskal's algorithm to find a minimum spanning tree of a connected graph G, where edge e weighs $w(e)$:

1. $F \leftarrow \emptyset$ (F is the set of edges that have been added)

2. For each edge $e \in E$, in order of increasing weight $w(e)$

   (a) If the endpoints of e are in different trees, then

      (i) $F \leftarrow F \cup \{e\}$

3. Return F as a minimum spanning tree.

Theorem 16.6. *Kruskal's algorithm creates a minimum spanning tree.*

Proof. The algorithm never joins vertices in the same tree, so it never creates a cycle. Since it tries every edge in G and G is connected, the final result is a spanning tree. It remains to be shown that the tree returned at the end has minimum weight.

We show that the forest F satisfies the following invariant predicate $P(F)$ at every stage of the execution of the algorithm: *F is a subgraph of a minimum spanning tree.*

$P(F)$ is plainly true at the beginning when F has no edges. If it is true at the end, then the tree returned is a minimum spanning tree. So we must show that if $P(F)$ is true before processing an edge, it will be true afterwards.

If $P(F)$ is true and e connects vertices in the same tree, then F does not change so the invariant is maintained. Now suppose that $P(F)$ is true, e has minimum weight among the remaining edges, and e connects vertices in two different trees, but $P(F + e)$ is false. We derive a contradiction from that assumption, which will establish that $P(F + e)$ is true and complete the proof.

Consider some minimum spanning tree T such that F is a subgraph of T (one must exist, since $P(F)$ is true). T does not include edge e, since $F \subseteq T$ but if $F + e$ were a subgraph of T then $P(F + e)$ would be true. So adding e to T creates a cycle by Theorem 16.4 (Figure 16.20). Then there is some other edge e' that is part of the cycle but not part of $F + e$ (otherwise $F + e$ would itself have contained a cycle). Now

- $w(e') \geq w(e)$, since the algorithm added e before adding e', but also

- $w(e') \leq w(e)$, since T, which contains e' instead of e, is a minimum-weight spanning tree.

(c)

(d)

Figure 16.18. At the next two steps, the edges of weight 2 and 3 are added. The forest now has two trees.

(e)

(f)

Figure 16.19. The next edge to be added is the edge of weight 4, but it would create a cycle and is skipped (e). The next edge is the one of weight 5, which completes the minimum spanning tree.

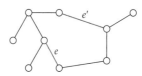

Figure 16.20. Crucial step in Kruskal's algorithm. F, the set of black edges, is a subgraph of some minimum spanning tree. The algorithm adds e, and $F + e$ is allegedly not a subgraph of any minimum spanning tree. But then some other edge e' is part of the minimum spanning tree including F, and there is a cycle including e and e'. Swapping e for e' cannot increase the weight of the tree.

So $w(e') = w(e)$, and replacing e' by e in T results in a spanning tree $T - e' + e$ that also has minimum weight, contradicting the assumption that $F + e$ was not a subgraph of any minimum spanning tree. ∎

Chapter Summary

- An *undirected graph*, or simply a *graph*, is composed of vertices and edges. Formally, a graph is an ordered pair $\langle V, E \rangle$ where V is the set of vertices and E is the set of edges. A edge is a set of two vertices, written $x{-}y$.

- Undirected graphs differ from digraphs in that their edges have no direction, and they cannot have self-loops.

- A *walk* in a graph is a series of vertices, where successive vertices are joined by an edge. The *length* of such a walk is the number of its edges. A *path* is a walk with no repeated edges. A *circuit* is a path in which the first and last vertices are the same, and a circuit is a *cycle* if it has no other repeated vertices.

- Two vertices are *connected* if there is a path between them. The equivalence classes of this relation are the graph's *connected components*. If a graph has only one connected component, it is *connected*.

- The connected components of a graph are the strongly connected components of its corresponding digraph. The *weakly connected components* of a digraph are the connected components of its *underlying* graph.

- The *degree* of a vertex is the number of edges *incident* on it.

- By *Euler's theorem*, a connected graph has an *Eulerian circuit* (a circuit that includes every edge exactly once) if and only if every vertex has even degree.

- Two graphs are *isomorphic* if they have the same structure, though their vertices' names may differ. If two graphs are isomorphic, they have the same number of edges, vertices, and vertices of each degree; but this is not sufficient for two graphs to be isomorphic.

- A *complete graph* has an edge between every pair of vertices.

- A *tree* is a connected acyclic graph. A tree has exactly one fewer edges than vertices. Removing any edge from a tree disconnects it, and adding any edge creates a cycle. Between any two vertices in a tree, there is exactly one path.

- A *forest* is an acyclic graph—that is, a graph in which the connected components are trees.

- The edges in a graph can be assigned *weights*. A *spanning tree* is a *subgraph* that is a tree, and includes every vertex of the original graph; a *minimum spanning tree* is one with minimum total weight.

■ *Kruskal's algorithm* creates a minimum spanning tree by adding edges in order of increasing weight, skipping any edge that would create a cycle.

Problems

16.1. What is the diameter of the graph of Figure 16.1? What is the length of the longest cycle in the graph?

16.2. Find an Eulerian circuit of the graph of Figure 16.21 by the method of Theorem 16.2, starting with the circuit $A-B-F-G-A$.

16.3. Prove that any graph with at least two vertices has two vertices of the same degree. Or to put it in terms of a social network: in any nontrivial social network, two people know the same number of people.

16.4. A *multigraph* is a generalization of a graph in which the same two vertices can be connected by multiple edges. In terms of the formalisms of this book, we can think of a multigraph as an undirected graph with edges labeled by positive integers to indicate their multiplicity. An edge adds its multiplicity to the degrees of both endpoints.

Show that Euler's theorem holds for multigraphs.

16.5. An *Eulerian walk* is a walk that traverses every edge exactly once (but does not necessarily end at its beginning). Show that a connected graph has an Eulerian walk if and only if it has at most two vertices of odd degree.

16.6. If $S_1, \ldots, S_n$ are subsets of a set U, the *intersection graph* of the collection $F = \{S_1, \ldots, S_n\}$ has vertices $v_1, \ldots v_n$ with an edge connecting $v_i - v_j$ if and only if $i \neq j$ and $S_i \cap S_j \neq \emptyset$.

(a) Draw the intersection graph for the sets of positive integers less than or equal to 10 that have a common divisor less than 10; that is,

$$\{\{i : 1 \leq i \leq 10 \text{ and } k \mid i\} : 1 \leq k \leq 10\}.$$

(b) Draw the intersection graph for the set of intervals of the natural numbers $\{[p, 3p] : p \text{ is a prime number less than } 10\}$, where $[a, b]$ denotes the set of numbers $\{x : a \leq x \leq b\}$.

(c) Show that every graph is an intersection graph. *Hint:* Think about the edges incident on each vertex.

(d) Find a graph that is not the intersection graph of any set of intervals like that of part (b). That is, find a graph that is not the intersection graph of any family of the form $\{[a_1, b_1], \ldots, [a_n, b_n]\}$, where the a_i and b_i are natural numbers.

16.7. *Prim's algorithm*[4] for finding a minimum spanning tree starts from an arbitrary vertex and grows a tree from that vertex. At each step, it adds the lowest-weight edge not yet in the tree that (1) is incident on one of the vertices already in the tree and (2) does not create a cycle.

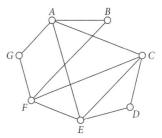

Figure 16.21. A graph with an Eulerian circuit.

[4]Named for the American computer scientist Robert Prim (born 1921), who discovered it in 1957 while working at Bell Labs, though it later emerged that the Czech mathematician Vojtěch Jarník had discovered it in 1930.

(a) Work out Prim's algorithm for the graph of Figure 16.16.

(b) Prove that Prim's algorithm produces a minimal spanning tree. *Hint:* Show by induction that as the tree grows, it is always a subtree of a minimal spanning tree.

16.8. Prove that the minimum spanning tree is unique if all edge weights are different.

16.9. Show that even when the minimum spanning tree of a graph G is unique, more than one spanning tree of G may have the second-lowest weight.

16.10. (a) Use Kruskal's algorithm to find a minimum spanning tree for the graph of Figure 16.22.

(b) Now replace each weight that is an odd number in Figure 16.22 by the next larger even number. Apply Kruskal's algorithm again and find two different minimum spanning trees.

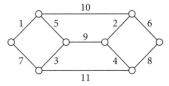

Figure 16.22. Find a minimum spanning tree of this graph.

Chapter 17

Connectivity

Graphs can be broken down into subgraphs representing disconnected parts. If a graph represents the social networks of two people who have almost no friends in common, only a few edges will connect their friendship networks, and the social groups might be disconnected if just a few friendships fade. Or if a graph represents a computer network, and there is a single vertex at the junction of two networks that are otherwise not connected to each other, then that critical vertex is a danger point—a malfunction or disabling attack there would sever communication between the subnetworks.

We have already seen certain ways of decomposing graphs and digraphs into disjoint subgraphs. A connected graph is one with only one connected component; if it has more than one connected component, those components have no paths between them. This is a coarse concept—either a graph is connected or it isn't. But connectivity comes in degrees. A graph may be connected, but vulnerable to disconnection by removal of only a few edges or vertices. We can refine the concept of connectivity to provide numerical measures of the connectivity of a graph.

The *edge connectivity* of a graph is the minimum number of edges that would have to be removed from the graph in order to disconnect it. The edge connectivity of a graph is never more than the minimum degree of the vertices of the graph, since removing all the edges incident on a vertex would disconnect that vertex from the rest of the graph.

Figure 17.1 shows a graph with edge connectivity 1, since removing the edge 3−4 disconnects it into two connected components, with vertices {1, 2, 3} and {4, 5, 6, 7, 8, 9}. After removing 3−4, the connected component on the right, consisting of vertices {4, 5, 6, 7, 8, 9}, has edge connectivity 2: removing a single edge cannot disconnect the graph, but the graph becomes disconnected if, for example, the edges 7−8 and 7−9 are both removed.

An edge of a graph is called a *bridge* if removing it increases the number of connected components. In graph G, 3−4 is a bridge.

The *vertex connectivity* of a graph is the minimum number of vertices with the property that removing them all results in a disconnected graph. The graph of Figure 17.1 has vertex connectivity 1, since removing any of 3, 4, or 7 disconnects it. The connected component in the lower right

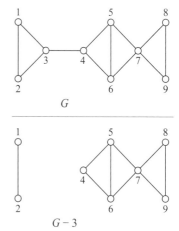

Figure 17.1. A graph with edge connectivity 1 and vertex connectivity 1.

of Figure 17.1 also has vertex connectivity 1, since removing vertex 7 disconnects it.

Strictly speaking, vertex connectivity is undefined for a complete graph K_n, which has an edge between every pair of distinct vertices (see Figure 16.11 on page 166). It is impossible to disconnect such a graph by removing vertices, but by convention, K_n has vertex connectivity $n - 1$, since removing $n - 1$ of its vertices leaves a single vertex.

If removal of a single vertex increases the number of connected components of a graph, that vertex is said to be an *articulation point* of the graph. So 3, 4, and 7 are all articulation points of the graph G of Figure 17.1. If a bridge connects nontrivial graphs, then its endpoints are both articulation points; but some articulation points, such as vertex 7 in Figure 17.1, are not the endpoints of bridges.

A graph is *k-connected* (where $k \geq 1$) if its vertex connectivity is at least k. So a 1-connected graph is connected. A 2-connected graph—that is, a connected graph with no articulation points—is said to be *biconnected*. A biconnected graph cannot be disconnected by removing any single vertex. In a computer or road network, biconnectedness is a minimal condition for survivability: if one junction is blocked, there is always another way for a pair of surviving vertices to communicate.

※

Let's imagine we are using a graph to represent the connections between two particular vertices, s and t. (Mnemonically, the "source" and the "target.") *Menger's theorem* relates the number of edges that have to be removed to disconnect s from t to the number of disjoint paths connecting them—that is, the number of paths from s to t that have no edges in common. For example, suppose we wanted to divide a social network into two parts, with given individuals s and t in different parts, in such a way as to minimize the number of friendship links that connect people in the part containing s with people in the part containing t. Menger's theorem says that the smallest number of edges that could be removed to separate s from t is the same as the largest number of edge-disjoint paths connecting s to t.

A few definitions first.

Let $G = \langle V, E \rangle$ be a finite graph, and let $s, t \in V$ be distinct vertices. An $\langle s, t \rangle$-*edge-cut* is a set of edges such that removing them from G leaves no path from s to t. The edge connectivity of a graph—defined earlier as the minimum number of edges such that removing them disconnects the graph—is then the minimum size of an $\langle s, t \rangle$-edge-cut for any pair of distinct vertices $s, t \in V$. Any superset of an $\langle s, t \rangle$-edge-cut is also an $\langle s, t \rangle$-edge-cut.

Two paths from s to t are said to be *edge-disjoint* if they share no edges. It is convenient to have a name for a set of edge-disjoint paths: an $\langle s, t \rangle$-*edge-connector* is a set of pairwise edge-disjoint paths from s to t—that is, all the paths connect s to t, and no two paths in the set share an edge. Think of an $\langle s, t \rangle$-edge-connector as a bundle of fibers with one end at s and the other at t. Any subset of an $\langle s,t \rangle$-edge-connector (even the empty set) is an $\langle s,t \rangle$-edge-connector.

With these definitions, we can state Menger's theorem more succinctly: the smallest edge-cut is the same size as the largest edge-connector. (See Figures 17.2–17.4 as an illustration.) To help prove the theorem,[1] we'll start with the following lemma:

Lemma 17.1. *Let $G = \langle V, E \rangle$ be a finite graph, and let s and t be distinct vertices in V. Let C be an $\langle s,t \rangle$-edge-cut of G of minimum size. Then there is an $\langle s,t \rangle$-edge-connector P such that C includes one edge of each path in P.*

For example, Figure 17.4 shows an $\langle s,t \rangle$-edge-connector of size 3. The $\langle s,t \rangle$-edge-cut of Figure 17.3 includes one edge from each of these paths.

Proof. A graph can be built up from a set of isolated vertices by adding edges, one at a time. The proof is by structural induction, with the constructor operation of adding an edge to G.

The base case is when there are no edges—the graph $G_0 = \langle V, E_0 \rangle$ where $E_0 = \emptyset$. Since $|E_0| = 0$, the empty set is an $\langle s,t \rangle$-edge-cut, and therefore the smallest one. The empty set is also an $\langle s,t \rangle$-edge-connector (in fact, the only one), and it's true that $C = \emptyset$ includes one edge of each $p \in P = \emptyset$.

So suppose Lemma 17.1 holds for all subgraphs of G with fewer edges than G. Let C be a smallest $\langle s,t \rangle$-edge-cut of G. (For example, suppose G is the graph in Figure 17.2, and $C = \{f, g, h\}$ as in Figure 17.3.) Consider any edge $e \in E$.

First, suppose the smallest $\langle s,t \rangle$-edge-cut of $G - e$ also has size $|C|$. (For example, take $e = i$ in Figure 17.2.) Then $e \notin C$, and C is also a smallest $\langle s,t \rangle$-edge-cut of $G - e$. By the induction hypothesis, $G - e$ has an $\langle s,t \rangle$-edge-connector P such that C includes one edge of every path in P. Then P is also an $\langle s,t \rangle$-edge-connector of G such that C includes one edge from each path in P.

Now, suppose instead that the minimum size of an $\langle s,t \rangle$-edge-cut of $G - e$ is less than $|C|$. (For example, take $e = f$ in Figure 17.2.) C was a smallest $\langle s,t \rangle$-edge-cut in G, so if a smaller $\langle s,t \rangle$-edge-cut exists in $G - e$, $C - \{e\}$ must be one such cut. Then G must have some path p from s to t that includes e, since $C - \{e\}$ is an $\langle s,t \rangle$-edge-cut of $G - e$ but not of G, but does not include any of the edges in $C - \{e\}$. By the induction hypothesis, $G - e$ has an $\langle s,t \rangle$-edge-connector P such that $C - \{e\}$ includes one edge of each path in P. We claim that p is edge-disjoint from every path in P: Take $q \in P$, where $e_q \in C - \{e\}$ is the edge that belongs to q. Suppose that p and q are not disjoint, sharing an edge e_s. Then $(C \cup \{e_s\}) - \{e, e_q\}$, which is smaller than C, is an $\langle s,t \rangle$-edge-cut of G, contradicting the assumption that C was a smallest $\langle s,t \rangle$-edge-cut of G. So in fact, p is edge-disjoint from all members of P, and therefore $P \cup \{p\}$ is an $\langle s,t \rangle$-edge-connector of G; and C includes one edge of each path in $P \cup \{p\}$. ∎

[1] The proof that follows is based on F. Göring, "Short Proof of Menger's Theorem," *Discrete Mathematics* 219 (2000): 295–6.

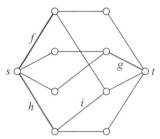

Figure 17.2. A graph with source s and target t identified.

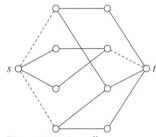

Figure 17.3. A smallest $\langle s,t \rangle$-edge-cut is of size 3 and includes the edges f, g, and h. Removing those three edges disconnects the graph.

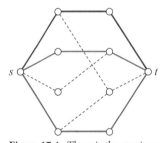

Figure 17.4. Three is the maximum size of an $\langle s,t \rangle$-edge-connector. Three pairwise edge-disjoint paths from s to t are shown—one in red, one in blue, and one in green. The red path includes edge f, the blue path includes edge g, and the green path includes edge h.

The theorem now follows:

Theorem 17.2. Menger's Theorem (point-to-point, edge version). *In any finite graph G and for any distinct vertices s and t of G, the minimum size of an ⟨s, t⟩-edge-cut is equal to the maximum number of edge-disjoint paths from s to t.*

Proof. Lemma 17.1 establishes that there exists a set of pairwise edge-disjoint paths from s to t of size equal to the minimum size of an ⟨s, t⟩-edge-cut. There can be no larger set of pairwise edge-disjoint paths, since that smallest ⟨s, t⟩-edge-cut would have to interrupt every path from s to t; but the paths share no edges, so each edge in the cut can interrupt only one path. ∎

Menger's theorem has many generalizations and variations. We state a few useful versions here, without proof.

Theorem 17.3. Menger's Theorem (set-to-set, edge version). *Let $G = ⟨V, E⟩$ be a graph and let S and T be disjoint subsets of V. Then the minimum size of an ⟨S, T⟩-edge-cut (that is, a set of edges such that removing them interrupts all paths from a vertex in S to a vertex in T) is equal to the maximum number of edge-disjoint paths from a vertex in S to a vertex in T.*

Theorem 17.2 is Theorem 17.3 with $S = \{s\}$ and $T = \{t\}$.

Theorem 17.4. Menger's Theorem (set-to-set, vertex version). *Let $G = ⟨V, E⟩$ be a graph and let S and T be disjoint subsets of V. Then the minimum size of an ⟨S, T⟩-vertex-cut (that is, a set of vertices such that removing them interrupts all paths from a vertex in S to a vertex in T) is equal to the maximum number of vertex-disjoint paths from a vertex in S to a vertex in T (that is, paths that share no vertices except their endpoints).*

A "point-to-point" version of Theorem 17.4, like Theorem 17.2, can be obtained by taking S and T to be singleton sets.

An extremely important generalization of Menger's theorem is a version that applies to weighted digraphs, in which edges are associated with weights that indicate the capacities of pipes, roadways, or segments of some other form of load-carrying network. In such a network, there is a natural notion of flow capacity: what is the maximum flow from a source s to a destination t, subject to the restrictions that (1) every edge can carry at most its capacity, and (2) at every vertex except s and t, the sum of the flows coming into a vertex must equal the sum of the flows leaving the vertex? The answer, called the *Max-Flow-Min-Cut Theorem*, is that the maximum flow from s to t is equal to the minimum capacity of the edges in an ⟨s, t⟩-edge-cut. The proof is constructive, and the algorithm that emerges from it, called the *Ford-Fulkerson Algorithm*, is one of the classic results of the theory of algorithms.

Chapter Summary

- A graph's *edge connectivity* is the minimum number of edges that would have to be removed to disconnect the graph.

- An edge in a graph is a *bridge* if its removal would increase the number of connected components.

- A graph's *vertex connectivity* is the minimum number of vertices that would have to be removed to disconnect the graph.

- A vertex in a graph is an *articulation point* if its removal would increase the number of connected components.

- A graph is *k-connected* if its vertex connectivity is at least k.

- *Menger's theorem* states that the number of edges that must be removed to disconnect two vertices is equal to the number of edge-disjoint paths between those vertices.

- The *Max-Flow-Min-Cut Theorem* states that, for a weighted digraph, the maximum load that can be carried from one vertex to another is equal to the minimum capacity of an *edge-cut* between them.

Problems

17.1. Let $K_n - e$ denote the result of removing any single edge from the complete graph K_n. (The resulting graph does not depend on which edge was removed, up to isomorphism.) What is the edge connectivity of $K_n - e$? The vertex connectivity?

17.2. Are two edge-disjoint paths necessarily also vertex-disjoint? Need two vertex-disjoint paths also be edge-disjoint? Prove, or disprove by counterexample.

17.3. Find a minimum-size $\langle s, t \rangle$-edge-cut of the graph of Figure 17.5, and an $\langle s, t \rangle$-edge-connector of the same size, such that each path includes one of the edges of the $\langle s, t \rangle$-edge-cut.

17.4. Prove that the vertex connectivity of any nontrivial graph is at most its edge connectivity.

17.5. Suppose G is k-connected, and a new vertex v is added to the graph, along with some edges connecting it to vertices of G. Show that the new graph is k-connected if and only if v is adjacent to at least k vertices of G.

17.6. Suppose graph G has n vertices. What is the minimum number of edges that G can have if it is connected? Biconnected? $(n-1)$-connected?

17.7. Prove that every two distinct vertices of a biconnected graph are on a cycle.

Figure 17.5. A graph with identified source and target vertices. What is the minimum size of an edge-cut and the maximum size of an edge-disjoint set of paths from s to t?

17.8. What is the maximum number of edges of an n-vertex graph that is *not* connected?

17.9. The *wheel* W_n on $n \geq 4$ vertices consists of a cycle with $n - 1$ vertices plus a vertex, called the "hub," with edges from the hub to each of the vertices on the cycle.

    (a) Show that $W_4 = K_4$ and any of the vertices can be chosen as the hub, but for W_n where $n > 4$, the hub is unique.

    (b) Show that W_n has vertex connectivity 3 for every $n \geq 4$.

17.10. Show that a graph is k-connected if and only if every pair of vertices is connected by at least k vertex-disjoint paths.

Chapter 18

Coloring

Six scientists—A, B, C, D, E, and F—are working in a lab with five pieces of very expensive equipment, 1, 2, 3, 4, and 5. Each scientist has a one-hour job to get done, but they can't all work at the same time, because several people need to use the same pieces of equipment and each piece can be used for only one job at a time. Figure 18.1 shows which scientists need which items of equipment to do their task—for example, A, B, and C all need to use 1.

If each person took a different time block, there would be no problem; they could each have the lab to themselves, and in six hours everyone would have finished. But of course, the lab itself is expensive to keep open, and there are strong incentives to find a more efficient way of scheduling the work.

Example 18.1. *What is the minimum number of hours into which the jobs of Figure 18.1 could all be scheduled?*

We know it is more than one—more than two, actually, since three different people need to use equipment item #1—and at most six. Is it three, four, five, or six?

This is a metaphor for many scheduling problems in which scarce resources must be allocated to meet competing demands. It doesn't sound like a problem in graph theory, but that is what it boils down to.

We can document the conflicts in a graph, where the vertices are the scientists and an edge between two of them indicates that they can't be working in the lab simultaneously. The conditions described above are captured in Figure 18.2.

We have colored the vertices in such a way that the endpoints of the edges are always different colors. We were able to color the graph in a way that maintains this property using only three colors—green, red, and blue. Returning to the motivating example, we can schedule all the work into just three one-hour time blocks, by having all the "green" scientists (A, D, and E) work simultaneously, and similarly with the other colors.

Situations of this kind arise in a great many different uses of graphs to model real-world phenomena. So let's be precise about the formalities.

Worker	1	2	3	4	5	Time
A	✓					GREEN
B	✓	✓	✓			RED
C	✓	✓	✓			BLUE
D			✓		✓	GREEN
E				✓	✓	GREEN
F				✓	✓	RED

Figure 18.1. Table showing which scientists need to use which equipment.

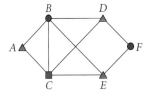

Figure 18.2. The conflict graph for the laboratory scenario.

A *coloring* of a graph $G = \langle V, E \rangle$ is an assignment of colors to vertices in such a way that adjacent vertices have different colors. That is, a coloring of G is a mapping $c : V \rightarrow C$, where C is a fixed finite set of colors, such that $c(x) \neq c(y)$ whenever the edge $x-y$ is in E. If G can be colored using k colors, then there exists a coloring c such that $|\{c(v) : v \in V\}| = k$, and c is said to be a k-coloring of G.

The graph of Figure 18.2 is not 2-colorable, because it contains a triangle—A, B, and C are all adjacent to each other.

✳

The minimum number of colors needed to color a graph G is called the *chromatic number* of G and is denoted $\chi(G)$. (That's a lowercase Greek "chi," the first letter of the word "chroma," or "color.") So if G is the graph of Figure 18.2, then $\chi(G) = 3$.

The chromatic number of a graph cannot be less than the chromatic number of any of its subgraphs. For example, another way of saying that the graph of Figure 18.2 is not 2-colorable is to note that it has the complete graph K_3 (a triangle) as a subgraph, so its chromatic number cannot be less than 3. Here are the chromatic numbers of a few important graphs and types of graphs:

- $\chi(K_n) = n$, where K_n is the complete graph on n vertices. If every vertex is connected to every other vertex, the graph can't be colored with fewer colors than there are vertices. More generally, a *clique* in a graph G is a subgraph of G that is a complete graph. If G has a k-clique, that is, a clique of k vertices, then the chromatic number of G cannot be less than k.

- A path with no branching can be colored with two colors by alternating the colors. So $\chi(P_n) = 2$, where P_n is the path connecting $n > 1$ distinct vertices by $n - 1$ edges. (See Figure 18.3.) The trivial graph P_1 has chromatic number 1.

- The chromatic number of any nontrivial tree is 2. (This generalizes the result that $\chi(P_n) = 2$.)

- If C_n is the cycle on n vertices, then $\chi(C_n)$ is 2 if n is even or 3 if n is odd (Figure 18.4).

- If the maximum degree of any vertex is k, then the chromatic number is at most $k + 1$.

- Say that a graph $G = \langle V, E \rangle$ is *bipartite* if V is the union of two disjoint subsets A and B, and every edge includes one endpoint from each subset. Any bipartite graph is 2-colorable: just color the vertices in A one color and the vertices in B the other color. (Figure 18.3 shows P_5 as a bipartite graph.) In fact, every graph that has chromatic number 2 is bipartite, the required subsets $A, B \subseteq V$ being the sets of vertices of each color.

Figure 18.3. A 2-coloring of P_5.

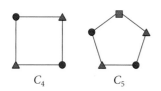

Figure 18.4. A 2-coloring of C_4 and a 3-coloring of C_5.

- As stated on page 165, a *planar* graph is one that can be drawn in the plane without any edges crossing. Planar graphs are closely related to two-dimensional maps: if each region of a map is represented by a vertex, then the corresponding graph has an edge between two vertices whenever the corresponding regions of the map share a border (Figure 18.5). It was a long-standing conjecture that four colors suffice to color any map in such a way that regions sharing a border have different colors, but every proposed proof was found to be flawed. What now can confidently be called the *Four-Color Theorem* was finally proved in 1976, with the aid of computers, by Kenneth Appel and Wolfgang Haken. It can be shown (see Problem 18.11) that the 4-colorability of all planar graphs is equivalent to the Four-Color Theorem for maps, so it is a consequence of the Four-Color Theorem that all planar graphs are 4-colorable.

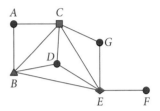

Figure 18.5. A map and the planar graph representing it.

Neither determining the chromatic number of a graph, nor finding a k-coloring given the knowledge that one exists, is in general a simple problem. (We discuss these problems further on page 226.) But checking whether a graph is 2-colorable is straightforward: the answer is no, unless all cycles are of even length.

We continue with two more examples of the usefulness of coloring.

Imagine you have a computer with a small number of fast-access registers $r_1, r_2, \ldots, r_k$ where data can be stored and manipulated. The basic operations of the computer are to perform an arithmetic operation on one or two registers and then store the result in a register; for example,

$$r_2 \leftarrow r_1 + r_3.$$

The registers within a given operation need not all be distinct; for example $r_2 \leftarrow -r_2$ is a valid instruction, which flips the sign of the number in the register. How many registers are needed to compute the value of the expression

$$\frac{(a+b) \cdot (a-b)}{b},$$

on the assumption that a and b are stored in two of the registers to start with?

The first column of Figure 18.6 shows the kind of code a compiler might generate to compute the result one step at a time—first computing $a + b$ and saving that value as c, then computing $a - b$ and saving it as d, and so on. As shown, the compiled code uses a different register for every intermediate

Step	a	b	c	d	e	f
Input a, b						
$c \leftarrow a + b$	✓	✓				
$d \leftarrow a - b$	✓	✓	✓			
$e \leftarrow c \cdot d$		✓	✓	✓		
$f \leftarrow e/b$		✓			✓	
Output f						✓

Figure 18.6. Table showing code to compute $\frac{(a+b)\cdot(a-b)}{b}$, and at which points in the execution of the program each variable is alive.

result, but that is wasteful. We can see that the value of a is not needed after $a - b$ has been computed, so the register assigned to a could be reassigned to hold some other value.

Example 18.2. *What is the minimum number of registers needed to compute the result in Figure 18.6?*

This is a graph coloring problem. The subsequent columns of Figure 18.6 show the points during the execution of this program at which the variables are *alive*; that is, a value has previously been assigned to them, and that value is needed now or will be needed in the future. If two checkmarks are in the same row, then those two variables cannot be stored in the same register. The *conflict graph* shown in Figure 18.7 captures all these conflicts. It has an edge between two vertices whenever a row has checkmarks in the columns for both vertices. Any coloring of this graph can safely be used to assign registers to variables, with all variables sharing the same color being assigned to the same register. Since the graph is 3-colorable, only one more register is needed in addition to the two in which a and b were initially stored. For example, if we assign the red values to r_1, green to r_2, and blue to r_3, then the code of Figure 18.6 is as shown in Figure 18.8. Many modern compilers incorporate a register allocation algorithm based on graph coloring.

As a final example, consider the problem of scheduling gates for arriving flights at an airport. For example, Figure 18.9 shows the arrival times for eight flights, A through H. Plainly D and E can't use the same gate because they arrive at the same time, but the airport actually has a stricter rule: two flights can't use the same gate if their arrival times are less than an hour apart.

We could assign the eight flights to eight different gates, but that's not practical—gates are extremely expensive, and they have to be used as efficiently as possible.

Example 18.3. *How few gates can accommodate all the flights of Figure 18.9?*

Let's start as we did in the register allocation example, by documenting all the conflicts—the flights that can't possibly use the same gate because they arrive within an hour of each other. That's a symmetric relation on the set of flights, and it can be represented by a graph in which the vertices are the flights and an edge connects two vertices if and only if those flights arrive less than an hour apart. The graph representing the conflicts in Figure 18.9 is shown in Figure 18.10.

It is immediately apparent that we need at least four gates, since flights B, C, D, and E all conflict with each other (that is, K_4 is a subgraph). We've colored those four vertices four different colors, and have colored the other vertices in such a way that no two adjacent vertices are the same color. The

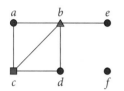

Figure 18.7. The conflict graph for intermediate values in the computation of Figure 18.6.

Input a, b into r_1, r_2
$r_3 \leftarrow r_1 + r_2$
$r_1 \leftarrow r_1 - r_2$
$r_1 \leftarrow r_3 \cdot r_1$
$r_1 \leftarrow r_1 / r_2$
Output r_1

Figure 18.8. Final code, after register allocation.

Flight	Arrives	Gate
A	1:30pm	GREEN
B	2:15pm	RED
C	2:20pm	ORANGE
D	3:00pm	GREEN
E	3:00pm	BLUE
F	3:17pm	RED
G	3:30pm	ORANGE
H	4:05pm	GREEN

Figure 18.9. Flights, arrival times, and gates.

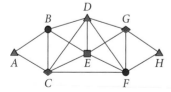

Figure 18.10. The conflict graph for the flights of Figure 18.9.

Gate column in Figure 18.9 is filled in according to these colors. Because the chromatic number of the graph is 4, four gates are both necessary and sufficient.

Chapter Summary

- A *coloring* of a graph is an assignment of colors to vertices such that any two vertices that share an edge have different colors. A coloring that uses k colors is called a *k-coloring*.

- The *chromatic number* of a graph G, denoted $\chi(G)$, is the minimum number of colors needed to color it.

- A complete graph on n vertices has chromatic number n. A nontrivial tree has chromatic number 2. An even-length cycle has chromatic number 2, and an odd-length cycle has chromatic number 3.

- A graph's chromatic number is at most one more than the maximum degree of any vertex.

- A graph is *bipartite* if its vertices can be divided into two sets such that every edge has one vertex in each set. A graph is bipartite if and only if it is 2-colorable, which is the case if and only if it has no odd-length cycles.

- A *planar* graph is one that can be drawn in two dimensions without any edges crossing. Any two-dimensional map has a corresponding planar graph, and all planar graphs are 4-colorable.

- Graph colorings are useful for problems that involve scheduling or resource-sharing.

Problems

18.1. Find a 4-coloring for the graph in Figure 18.11. Find a 3-coloring, or explain why no 3-coloring can exist. Find a 2-coloring, or explain why no 2-coloring can exist.

18.2. What is the chromatic number of $K_n - e$, the result of removing a single edge from K_n? Why?

18.3. What is the chromatic number of an $n \times n$ grid; that is, a graph with vertices $\{\langle i, j \rangle : 1 \le i, j \le n\}$ and edges

$$\{\langle i,j \rangle - \langle i+1,j \rangle : 1 \le i < n, 1 \le j \le n\}$$

$$\cup \{\langle i,j \rangle - \langle i,j+1 \rangle : 1 \le i \le n, 1 \le j < n\}?$$

18.4. Let C'_n be the cycle of n vertices ($n \ge 3$) with additional edges connecting each vertex to the vertex two positions later in the cycle (see Figure 18.12).

- (a) Using the Four-Color Theorem, show that for even n, C'_n is 4-colorable.
- (b) What is $\chi(C'_n)$ for odd n?

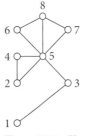

Figure 18.11. How many colors are needed?

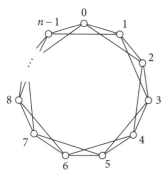

Figure 18.12. C'_n, the cycle with additional edges between vertices that are separated by 2 positions.

18.5. Andre can't work with Bridget, Bridget can't work with Chloe, Chloe can't work with Daniel, Daniel can't work with Eloise, and Eloise can't work with Andre. Fabio can't work with either Andre or Daniel. If each person must be able to work with everyone who shares their work room, how many rooms are needed? Solve this as a graph coloring problem.

18.6. Using the machine model of Example 18.2, find an allocation of registers that uses as few registers as possible to compute

$$\frac{a \cdot (a-b)^2 \cdot (a-c)}{(a+1)^2}.$$

18.7. Amal needs to use a computer for two hours starting at noon. Bianca needs one for four hours starting at 3pm. Chung needs one for two hours starting at 6pm. Diego needs one for four hours starting at 1pm. Elijah needs one for an hour starting at 4pm. Frances needs one for three hours starting at 2pm. How many computers are needed?

18.8. Consider a formula of propositional logic consisting of a conjunction of clauses of the form $(\pm p \oplus \pm q)$, where p and q are propositional variables (not necessarily distinct) and $\pm p$ stands for either p or $\neg p$. Consider the graph in which the vertices include p and $\neg p$ for all propositional variables p appearing in the formula, and in which there is an edge (1) connecting p and $\neg p$ for each variable p, and (2) connecting two literals if their exclusive-or is a clause of the formula. Prove that the formula is satisfiable if and only if the graph is 2-colorable.

18.9. Five senators serve on six committees as shown below. The committees meet once every week. What is the minimum number of weekly meeting times needed to ensure there are no scheduling conflicts for any of the senators? (Multiple meetings can be run at the same meeting time as long as there aren't any senators who would need to be in two meetings at once.)

Athletics: Ava, Bill, Cho

Budget: Bill, Dara, Ella

Compensation: Ava, Cho, Ella

Diversity: Cho, Dara, Ella

Education: Ava, Bill

Football: Bill, Cho, Ella

18.10. A college waits to schedule classes until it knows what courses students want to take. It then creates a graph with courses as vertices and edges between courses that have at least one student in common. Explain how graph coloring could be used to slot these courses into as few time periods as possible.

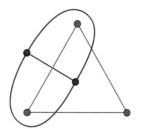

Figure 18.13. A graph G (in blue) and its dual (in red). The dual has three edges corresponding to the three edges of G, and two vertices corresponding to the two regions defined by G, the inside and outside of the triangle.

18.11. Show that every map can be colored with 4 colors if and only if every planar graph is 4-colorable. *Hint:* In Figure 18.5, the planar graph constructed from the map can also be viewed as another map. It divides the plane into six regions, one of which is the exterior region. This process of constructing a so-called *dual* of a map or a planar graph, where the dual has as many vertices as the original had regions and as many edges as the original had borderlines between regions, is used in both halves of the proof. But the dual may not be a graph; it may have a loop from a vertex to itself and may have multiple edges between two vertices. For example, a map for which the planar graph consists of a single triangle would yield a dual with two vertices and three lines connecting those two vertices, not crossing each other but with one crossing each of the border lines of the triangle. See Figure 18.13.

Chapter 19

Finite Automata

In Chapter 15, we showed how to model the behavior of a computer program as a directed graph, with the vertices representing states. In this chapter, we focus on a very simple model for a computer, the *finite automaton*, also known as the *finite state machine*. The finite automaton is an unrealistic model for an entire computer system, but an extremely useful model for small parts of a computer system.

Let's begin with an example. Suppose we wanted to design a computer to decide whether an input string of symbols from the two-symbol alphabet $\Sigma = \{a, b\}$ has an odd number of occurrences of a and an odd number of occurrences of b. We could use a device with the state space of Figure 15.5 (page 153), by simply counting up the numbers of occurrences of a and b and then checking at the end to see whether those numbers are even or odd. But that would be overkill. It would require registers capable of holding integers of arbitrarily large size, when just two bits of information need to be remembered. It suffices to know whether the number of occurrences of a seen so far is even or odd, and whether the number of occurrences of b seen so far is even or odd. This can be done in a state space with only four states, as shown in Figure 19.1. We have slightly embellished the notion of state space as previously defined, numbering the states 0, 1, 2, and 3, designating state 0 as the *start state* by marking it with the > character, and double-circling state 3 to indicate that it is what we will call a "final" or "accepting" state. A finite automaton "accepts" an input string by reaching a final state after starting in the start state and reading the entire string.

Figure 19.1 is a "folded" version of Figure 15.5. State 0 of Figure 19.1 corresponds to all states of Figure 15.5 after an input with an even number of occurrences of a and an even number of occurrences of b, and the other three states have similar characterizations. Let's be precise about exactly what a finite automaton is, and how their operations relate to the notion of invariants of a state space.

A *finite automaton* is a quintuple $\langle K, \Sigma, \Delta, s, F \rangle$, where

> K is a finite set of states,   (19.1)
>
> Σ is an alphabet,

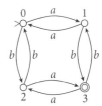

Figure 19.1. A finite automaton accepting all strings with an odd number of occurrences of a and an odd number of occurrences of b.

$$\Delta \subseteq K \times (\Sigma \cup \{\lambda\}) \times K \text{ is a set of labeled arcs,} \tag{19.2}$$

$$s \in K \text{ is the } \textit{start state,} \text{ and}$$

$$F \subseteq K \text{ is the set of } \textit{final states.} \tag{19.3}$$

So if $M = \langle K, \Sigma, \Delta, s, F \rangle$ is a finite automaton, then $\langle K, \Delta \rangle$ is a state space as defined on page 152. In addition,

- K is finite (that is important!);
- the labels can be either individual symbols from the finite alphabet Σ, or the empty string λ;
- one of the states is designated as the start state; and
- some of the states (possibly none, possibly all) are designated as final or accepting states.

In the finite automaton of Figure 19.1,

$$K = \{0, 1, 2, 3\},$$

$$\Sigma = \{a, b\},$$

$$\Delta = \{0 \xrightarrow{a} 1, 1 \xrightarrow{a} 0, 0 \xrightarrow{b} 2, 2 \xrightarrow{b} 0, 2 \xrightarrow{a} 3, 3 \xrightarrow{a} 2, 1 \xrightarrow{b} 3, 3 \xrightarrow{b} 1\},$$

$$s = 0, \text{ and}$$

$$F = \{3\}.$$

The labeled arcs of a finite automaton are known as its *transitions*. A transition $x \xrightarrow{a} y$, where $a \in \Sigma$, is meant to suggest that a transition from state x to state y may occur in response to input a. A transition $x \xrightarrow{\lambda} y$ signifies that the system can jump directly from state x to state y without any input; we shall shortly see the usefulness of these λ-*transitions*.

A diagram like that of Figure 19.1 succinctly represents all the same information as the quintuple $\langle K, \Sigma, \Delta, s, F \rangle$. It's often easier to reason about a finite automaton in diagram form. For example, if we double-circled state 0 as well as state 3, so the set of final states became $\{0, 3\}$ rather than $\{3\}$, then the resulting automaton would identify all strings in which the numbers of occurrences of a and b have the same *parity* (both even or both odd). If that was actually the task at hand, however, the simpler finite automaton of Figure 19.2 would do the trick, since these are simply the even-length strings! In the diagram, we have written $0 \xrightarrow{a,b} 1$ as a shorthand for the two labeled arcs $0 \xrightarrow{a} 1$ and $0 \xrightarrow{b} 1$.

The finite automata illustrated in Figures 19.1 and 19.2 share an important characteristic: they are deterministic. This means that from every state, in response to every input symbol, there is one and only one possible transition; and there are no λ-transitions. Consequently, the behavior is completely determined by the structure of the automaton and by its input. Formally, we say that a finite automaton $M = \langle K, \Sigma, \Delta, s, F \rangle$ is *deterministic*

Figure 19.2. A finite automaton to recognize even-length strings.

if and only if Δ can be described by a function; that is, there is a function $\delta : K \times \Sigma \to K$ such that

$$\Delta = \{q \overset{\sigma}{\to} \delta(q, \sigma) : q \in K, \sigma \in \Sigma\}.$$

In the state diagram for a *deterministic finite automaton*, there is exactly one arc out of every state labeled with each symbol of the alphabet. The function δ for a deterministic finite automaton is called its *transition function*. For example, the transition function for the finite automaton of Figure 19.1 is shown in Figure 19.3.

A finite automaton is sometimes called a *nondeterministic finite automaton* to emphasize the fact that it may not be deterministic. But every deterministic finite automaton is also a nondeterministic finite automaton: the DFAs, as they are called, are a proper subset of the NDFAs.

✳

To see the usefulness of nondeterminism, consider the finite automaton of Figure 19.4.

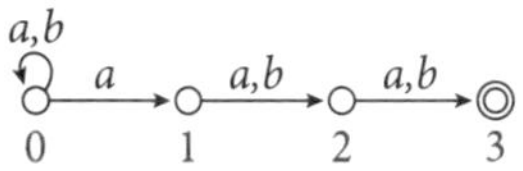

Here the objective is to recognize input strings from the alphabet $\Sigma = \{a, b\}$ with length at least 3 and with third-to-last symbol a. For example, *aaaa* and *bbabb* should be accepted, but *aabaa* and *aa* should not. The easiest way to do this, of course, would be to get the whole input string first and then go through it backward starting from the end rather than the beginning. But that is not an option, because the automaton must be in the appropriate state as each input symbol is processed.

The NDFA of Figure 19.4 seems to play a guessing game. It stays in the start state (state 0) for an arbitrary number of input symbols, whether they are occurrences of a or of b. Then, on an input of a, it nondeterministically transitions to state 1 rather than staying in state 0. The automaton then transitions from state 1 to state 2 and from state 2 to state 3 on reading the next two symbols, whatever they are, entering the unique final state. The diagram suggests that if the automaton has not actually reached the end of the input when it reaches state 3, it is stuck there and cannot process any more input symbols.

The two transitions on symbol a out of state 0, $0 \overset{a}{\to} 0$ and $0 \overset{a}{\to} 1$, are one reason that this automaton is not deterministic. It also fails to be a DFA because there are no transitions out of state 3, while a DFA must have a defined action for every state and every symbol.

We will shortly make the rules of operation completely precise; for now we explain them informally. A nondeterministic finite automaton is said to accept an input string if there is *some* set of choices it could make on reading an input string that will drive it from the start state to a final state while reading the entire input. It fails to accept an input string if and only if there is *no* set of choices it could make that will drive it from the start state to a final state, reading the entire input.

q	$\delta(q,a)$	$\delta(q,b)$
0	1	2
1	0	3
2	3	0
3	2	1

Figure 19.3. The transition function for the deterministic finite automaton of Figure 19.1.

Figure 19.4. A finite automaton to recognize strings with the property that the third-to-last symbol is a.

So for example, the NDFA of Figure 19.4 accepts *bbabb* since, starting from state 0, the automaton could remain in state 0 while reading the first two occurrences of *b*, then transition to state 1 while reading the *a* (rather than remaining in state 0, which it also had the option of doing), and then transitioning to states 2 and 3 while reading the final two occurrences of *b*. There are other choices it could have made which would not have resulted in it reaching the final state after reading the five input symbols, but it suffices that there be at least one set of choices that drives it to the final state. With input *aabaa*, by contrast, no sequence of choices would cause the automaton to reach the final state after reading the entire string.

It is an odd sort of computational model that seems to rely on guess-work to get the right answer. To solve the same problem with a deterministic device is possible, but requires a more complicated construction. Figure 19.5 shows a deterministic finite automaton that does the trick.

Figure 19.5 looks like a mess of spaghetti code. It is not at all obvious that it works, why it works, or how even a clever programmer could have come up with it. We shall shortly relieve the mystery of how it was derived, but for the moment we'll simply suggest how one might go about proving its correctness. A programmer might try to solve this problem by remembering, in a three-symbol buffer, the last three symbols that have been seen. When a new symbol is read, what had been the third symbol back is now the fourth symbol back and is irrelevant, so it is forgotten. The new symbol is retained in the rightmost position, what had been the rightmost symbol moves to the middle position, and what had been in the middle position moves to the leftmost. Then the final states would be those in which there is an *a* in the leftmost position in the buffer. Figure 19.5 does something like this, appropriately accounting for the situations in which fewer than three symbols have been read.

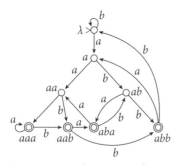

Figure 19.5. A deterministic finite automaton to recognize strings in which the third-to-last symbol is *a*.

The names of the states suggest the invariants that are maintained at those states. We'll just give two examples. In state *aab*, the invariant is that *at least three symbols have been read, and the last three symbols read were a, a, and b* (which were read, respectively, two steps back, one step back, and on the most recent step). Now the transition *aab* $\xrightarrow{b}$ *abb* makes sense: from state *aab* on reading a *b*, that *b* becomes the third buffered symbol, and the *ab* that used to be the second and third symbols retained now become the first and second symbols. The invariant for state *a*, on the other hand, is that *at least one symbol has been read, the symbol most recently read was a, and the second and third most recently read symbols, if any, were b.* So this is the state the automaton would reach on reading strings such as *a, ba, bba,* or *ababbabbaaaabba.*

It is possible to write such an invariant for each of the eight states and then to prove by induction that all are maintained by state transitions. But that would still leave the mystery of how this automaton was discovered in the first place. The answer is: by applying a straightforward construction, called the *subset construction*, to the nondeterministic finite automaton of

Figure 19.4. In the language of computer science, the subset construction compiles an input NDFA and produces an equivalent DFA as output, just as a software compiler produces object code from source code in a different language.

✳

It is time now to be more precise about what exactly we mean by equivalence between finite automata.

First, we need to state formally what it means for a finite automaton (deterministic or nondeterministic) to accept a string. We have explained this in words; to explain it in math, we need to start with a bit of notation about strings and languages.

A *language* is a set of strings. A language with alphabet Σ is any subset of Σ^*. Other examples of languages are $\emptyset$ (a language with zero elements), $\{a\}$ (a language containing just one string, which happens to have length 1), and $\{\lambda\}$ (a language containing just one string, which happens to have length 0). Another language is the one recognized by the automata of Figures 19.4 and 19.5; that is, the set of all strings of length at least 3 in which the third symbol from the right end is an a.

Since languages are sets, they can be combined using the operations of set theory. The union or intersection of two languages is a language, for example. In addition, there are two operations that are specific to languages.

If L_1 and L_2 are languages, then the *concatenation* of L_1 and L_2 is the language containing all concatenations of a string from L_1 with a string from L_2. That is, $L_1 L_2 = \{xy : x \in L_1, y \in L_2\}$. For example, $\{a, aa\}\{a, aa\} = \{aa, aaa, aaaa\}$ (the string aaa can be gotten in two different ways). $L\{\lambda\} = L$ for any L, and $L\emptyset = \emptyset$ for any L since there is no way to take one string from each set.

If L is any language, then L^*, the *Kleene star* of L, is the result of concatenating any number of strings from L, each of which can be used any number of times, in any order. For example, if $L = \{aa, b\}$ then L^* contains all strings made up of the symbols a and b in which occurrences of a always appear two at a time. Formally,

$$L^* = \{w_1 \ldots w_n : n \geq 0, w_1, \ldots, w_n \in L\}.$$

In particular, when $n = 0$, $w_1 \ldots w_n = \lambda$, so $\lambda \in L^*$ for any[1] L.

This definition of L^* is consistent with our previous definition of Σ^* (page 81) as the set of all strings of symbols from Σ: if L is the language consisting of just single-character strings, one for each symbol of Σ, then L^* and Σ^* describe the same set of strings.

Now let $M = \langle K, \Sigma, \Delta, s, F \rangle$ be a finite automaton. The *configurations* of M are the members of $K \times \Sigma^*$. A configuration $\langle q, w \rangle$ represents everything that can affect the future behavior of the automaton: q is its current state, and w is the part of the input that has not yet been read. So at the beginning

[1] This is true even for $L = \emptyset$. In this case L^* contains just the empty string: $\emptyset^* = \{\lambda\}$.

of the computation, q is the start state, and w is the entire input string. If the entire input has been read, $w = \lambda$.

The binary relation $\vdash_M$ between configurations of M describes whether one configuration can result in another after a single computational step. For example, if M is the NDFA of Figure 19.4, we can say $\langle 0, aba \rangle \vdash_M \langle 1, ba \rangle$ since the machine can pass from state 0 to state 1 via the transition between those states labeled with an a.

To be precise, $\langle q, w \rangle \vdash_M \langle r, u \rangle$ if and only if there is a $\sigma \in \Sigma \cup \{\lambda\}$ such that

$$w = \sigma u, \text{ and}$$

$$q \xrightarrow{\sigma} r \in \Delta.$$

That is, u is the result of removing σ from the left end of w (this doesn't change w if $\sigma = \lambda$), and the state changes from q to r via a transition with label σ. The symbol "$\vdash$" is read as "yields in one step"; if the machine M is understood from context, then we can write "$\vdash$" instead of "$\vdash_M$."

We use $\vdash_M^*$ to denote the reflexive, transitive closure of $\vdash_M$. ("$\vdash^*$" is pronounced "yields in zero or more steps," or just "yields.") Then finite automaton M *accepts* string $w \in \Sigma^*$ if and only if

$$\langle s, w \rangle \vdash_M^* \langle f, \lambda \rangle$$

for some final state $f \in F$. That is, an input string w is accepted if in the digraph of configurations, a configuration with a final state and an empty input is reachable from the initial configuration $\langle s, w \rangle$. A series of configurations leading from the initial configuration to a final configuration, each related to the next by $\vdash_M$, is called a *computation*. Again using the automaton M from Figure 19.4 as an example, the string aba is accepted because $\langle 0, aba \rangle \vdash_M^* \langle 3, \lambda \rangle$, via the computation

$$\langle 0, aba \rangle \vdash_M \langle 1, ba \rangle \vdash_M \langle 2, a \rangle \vdash_M \langle 3, \lambda \rangle.$$

$\mathcal{L}(M)$ is the language accepted by M. For example, if M is the DFA of Figure 19.2, which accepts even-length strings of the symbols a and b, then

$$\mathcal{L}(M) = \{w \in \{a, b\}^* : |w| \bmod 2 = 0\}$$

Two finite automata are *equivalent* if they accept the same language. For example, the finite automata of Figures 19.4 and 19.5 are equivalent.

✳

Theorem 19.4. *Any language accepted by a nondeterministic finite automaton is accepted by some deterministic finite automaton.*

Proof. The proof is constructive. We will show how to start with any finite automaton $M = \langle K, \Sigma, \Delta, s, F \rangle$ and construct a deterministic finite automaton $M' = \langle K', \Sigma, \Delta', s', F' \rangle$ that is equivalent to M.

The basic idea is that the state of M' after reading an input string w "remembers" what states M could have reached after reading the same input. So the states of M' correspond to sets of states of M. That increases the size of the state space exponentially, but the constructed automaton is still finite. Intuitively, a transition on symbol σ out of a state $Q \in K'$, which is a *set* of states of M, goes to the set containing all states of M to which there is a transition on σ from any state in Q.

To simplify matters, let's first lay out the construction for the case in which M has no λ-transitions, and then explain how to modify the proof to be fully general. The components of M' are as follows:

$$K' = \mathcal{P}(K), \text{ the power set of } K \tag{19.5}$$

$$\Delta' = \{P \xrightarrow{\sigma} Q : P \in K', \sigma \in \Sigma, \text{ and}$$

$$Q = \{q : p \xrightarrow{\sigma} q \text{ for some } p \in P\}\}$$

$$s' = \{s\}$$

$$F' = \{Q \in K' : Q \cap F \neq \emptyset\}. \tag{19.6}$$

That is, the transition function (let's call it δ') of M' has $\delta'(P, \sigma) = Q$, for a state $P \in K'$ and $\sigma \in \Sigma$, where Q is the set of all states q such that there is a transition of M on input σ from some state $p \in P$ to q. The final states of M' are those containing any of the final states of M, since reaching such a state of M' means that there was some way to reach one of the final states of M.

The effect of these definitions is that

$$\text{if there is a path } s = q_0 \xrightarrow{\sigma_1} q_1 \xrightarrow{\sigma_2} \ldots \xrightarrow{\sigma_n} q_n \in F \tag{19.7}$$

$$\text{then there is a path } s' = Q_0 \xrightarrow{\sigma_1} Q_1 \xrightarrow{\sigma_2} \ldots \xrightarrow{\sigma_n} Q_n \in F' \tag{19.8}$$

where $q_i \in Q_i$ for each $i \geq 0$. Conversely, if there is a path (19.8) in M', then there exist $q_0, \ldots, q_n \in K$ such that $q_i \in Q_i$ for each i and there is a path (19.7) in M. Together these show that $\langle s, w \rangle \vdash_M^* \langle f, \lambda \rangle$ if and only if $\langle s', w \rangle \vdash_{M'}^* \langle Q, \lambda \rangle$ for some $Q \in F'$, or in other words, that $\mathcal{L}(M') = \mathcal{L}(M)$.

This argument is easily modified to cover the case in which M has λ-transitions. Let's say that $p \xrightarrow{\lambda}{}^* q$ if q is reachable from p via a series of zero or more λ-transitions, using * in the usual way to denote "zero or more times." It is easy to determine, for any p and $q \in K$, whether $p \xrightarrow{\lambda}{}^* q$, by simply following arcs labeled with λ. If M has λ-transitions, the automaton defined by the following alternatives to (19.5)–(19.6) correctly takes account of the need to traverse those labeled arcs:

$$K' = \mathcal{P}(K)$$

$$\Delta' = \{P \xrightarrow{\sigma} Q : P \in K', \sigma \in \Sigma, \text{ and}$$

$$Q = \{q \in K : \text{for some } p \in P \text{ and } r \in K, p \xrightarrow{\lambda}{}^{*} r \xrightarrow{\sigma} q\}\}$$

$$s' = \{q \in K : s \xrightarrow{\lambda}{}^{*} q\}$$

$$F' = \{Q \in K' : \text{there is a } q \in Q \text{ and an } f \in F \text{ such that } q \xrightarrow{\lambda}{}^{*} f\}.$$

That is, $P \xrightarrow{\sigma} Q$ if from a state in P, M could follow zero or more λ-transitions to get to some state r, from which there is a single transition on input σ to a state in Q. The last transition in this sequence is always a transition on a symbol in Σ, not on λ. To take account of the fact that it may take some additional λ-transitions to reach a final state, we define F', the set of final states of the constructed DFA, to include all states of M from which a final state of M can be reached by a series of zero or more λ-transitions. ■

Theorem 19.4 is a remarkable result. Because the proof is constructive—the proof tells us exactly how to build M' from M—we are free to use nondeterministic finite automata any time we want, secure in the knowledge that a purely mechanical compilation process can be invoked, if necessary, to produce an equivalent deterministic automaton.

Let us see how the construction "determinizes" the nondeterministic automaton of Figure 19.4. The original machine has four states: 0, 1, 2, and 3. So the states of the deterministic machine will be subsets of $\{0, 1, 2, 3\}$. The table of Figure 19.6 was constructed by starting with the start state $\{0\}$, tracing the states that can be reached from each state. (Ignore the leftmost column the first time you go through this table.) When a new set of states is discovered it is added to the table. For example, from state 0 on reading an a, M could go to either state 0 or state 1, so $\delta'(\{0\}, a) = \{0, 1\}$. State $\{0, 1\}$ of M' presents the possibility that M might be in state 0 or in state 1. So to calculate $\delta'(\{0, 1\}, a)$, we note that from state 0 on reading an a, M could go to either state 0 or state 1, and from state 1 on reading an a, M could go to state 2, so from state $\{0, 1\}$ on reading an a, M' goes to state $\{0, 1, 2\}$.

The leftmost column shows the names of the states as we assigned them in Figure 19.5. We assigned the "buffer" contents as names to help explain what the automaton is doing, but the automaton itself was the output of the purely mechanical construction process of Theorem 19.4. The subset construction "discovered" the buffering idea!

It turned out that in this example, only eight of the sixteen possible states were actually reachable from the start state. So this example illustrates an important practical improvement in the general algorithm described in the proof of Theorem 19.4, which describes how to construct from an NDFA with n states an equivalent DFA with 2^n states. While in the worst case that

buffer	Q	$\delta'(Q, a)$	$\delta'(Q, b)$
λ	$\{0\}$	$\{0, 1\}$	$\{0\}$
a	$\{0, 1\}$	$\{0, 1, 2\}$	$\{0, 2\}$
aa	$\{0, 1, 2\}$	$\{0, 1, 2, 3\}$	$\{0, 2, 3\}$
ab	$\{0, 2\}$	$\{0, 1, 3\}$	$\{0, 3\}$
aaa	$\{0, 1, 2, 3\}$	$\{0, 1, 2, 3\}$	$\{0, 2, 3\}$
aab	$\{0, 2, 3\}$	$\{0, 1, 3\}$	$\{0, 3\}$
aba	$\{0, 1, 3\}$	$\{0, 1, 2\}$	$\{0, 2\}$
abb	$\{0, 3\}$	$\{0, 1\}$	$\{0\}$

Figure 19.6. The construction of a DFA from the NDFA of Figure 19.4.

exponential blowup may be necessary, the DFA actually constructed may be much smaller if only reachable states are created. In particular, if the original NDFA was actually deterministic, the DFA constructed by creating only reachable states is identical to the original automaton!

Recall that the empty set is a subset of every set—so the subset construction may result in the empty set appearing as a state of the constructed DFA. This represents the scenario where no state of the original NDFA corresponds to the input read so far, and the string gets "stuck."

For example, consider the NDFA M of Figure 19.7, which nondeterministically recognizes strings that start with any number of occurrences of a and then end with ab. The subset construction results in the table of Figure 19.8 and its corresponding DFA M' shown in Figure 19.9: Beginning in state $\{0\}$, M' transitions to $\{0, 1\}$ on input a, or to $\emptyset$ on input b, since there is no transition on b out of state 0 in M. Then from state $\{0, 1\}$, M' transitions back to $\{0, 1\}$ on input a, or to state $\{2\}$ on input b. From state $\{2\}$, M' transitions to $\emptyset$ on any input, since there are no transitions out of state 2 in M.

Once M' enters the state $\emptyset$, it returns there on every transition for the rest of the computation. This state means "there is no state that M could be in after reading the input so far," and that remains true no matter what symbols follow in the remainder of the input.

As a final example, consider the NFA of Figure 19.10(a), which is "almost" deterministic. It accepts the language $\{a\}$, and fails to be deterministic only because it is missing some transitions: on b out of state 0 and on either a or b out of state 1. The subset construction (Figure 19.11) creates a new state and the needed transitions, and results in the DFA of Figure 19.10(b).

∗

It is easy to construct finite automata to recognize certain simple patterns. For example, the automaton of Figure 19.12 accepts all and only those input strings that have no more than two occurrences of a in a row.

The automaton of Figure 19.13 determines whether a string of bits (read from most significant to least significant) is a binary numeral representing a number n with the property that $n \bmod 3 = 2$. The invariants for the states 0, 1, and 2 are that the automaton reaches state i if and only if $n \bmod 3 = i$, where n is the value of the binary numeral that has been read so far. Note that reading a 0 doubles the current value, while reading a 1 doubles it and adds 1. For example, if the automation is in state 1, so the number read so far is equal to 1 mod 3, and the next symbol is a 0, the automaton moves to state 2: if $n \bmod 3 = 1$ then $2n \bmod 3 = 2$.

On the other hand, finite automata are severely limited by their, well, finiteness! For example, a simple pigeonhole argument shows that no finite automaton can determine whether its input string has the same number of occurrences of a and b.

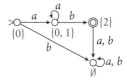

Figure 19.7. An NDFA to recognize strings composed of any number of occurrences of a followed by ab.

Q	$\delta'(Q, a)$	$\delta'(Q, b)$
$\{0\}$	$\{0, 1\}$	$\emptyset$
$\{0, 1\}$	$\{0, 1\}$	$\{2\}$
$\{2\}$	$\emptyset$	$\emptyset$
$\emptyset$	$\emptyset$	$\emptyset$

Figure 19.8. The construction of a DFA from the NDFA of Figure 19.7.

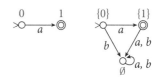

Figure 19.9. A DFA equivalent to the NDFA of Figure 19.7.

Figure 19.10. (a) An NFA that is a few transitions short of being a DFA; it would get "stuck" on reading input b. (b) An equivalent DFA, the result of the subset construction illustrated in Figure 19.11.

Q	$\delta'(Q, a)$	$\delta'(Q, b)$
$\{0\}$	$\{1\}$	$\emptyset$
$\{1\}$	$\emptyset$	$\emptyset$
$\emptyset$	$\emptyset$	$\emptyset$

Figure 19.11. The construction of a DFA from the NDFA of Figure 19.10(a).

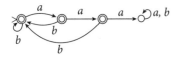

Figure 19.12. A finite automaton to recognize strings with no more than two consecutive occurrences of a.

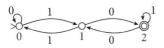

Figure 19.13. A finite automaton to recognize binary numerals representing values n such that $n \bmod 3 = 2$.

Theorem 19.9. *There is no finite automaton M such that $\mathcal{L}(M)$ is exactly the set of strings in $\{a, b\}^*$ with the same number of occurrences of a and b.*

Proof. Suppose M were a finite automaton that accepts all and only those strings with the same number of occurrences of a and b. We prove by contradiction that it does not, in fact, behave as advertised.

M has a finite number of states, say k. Consider the string $a^k b^k$, which does indeed have the same number of occurrences of a and b. The computation of M on this input looks like this:

$$\langle q_0, a^k b^k \rangle \vdash_M \langle q_1, a^{k-1} b^k \rangle \vdash_M \langle q_2, a^{k-2} b^k \rangle \vdash_M \ldots \vdash_M \langle q_k, b^k \rangle \vdash_M^* \langle q_{2k}, \lambda \rangle$$

where q_0 is the start state and q_{2k} is a final state since the string is accepted. Not all of the $k + 1$ states $q_0, \ldots, q_k$ can be different from each other, since M has only k states. So there must be i and j, where $0 \leq i < j \leq k$, such that $q_i = q_j$. But then M will also accept the string $a^{k+j-i} b^k$, which has unequal numbers of occurrences of a and of b since $i < j$. In essence, M gets into a cycle and loses track of whether it has been around that cycle already or not. So $\mathcal{L}(M)$ is not in fact the set of strings with equal numbers of occurrences of a and of b. ∎

There is something remarkable about this argument that should not be lost in the tangle of symbols. It proves an impossibility: in all the infinite universe of all possible finite automata, not one can, for all possible inputs, carry out the simple task of determining whether the input has equal numbers of a and b symbols. We first saw an impossibility argument in Chapter 2 (where we showed that $\sqrt{2}$ is not the ratio of any two integers), and saw another in Chapter 7 (where we showed by diagonalization that there can be no bijection between $\mathbb{N}$ and $\mathcal{P}(\mathbb{N})$). The proof method used here is called *pumping*—it argues that any finite automaton accepting all the strings it is required to accept would also have to accept others it should not (strings with an extra $j - i$ occurrences of a "pumped" in). While computer scientists are naturally most interested in how to devise clever solutions to solve difficult problems, it has been one of the great achievements of the field (though rooted, as the $\sqrt{2}$ example shows, in ancient history) to develop ways of proving irrefutably that some problems simply can't be solved using specified methods. To make progress, either the methods have to change, or the problem has to be changed!

Chapter Summary

■ A *finite automaton* or *finite state machine* is a simplified model of a computer.

■ A finite automaton is formally designated as a finite set of states, an alphabet of available symbols, a set of labeled arcs, a single *start state* (in a diagram, marked by a >), and a set of *final states* (each marked by a double circle).

- A labeled arc represents a *transition* from one state to another in response to the specified input symbol. A λ-*transition* is a transition that consumes no input.

- A finite automaton is *deterministic* (a "DFA") if there is exactly one transition from each state in response to every symbol, and there are no λ-transitions: its behavior for any given input is fully determined. The function that maps a state and symbol to the corresponding next state is the DFA's *transition function*.

- A finite automaton is sometimes called a *nondeterministic finite automaton* (an "NDFA") to emphasize that it is not necessarily deterministic. However, an NDFA may be deterministic—the DFAs are a proper subset of the NDFAs.

- An NDFA can "guess" which of multiple paths to take; it accepts an input if there is any path from the start state to a final state that consumes the entire input.

- Any NDFA is *equivalent* to some DFA, and such a DFA can be found using the *subset construction*.

- A *language* is a set of strings. Two languages can be combined via union, intersection, or concatenation to form another language. The *Kleene star* of any language (denoted by *) is also a language.

- A *configuration* of an automaton specifies its current state and the remaining input.

- A configuration *yields* another configuration (denoted ⊢*) if the automaton can proceed from the first to the second in any number of steps. The sequence of steps between an initial configuration and a final one is a *computation*.

Problems

19.1. Draw a deterministic finite automaton that accepts all and only strings from the alphabet $\Sigma = \{a, b\}$, of length at least 3, in which every third symbol is a *b*, starting with the symbol in the third position. For example, the automaton should accept *abb* and *aaba* and *bbbbbbb*, but not *ab* or *bba* or *aabaaa* or *abaab*.

19.2. (a) Show the full computation of the DFA of Figure 19.14, for the alphabet $\Sigma = \{a, b\}$, on input *ababaa*.

(b) Describe in English the language accepted by this DFA.

(c) For each state, write an invariant describing the input strings that drive the automaton to that state. Prove the invariants, and conclude that the language accepted is what you claim.

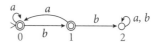

Figure 19.14. A finite automaton.

19.3. For each of the NDFAs of Figure 19.15, determine which strings are accepted. For any string that is accepted, show an accepting computation.

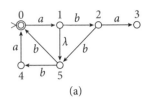

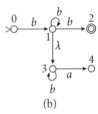

(a)

(b)

Figure 19.15. Two nondeterministic finite automata.

(a)   *aa*

aba

abb

ab

abab

(b)   *ba*

ab

bb

b

bba

19.4. Draw deterministic finite automata that accept the following languages, using the alphabet $\Sigma = \{a, b\}$.
  (a) Strings with *abab* as a substring.
  (b) Strings with neither *aa* nor *bb* as a substring.
  (c) Strings with an odd number of occurrences of *a* and a number of occurrences of *b* that is divisible by 3.
  (d) Strings with both *ab* and *ba* as substrings.

19.5. Let *M* be a DFA. Under what conditions is it true that
  (a) $\lambda \in \mathcal{L}(M)$?
  (b) $\mathcal{L}(M) = \emptyset$?
  (c) $\mathcal{L}(M) = \{\lambda\}$?

19.6. Let *M* be an NDFA. Reanswer the questions of Problem 19.5: under what conditions is it true that
  (a) $\lambda \in \mathcal{L}(M)$?
  (b) $\mathcal{L}(M) = \emptyset$?
  (c) $\mathcal{L}(M) = \{\lambda\}$?

19.7. Consider the NDFA of Figure 19.16, for the alphabet $\Sigma = \{a, b\}$.
  (a) Show all possible computations by this automaton on the input string *ababaa*, including computations that stop short of the end of the string.
  (b) Use the subset construction to find a DFA equivalent to the NDFA of Figure 19.16.

19.8. Show that equivalence of finite automata is an equivalence relation.

Figure 19.16. Convert this NDFA to a DFA.

19.9. Consider the language accepted by the automata of Figures 19.4 and 19.5. Show that no deterministic finite automaton with fewer than eight states accepts this language. *Hint:* Consider the eight strings of length three, and apply the Pigeonhole Principle. You can then conclude that an automaton must reach the same state after reading two different strings from among these eight, but that doesn't quite complete the argument.

19.10. Show that the subset construction is at least nearly optimal, in the sense that for every n, there is an n-state nondeterministic finite automaton that is equivalent to no deterministic finite automaton with fewer than 2^{n-1} states.

19.11. In Figure 19.13, explain why the transition from state 2 on input 1 is to state 2.

Chapter 20

Regular Languages

Those languages that are accepted by finite automata are ordinarily called *regular languages*. The term "regular language" comes from another characterization of these languages: they are the languages that can be described by what are known as "regular expressions." In this chapter, we explain the regular languages and show the way in which they are equivalent to finite automata.

The table below shows a few regular expressions and the languages they denote. Ignore the first column for the moment.

Formal	Informal	Language
a^*	a^*	Strings of any number of occurrences of a
$(a^* \cup b^*)$	$a^* \cup b^*$	Strings consisting entirely of the symbol a or entirely of the symbol b
$(a^* b^*)$	$a^* b^*$	Strings in which all occurrences of a precede any occurrence of b
$((a \cup b)^*(a(b(ab))))$	$(a \cup b)^* abab$	Strings ending with $abab$
$\emptyset$	$\emptyset$	The empty language
$\emptyset^*$	$\emptyset^*$	$\{\lambda\}$
$(a^*(b(a^*(ba^*)))^*)$	$a^*(ba^* ba^*)^*$	Strings with an even number of occurrences of b

Regular expressions are *strings* that denote *languages*. Regular expressions are made up of symbols such as a and b, plus the symbol $\emptyset$, parentheses, the cup symbol $\cup$, and the star symbol $*$. Certain strings of these symbols make up the *metalanguage* used to describe sets of strings in the *object language*.[1] In this case the object language strings are made up of the symbols a and b.

As a rough analogy, consider a textbook for the Latin language written in English. The subject it describes is correct expressions (phrases and sentences) of the Latin language, so Latin is the object language. The language used to describe which expressions are proper Latin is English, so English is the metalanguage. In the same way, we use a metalanguage consisting of strings of cups, stars, et cetera to describe an object language consisting of certain strings of the letters a and b.

We saw the metalanguage symbols "$\emptyset$," "$\cup$," and "$*$" in our discussion of sets (pages 49, 52, and 81). To parse just the last of the examples in the table,

[1] The ideas of object language and metalanguage were also used in our discussion of propositional logic, on page 93.

a string has an even number of occurrences of b if and only if it can be written as a string of zero or more consecutive occurrences of a, followed by any number of blocks, each consisting of a single b followed by some number of occurrences of a followed by a single b followed by some number of occurrences of a. So the string $aaabaaaaabbaaabaa$ fits the pattern, if we imagine it grouped as $a^3(ba^5ba^0)(ba^3ba^2)$, where we have used a^3, for example, as an abbreviation for aaa.

Let's now be more precise with the notation. Since expressions like these are commonly processed by computer, it is important that they be unambiguous. For example, in the second line, the expression $a^* \cup b^*$ could conceivably mean $(a^* \cup b)^*$, though by convention * "binds more tightly" than $\cup$. The "fully parenthesized" versions in the first column are what we officially call *regular expressions*.

For the time being, we will be extremely precise in distinguishing the symbols in regular expressions from the ordinary "mathematical English" we are using to talk about languages. We'll be able to blur the distinction once we have been clear about it, but for the moment, all the symbols in the metalanguage—that is, the symbols appearing in the regular expressions themselves—will be written in blue. Using that convention, we can define the class of regular expressions—a set of blue strings—by structural induction as follows.

Let's consider the alphabet $\Sigma = \{a, b\}$ for simplicity. Regular expressions are written using an alphabet that includes "blue" versions of the symbols of Σ, along with five others—so in this case, seven symbols in all:

$$a \quad b \quad \emptyset \quad \cup \quad ^* \quad (\quad)$$

and are constructed according to the following rules:

1. $\emptyset$, a, and b are regular expressions.

2. If α and β are regular expressions, then so are

   (a) $(\alpha \cup \beta)$,

   (b) $(\alpha\beta)$, and

   (c) α^*.

So strictly speaking, ab is not a regular expression, but (ab) is. The justification for dropping parentheses comes from the intended meaning of regular expressions, formalized in a mapping $\mathcal{L}$ from regular expressions to languages.

1. $\mathcal{L}(\emptyset) = \emptyset$, $\mathcal{L}(a) = \{a\}$ and $\mathcal{L}(b) = \{b\}$.

2. (a) $\mathcal{L}((\alpha \cup \beta)) = \mathcal{L}(\alpha) \cup \mathcal{L}(\beta)$.

   (b) $\mathcal{L}((\alpha\beta)) = \mathcal{L}(\alpha)\mathcal{L}(\beta)$.

   (c) $\mathcal{L}(\alpha^*) = \mathcal{L}(\alpha)^*$.

The arguments of $\mathcal{L}(\cdot)$ on the left are regular expressions; that is, individual strings of "blue" symbols. The values on the right are languages; that is, subsets of Σ^*. To name the function that associates regular expressions with the languages they represent, we use the same symbol $\mathcal{L}$ as was used in the previous chapter to denote the function that associates finite automata with the languages they accept. In each case, the argument of $\mathcal{L}(\cdot)$ is a finite object and its value could be an infinite set.

Because set union and language concatenation are associative operations, we are justified in dropping parentheses in practice as was done in the table above. For example, $\mathcal{L}(((ab)a)) = \mathcal{L}((a(ba))) = \{aba\}$; these two distinct regular expressions have the same meaning. We use the "blue symbols" only to stress that regular expressions have a very specific syntax. In our more relaxed moments, we'll drop the parentheses and write in black, eliding the distinction between the expression and the set it represents. So for example,

$$((ab)a) \neq (a(ba)),$$

because the two expressions are different (the one on the right ends with two right parentheses, the one on the left with only one). But they are equivalent:

$$\mathcal{L}(((ab)a)) = \mathcal{L}((a(ba))) = \{aba\},$$

a fact we will ordinarily state by writing

$$((ab)a) \equiv (a(ba)),$$

or even more simply,

$$((ab)a) = (a(ba)) = aba.$$

Here we have blurred the distinction between blue and black symbols, and also the difference between a string and the language consisting of that one string.

Now that we know exactly what is and is not a regular expression (note that $\cap$ is not an operator used in regular expressions, for example), we can pin down the connection between finite automata and regular expressions.

Theorem 20.1. *For every regular expression, there is an equivalent finite automaton. That is, for any regular expression α, there is a finite automaton M such that $\mathcal{L}(M) = \mathcal{L}(\alpha)$.*

Proof. By structural induction on the construction of the regular expression. Figure 20.1 shows finite automata accepting the languages $\emptyset$, $\{a\}$, and $\{b\}$, respectively. If M_1 and M_2 are arbitrary finite automata, then Figures 20.2

Figure 20.1. Base case of the induction: finite automata that accept the languages $\emptyset$, $\{a\}$, and $\{b\}$.

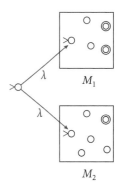

Figure 20.2. Constructor case: a finite automaton that accepts the union of the languages accepted by M_1 and M_2.

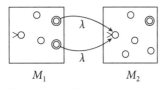

Figure 20.3. Constructor case: a finite automaton that accepts the concatenation of the languages accepted by M_1 and M_2.

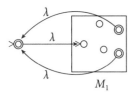

Figure 20.4. Constructor case: a finite automaton that accepts the Kleene star of the language accepted by M_1.

and 20.3 depict automata accepting the union and concatenation of the languages accepted by M_1 and M_2, respectively; and Figure 20.4 depicts an automaton accepting the Kleene star of the language accepted by M_1.

We show the original machines in black, with their start and final states, as just a suggestion of their internal structure. The constructed machines include the additional transitions and (for the union and Kleene star) the additional states, shown in red.

For example, the automaton for the union (Figure 20.2) has a new start state, and λ-transitions from that state to the start states of the two component machines (which are not the start state of the new machine, but would be final states of the new machine if they were final states of the component machines). Formally, if $M_i = \langle K_i, \Sigma, \Delta_i, s_i, F_i \rangle$ for $i = 1, 2$, and $K_1 \cap K_2 = \emptyset$ (that is, the states of the two machines have different names), then $\mathcal{L}(M) = \mathcal{L}(M_1) \cup \mathcal{L}(M_2)$, where M is as follows:

$$M = \langle K, \Sigma, \Delta, s, F \rangle, \text{ where}$$
$$s \text{ is a new state,}$$
$$K = K_1 \cup K_2 \cup \{s\},$$
$$F = F_1 \cup F_2, \text{ and}$$
$$\Delta = \Delta_1 \cup \Delta_2 \cup \{s \xrightarrow{\lambda} s_1, s \xrightarrow{\lambda} s_2\}. \qquad \blacksquare$$

This proof illustrates the usefulness of nondeterminism and λ-transitions, and the power of the subset construction. Even a simple regular expression would turn into a rather complicated and highly nondeterministic finite automaton, but one that could be mechanically transformed into a deterministic automaton if desired.

Remarkably, the converse is also true. It is always possible to describe the language accepted by a finite automaton using a regular expression, and the transformation from automaton to regular expression is mechanical.

Theorem 20.2. *The language accepted by any finite automaton can be described by a regular expression. That is, for any finite automaton M, there is a regular expression α such that $\mathcal{L}(\alpha) = \mathcal{L}(M)$.*

Like the proof of Theorem 20.1, the proof of this theorem is constructive. The proof shows that in the finite automaton model, there is a mechanical procedure for turning a snare of intersecting loops—a mess of spaghetti code—into a set of neatly nested Kleene-star expressions.

Proof. To prove this result, it will be convenient to extend our concept of a finite automaton beyond the nondeterministic model. In a finite automaton, transitions of two kinds are allowed: $p \xrightarrow{a} q$, where a is a single symbol, and $p \xrightarrow{\lambda} q$, indicating that a transition is possible without reading any input. In a *generalized finite automaton*, we will allow any regular expression on an arc. For example, $p \xrightarrow{ba} q$ means that it is possible to transition from state p to state q while consuming the two input symbols ba (first b, and then a); $p \xrightarrow{a \cup b} q$ means reading only one symbol, but it can be either an a or a b; and $p \xrightarrow{a^*} q$ means that a transition from p to q is possible while reading any string consisting entirely of occurrences of the symbol a (including the empty string). To be precise, a generalized finite automaton is defined exactly as a finite automaton is defined by (19.1)–(19.3), except that there can be only one transition between two states, and (19.2) is replaced by

$$\Delta \subseteq K \times R \times K,$$

where R is the set of all regular expressions on alphabet Σ. (The effect of having two transitions between p and q, say $p \xrightarrow{\alpha} q$ and $p \xrightarrow{\beta} q$, can be achieved by the single transition $p \xrightarrow{\alpha \cup \beta} q$.) Note that Δ is finite even though R is not, since there can be only one transition between any pair of states, and the set of states is finite.

Starting with a finite automaton $M = \langle K, \Sigma, \Delta, s, F \rangle$, we will transform M into a series of generalized finite automata, each accepting the same language as the previous. The last in the series will have only two states and a single transition $s \xrightarrow{\alpha} f$, where s is the start state and f is the one final state. It will then follow that the language accepted by M is $\mathcal{L}(\alpha)$.

By way of preparation, add to M a new start state s and a new final state f, and λ-transitions from s to the start state of M and from each of the final states of M to f. This generalized finite automaton has no transitions into its start state or out of its unique final state. Figure 20.5 shows the result of so modifying the finite automaton of Figure 19.13 (page 196), which accepts all binary numerals representing numbers that leave a remainder of 2 when divided by 3.

Now for any two states p and q, we'll create a single regular expression labeling the transition from p to q, which we will denote α_{pq}. If there was originally exactly one transition from p to q, then α_{pq} is the original label; if there were multiple transitions from p to q, then α_{pq} is the union of the expressions labeling all the original transitions; and if there was no such transition, $\alpha_{pq} = \emptyset$. So for any states p and q (including $p = q$), α_{pq} is a regular expression. We have in effect added dummy transitions anywhere there were none, labeled with $\emptyset$—no such transition can actually be traversed, since the set of strings in the label is empty.

We now remove the original states of M one at a time until only s and f are left. When we remove a state and the transitions going into and out of it, we repair other transitions to account for the damage done. Formally, to convert generalized finite automaton $M = \langle K, \Sigma, \Delta, s, \{f\} \rangle$ into an equivalent regular expression:

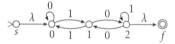

Figure 20.5. The result of preparing the finite automaton of Figure 19.13 as a generalized finite automaton for conversion to a regular expression.

1. While $|K| > 2$

   (a) Pick any state q to eliminate, other than s and f.

   (b) Simultaneously for each pair of states $p, r \in K$ such that $p \neq q$, $p \neq f$, $r \neq q$, and $r \neq s$, replace the label α_{pr} on the transition from p to r by the label

   $$\alpha_{pr} \cup \alpha_{pq}\alpha_{qq}^*\alpha_{qr}.$$

   (c) Remove state q.

2. Only two states remain, and they must be s and f. Then M accepts the language α_{sf}.

The regular expression $\alpha_{pq}\alpha_{qq}^*\alpha_{qr}$ describes the strings that could be read by passing from p to r through q, and perhaps looping at q any number of times. There was already a transition from p directly to r, so the label on that transition must be unioned with the new regular expression. ■

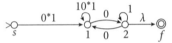

Figure 20.6. The result of eliminating state 0.

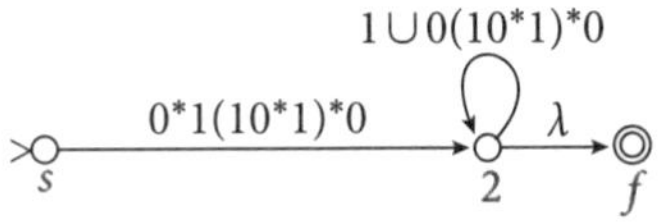

Figure 20.7. The result of eliminating state 1.

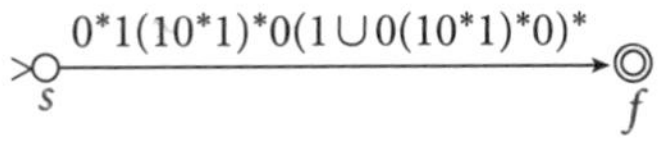

Figure 20.8. The result of eliminating state 2.

Figures 20.6–20.8 show the results of eliminating states 0, 1, and 2 in succession. For example, in Figure 20.6, state 0 has been eliminated. Since there were transitions $s \xrightarrow{\lambda} 0$, $0 \xrightarrow{0} 0$, and $0 \xrightarrow{1} 1$, the new transition from s to 1 gets the label $\lambda 0^*1$, which we simplify to the equivalent expression 0^*1. Similarly, a loop with the label 10^*1 appears at state 1, the result of taking into account the transitions from 1 to 1 by passing through (and perhaps looping at) state 0. (Technically, the new label on the loop at 1 is $\emptyset \cup 10^*1$, since we had set $\alpha_{11} = \emptyset$. Again, we simplified the expression to a shorter equivalent one.)

The final result is a regular expression that certainly would have been hard to come up with just by thinking about it—it has a starred expression within a starred expression within a starred expression! Whether or not there is another, simpler regular expression for this language, this one is correct.

✳

Theorems 20.1 and 20.2 establish that *a language is accepted by some finite automaton if and only if it is represented by some regular expression*. These languages are called the *regular languages*, and they have many beautiful properties, some of which are explored in the exercises. For example, it is obvious from the definition of regular expressions that the union or concatenation of any two regular languages is regular. It is less obvious that the complement of any regular language is regular.

Theorem 20.3. *The complement $\Sigma^* - L$ of any regular language $L \subseteq \Sigma^*$ is regular. Also, the union, concatenation, and intersection of any two regular languages are regular.*

Proof. It is hard to imagine how to turn a regular expression into another regular expression representing the complementary language. But we can convert a regular expression into a finite automaton, and we can transform that finite automaton into an equivalent deterministic finite automaton. For a *deterministic* finite automaton, if the final states are changed to nonfinal and the nonfinal states to final, the result is a deterministic automaton accepting the complementary language. We could then convert that finite automaton into a regular expression. The net effect is that for any regular expression α, we can construct another regular expression α' such that $\mathcal{L}(\alpha') = \Sigma^* - \mathcal{L}(\alpha)$.

It is clear that the union of two regular languages is regular, since if they are represented by regular expressions α and β, then their union is represented by the regular expression $\alpha \cup \beta$; similarly, their concatenation is represented by the regular expression $\alpha\beta$.

Finally, if L_1 and L_2 are regular, then so is $L_1 \cap L_2$, since

$$L_1 \cap L_2 = \overline{\overline{L_1} \cup \overline{L_2}}$$

by De Morgan's law. ■

Chapter Summary

- A *regular expression* is a finite string that denotes a possibly infinite language.

- Regular expressions are composed of symbols from an alphabet, as well as the symbols $\emptyset$, (,), $\cup$, and *.

- Regular expressions can be constructed inductively: $\emptyset$ and any single symbol from the alphabet are themselves regular expressions. The union or concatenation of two regular expressions is a regular expression, and the Kleene star of any regular expression is a regular expression.

- A language can be represented by a regular expression if and only if there exists a finite automaton that accepts the language. Such a language is called a *regular language*.

- A *generalized finite automaton* allows transitions on any regular expression, rather than on just a single symbol. All transitions between one state and another are combined into a single regular expression, so that there is at most one transition between the two states.

- The complement of a regular language is a regular language; and for any two regular languages, their union, concatenation, and intersection are all regular languages.

Problems

20.1. What are these languages, in simpler terms? Explain.

(a) $(a \cup aa)^*$

(b) $(aa \cup aaa)^*$

(c) $\{\lambda\}^*$

(d) $\emptyset^*$

(e) $(L^*)^*$, where L is an arbitrary language

(f) $(a^*b)^* \cup (b^*a)^*$

20.2. Write regular expressions for the following languages. Assume $\Sigma = \{a, b\}$.

(a) Strings with no more than three occurrences of a.

(b) Strings with a number of occurrences of a that is divisible by 4.

(c) Strings with exactly one occurrence of the substring bbb.

20.3. For any language L, let L^R, the *reversal* of L, be the set of reversals of strings in L (see page 87).

(a) Prove by structural induction that the reversal of a regular language is regular.

(b) Sketch how you could prove the same fact starting with a deterministic finite automaton accepting L.

20.4. A language L is said to be *definite* if there is a number k such that whether or not a string w is in L depends only on the last k symbols of w.

(a) State this definition more precisely.

(b) Show that every definite language is regular.

(c) Show that the union of definite languages is regular.

(d) Show that the concatenation of two definite languages need not be definite.

20.5. Write a regular expression for the language consisting of all strings with an even number of occurrences of a and an even number of occurrences of b. *Hint:* You could use Theorem 20.2, applied to a four-state automaton accepting this language. Another way, which yields a simpler result, is to note that any such string has even length. Then, a string of length $2n$ can be viewed as a sequence of n two-symbol blocks. Each block is either aa or bb (we'll call these "even") or ab or ba (let's call those "odd"). Then a string with an even number of occurrences of a and an even number of occurrences of b must have an even number of odd blocks, and this condition can be expressed with the technique we used for strings with an even number of occurrences of b (in the table at the beginning of this chapter).

20.6. We showed using De Morgan's law that the intersection of regular languages is regular.

(a) Suppose we start with finite automata M_1 and M_2 and construct a finite automaton M that accepts $\mathcal{L}(M_1) \cap \mathcal{L}(M_2)$ using the constructions for the union and complement of languages accepted by finite automata. If the state set of M_1 is of size n_1 and the state set of M_2 is of size n_2, estimate the size of the state set of M.

(b) Give a more direct construction of a DFA accepting $\mathcal{L}(M_1) \cap \mathcal{L}(M_2)$ by taking the cross product of the state sets of M_1 and M_2. How large is the resulting machine?

20.7. An *arithmetic progression* is a set of numbers $\{j + n \cdot k : n \geq 0\}$ for some natural numbers j and k. So, for example, $\{7, 11, 15, 19, \dots\}$ is an arithmetic progression, with $j = 7$ and $k = 4$. Define the set of *word lengths* for any language L to be the set of lengths of words in L; that is, $\{|w| : w \in L\}$. For example, the set of word lengths of $(ab)^*$ is the set of even numbers.

Prove that a set of natural numbers is a finite union of arithmetic progressions if and only if it is the set of word lengths of some regular language. *Hint:* The "only if" direction is easier; given a finite union of arithmetic progressions, you need to construct a regular language in which the word lengths are exactly the numbers in the set. For the "if" direction, start by assuming an alphabet with only one symbol, and then generalize.

20.8. (a) Describe the language accepted by the DFA of Figure 20.9.

(b) Mechanically convert this DFA into a regular expression using the method described in the proof of Theorem 20.2.

20.9. Using the method of Theorem 20.1, systematically build nondeterministic finite automata equivalent to these regular expressions.

(a) $(ba \cup b)^* \cup (bb)^*$

(b) $((ab)^*(ba)^*)^*$

20.10. Here we will establish that there is a version of "disjunctive normal form" for regular expressions.

(a) Prove that for any regular expressions α, β, and γ,

$$(\alpha \cup \beta)\gamma \equiv (\alpha\gamma) \cup (\beta\gamma), \text{ and}$$

$$\gamma(\alpha \cup \beta) \equiv (\gamma\alpha) \cup (\gamma\beta).$$

(b) Prove that for any regular expressions α and β,

$$(\alpha \cup \beta)^* \equiv \alpha^*(\beta\alpha^*)^*.$$

(c) Prove that any regular expression is equivalent to one of the form $(\alpha_1 \cup \dots \cup \alpha_n)$, where $\alpha_1, \dots, \alpha_n$ use only concatenation and Kleene star (and not union).

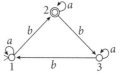

Figure 20.9. A DFA to compile into a regular expression.

Chapter 21

Order Notation

Even if something is in theory possible to compute, it may be practically infeasible to carry out the computation. For example, suppose we were asked to list all sets of days of the year. The listing might begin

> {January 2, February 13, March 15, April 19},
> {April 21, December 2},
> . . . ,

with many more sets to follow. How long would it take to list them all if we could list one set in, say, a nanosecond? Well, there are 365 days in a year, so there are 2^{365} sets of days, without even taking February 29 into account. But 2^{365} nanoseconds is more than 10^{93} years, and the age of the universe is less than $1.4 \cdot 10^{10}$ years! Glad we figured that out before we got too far in the enumeration.[1]

How can we determine whether it is worth actually implementing an algorithm? Or, given two approaches to solving a problem, how can we decide which one is better?

Time is one factor that may limit the usefulness of an algorithm; another potential constraint is the amount of memory it uses. For us, memory is of secondary importance, since an algorithm can only make use of memory that it takes the time to access. Therefore, if the units are chosen appropriately, any algorithm uses at least as much time as memory. Memory is often referred to as *space*, a more abstract measure that might correspond to the number of bits of storage in the RAM chips of a digital computer or the number of digits written down by a human being during a paper-and-pencil computation.[2]

To formalize these notions, we can express the time (or space) requirements of an algorithm as a function of the size of input. So we need some measure of the time taken in the execution of a computer program. That should not be microseconds, or any other measure of clock time, since our results should not be invalidated when a new, faster chip set is released. So we will assume some notion of what comprises a step of a computation, and use step count as our measure of time. A step might be the execution of a

[1] In a similar vein, we noted on page 94 that just writing down the full truth table for a formula of propositional logic with 300 variables would take more lines than there are particles in the universe.

[2] The word *computer*, as used during the 1950s, referred almost exclusively to the people, mostly women, who carried out mathematical calculations in science and engineering.

single machine instruction, or of a single line of code in a higher-level language, or writing down a single digit during a multiplication by hand. For our purposes, the details of the architecture of the computer won't matter, since we will be measuring only how the time taken by a given machine or procedure increases as the size of the input increases.

Let's review a few basic notions. An *algorithm* is a fully specified computational method for producing outputs from inputs. Given any input, an algorithm will eventually stop computing and deliver an answer, and given the same input on some subsequent occasion, it will produce the same output. That is, the relation between input and output of an algorithm is a function: uniquely defined, and defined for all values in the domain.

Now for the measure of the runtime of an algorithm: an algorithm A has *runtime* function $f : \mathbb{N} \to \mathbb{N}$ if $f(n)$ is the maximum number of steps that A takes on any input of size n. So we are adopting the *worst-case* view of the runtime of an algorithm. Worst-case analysis makes sense because if we know the size of the input, the result of our analysis allows us to give iron-clad guarantees about how long we might have to wait for a computation to end. Worst-case analysis is not the only possibility, however. Under some circumstances we might be more interested in an *average-case* analysis—for example, if we expect to run the algorithm on many different inputs and prefer an approach that is faster overall, though it may be slower for some individual inputs. An average-case analysis requires knowing something about what the "average case" of the possible inputs would actually be, and therefore tends to be trickier and less widely applicable than worst-case analysis. We will return to average-case analysis in Chapter 29.

Analysis of an algorithm may reveal that its runtime is a familiar function, for example:

- A *linear* function is of the form $f(n) = a \cdot n + b$, for some constants $a > 0$ and b.

- A *quadratic* function is of the form $f(n) = a \cdot n^2 + b \cdot n + c$, for some constants $a > 0$, b, and c.

- A *polynomial* function is of the form $f(n) = \sum_{i=0}^{d} a_i n^i$, for some constants $a_0, \ldots, a_d$. Provided that $a_d > 0$, so that d is the largest exponent among the nonzero terms, f is said to have *degree* d. So linear functions are degree-1 polynomials, quadratic functions are degree-2 polynomials, and *cubic* functions are degree-3 polynomials. The degree of a polynomial need not be an integer—we will see a $\log_2 7 \approx 2.81$-degree polynomial in Chapter 25. The degree d may even be less than 1, as long as $d > 0$. For example, $f(n) = \sqrt{n} = n^{0.5}$ is a degree-0.5 polynomial. Polynomials with degree $0 < d < 1$ are also called *fractional power* functions.

- A *constant* function is one for which the value does not depend on the argument—that is, a constant function has the form $f(n) = c$, for some constant c not depending on n. We sometimes think of constant functions as degree-0 polynomials.

- An *exponential* function is of the form $f(n) = c^n$, for some constant $c > 1$. The constant c is called the *base* of the exponential function.

In general, some of these functions may take negative arguments or may have negative values. (For example, $f(n) = -n^2$ is a polynomial with negative values for positive arguments.) However, in this chapter we are assuming that all functions under consideration have domain and codomain $\mathbb{N}$, so all values are nonnegative. Usually, as in each of these examples except for the constant functions, runtimes are *monotonic*: $f(n) > f(m)$ if $n > m$; or in other words, the larger the input size n, the longer the runtime.

Runtime cannot be defined for a computation that may not halt, since the runtime would then not be a function. (For example, consider a program that takes an integer as input, and then counts upward from 1 until reaching the input number. If the input is 0, the program will never halt.)

One more note to set the stage. We have said that we will measure the runtime of an algorithm as a function f from $\mathbb{N}$ to $\mathbb{N}$, where $f(n)$ is the number of steps taken, in the worst case, for any input of size n. But what is the "size" of an input? The *size* of a data item is some measure of the amount of space it takes to write it down. So the size of a string of symbols $w \in \{a, b\}^*$ is just its length $|w|$, and the size of an $n \times n$ array of bits is n^2. In the case where the input is a number, note that its size is *not* equal to its value. If an algorithm takes a natural number n as input, for example, and computes its square root $\sqrt{n}$, what is the size of the input? The answer is $\log_2 n$ (commonly written as $\lg n$)—or to be precise, $\lceil \lg n \rceil$, since that is the number of bits in the binary representation of n (Figure 21.1).

Let's take a simple example. Consider an algorithm for determining how many 1s there are in the binary representation of n, by simply scanning that representation and returning the total number of 1s that are encountered. We would describe this as a *linear*-time algorithm, not a logarithmic algorithm: the time taken is proportional to the size of the representation of n, though it is logarithmic in the numerical value of n itself.

In essence, the size of an input or output of an algorithm is the amount of paper it takes to write it down. The choice of notational system has only a constant-factor effect on the size of the data representation. So, for example, a string of n ASCII characters could be described as being n characters long or $8n$ bits long; these differ by a factor of 8. The representation of a number n could be described as being $\lceil \lg n \rceil$ bits or $\lceil \log_{10} n \rceil$ decimal digits, but these differ by a factor of roughly $\log_2 10$, or about 3.32.

✳

Our analysis of runtime will require that we define a way of comparing two functions. One natural method of comparison is to take the ratio of their values: for example, if $f(n) = n$ and $g(n) = 2n$, then

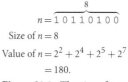

$$n = \overbrace{1\,0\,1\,1\,0\,1\,0\,0}^{8}$$

Size of $n = 8$

Value of $n = 2^2 + 2^4 + 2^5 + 2^7$
$$= 180.$$

Figure 21.1. The size of a number is the amount of paper needed to write it down, not its value, which in general is exponentially larger.

$$\frac{f(n)}{g(n)} = \frac{n}{2n} = \frac{1}{2}.$$

The value of f is half as large as the value of g, for all inputs n. But the ratio between two functions need not be a constant. For example, if $h(n) = n + 2$ and $k(n) = 2n + 1$, the ratio depends on what input n we choose: for $n = 1$,

$$\frac{h(1)}{k(1)} = \frac{1+2}{2+1} = 1,$$

but for $n = 100$,

$$\frac{h(100)}{k(100)} = \frac{100+2}{200+1} \approx 0.51.$$

In general, we care about the ratio when the input n is large, since that is when the difference between the functions is most significant. For example, if the functions $h(n) = n + 2$ and $k(n) = 2n + 1$ are runtime functions of two different algorithms, then both algorithms take 3 steps for an input of size 1. But for an input of size 100, the algorithm with runtime function k takes 99 more steps—and for an input of size 1000, takes 999 more steps.

Instead of picking a particular number n at which to evaluate the ratio, we can consider the *limit* of the ratio as n approaches infinity, written $\lim_{n \to \infty}$. Limits are an important topic in calculus, but for our purposes only a few facts will be needed.

Theorem 21.1. Basic Limits.

$$\lim_{n \to \infty} c = c, \text{ for any constant } c$$

$$\lim_{n \to \infty} n = \infty$$

$$\lim_{n \to \infty} \frac{1}{n} = 0.$$

In the first rule, the expression inside the limit doesn't refer to n, so the value of n is irrelevant.

The second rule simply states that as n approaches infinity, the value of n grows without bound. In general, we will write $\lim_{n \to \infty} f(n) = \infty$ (or $\lim_{n \to \infty} f(n) < \infty$) to mean that $f(n)$ grows without bound (or converges to a finite nonnegative constant).

The third rule states that as n approaches infinity, the reciprocal of n approaches 0. Though $\frac{1}{n}$ is never actually equal to 0, it comes arbitrarily close: for any tiny value $\varepsilon > 0$, there exists some number (to be precise, $\frac{1}{\varepsilon}$) above which the value of $\frac{1}{n}$ is invariably less than ε.

The limits of more complex expressions can be evaluated with the help of a few additional rules. First, in arithmetic expressions, "infinity" behaves as an indefinitely large number, so $\infty + \infty = \infty \cdot \infty = \infty$, and for that matter $3 \cdot \infty = \infty$. Second, in general, the limit of the sum, difference, product, or quotient of two functions is the sum, difference, product, or quotient of the limits of those two functions. For example, to evaluate $\lim_{n \to \infty} n^2$, since we know that $\lim_{n \to \infty} n = \infty$, we can calculate

$$\lim_{n \to \infty} n^2 = \lim_{n \to \infty} (n \cdot n) = (\lim_{n \to \infty} n) \cdot (\lim_{n \to \infty} n) = \infty \cdot \infty = \infty.$$

Here are the details:

Theorem 21.2. Properties of Limits. *For any nonnegative functions $f(n)$ and $g(n)$,*

$$\lim_{n \to \infty} \big(f(n) + g(n) \big) = \left(\lim_{n \to \infty} f(n) \right) + \left(\lim_{n \to \infty} g(n) \right).$$

Similar rules apply for evaluating the limit, as $n \to \infty$, of

$$c \cdot f(n), \text{ for any constant } c,$$
$$(f(n))^c, \text{ for any constant } c,$$
$$f(n) - g(n),$$
$$f(n) \cdot g(n), \text{ and}$$
$$\frac{f(n)}{g(n)},$$

provided that *calculating the power, difference, product, or quotient does not require finding a value for the* indeterminate forms 0^0, ∞^0, $\infty - \infty$, $0 \cdot \infty$, $\infty \cdot 0$, $\frac{0}{0}$, $\frac{\infty}{\infty}$, or $\frac{L}{0}$ for any L.

The exceptions are important. We know that $\lim_{n \to \infty} n = \infty$ and that $\lim_{n \to \infty} \frac{1}{n} = 0$, but $\lim_{n \to \infty} (n \cdot \frac{1}{n}) = 1$, not $\infty \cdot 0$. Calculating a value for $\infty \cdot 0$ is prohibited. (Though whoever wrote the 1933 epigram appearing at the beginning of this book might have had such a product in mind!)

In the case of limits that take the indeterminate forms $\frac{0}{0}$ or $\frac{\infty}{\infty}$, *l'Hôpital's rule* from calculus can help, and we will use it occasionally. When $\lim_{n \to \infty} f(n) = \lim_{n \to \infty} g(n) = 0$, or $\lim_{n \to \infty} f(n) = \lim_{n \to \infty} g(n) = \infty$, l'Hôpital's rule states that

$$\lim_{n \to \infty} \frac{f(n)}{g(n)} = \lim_{n \to \infty} \frac{f'(n)}{g'(n)},$$

where $f'(n)$ and $g'(n)$ are the derivatives of $f(n)$ and $g(n)$, provided that these derivatives exist and that $g'(n) \neq 0$ for all n.

✳

When analyzing the runtime of an algorithm, we generally don't care about the *exact* number of steps in the worst case, which may be very difficult to determine precisely. It is usually sufficient to find a simple function that is roughly equivalent to the runtime, rather than a more complex function that gives a more precise value. One version of "rough equivalence" is as follows: two functions $f(n)$ and $g(n)$ are said to be *asymptotically equivalent*, denoted $f(n) \sim g(n)$, if and only if

$$\lim_{n \to \infty} \frac{f(n)}{g(n)} = 1.$$

See Figure 21.2. For example, the functions $f(n) = n^2 + 1$ and $g(n) = n^2$ are asymptotically equivalent:

$$\lim_{n \to \infty} \frac{n^2 + 1}{n^2} = 1 + \lim_{n \to \infty} \frac{1}{n^2} = 1.$$

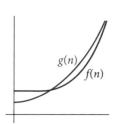

Figure 21.2. Asymptotically equivalent functions. The ratio $f(n)/g(n)$ tends to 1 as $n \to \infty$.

As its name suggests, asymptotic equivalence is an equivalence relation (Problem 21.4).

An important notational convention is worth mentioning here. The statement $f(n) \sim g(n)$ is an assertion about the functions f and g in their entirety, not about any particular value of n. We might more correctly write this simply as $f \sim g$, but it is conventional to include the "dummy" argument n to stress that f and g are functions of one variable. This convention also allows us to write statements such as

$$n^2 + 1 \sim n^2 + n,$$

without introducing names for the two functions of n that are being compared. This particular statement is true, since

$$\lim_{n \to \infty} \frac{n^2 + 1}{n^2 + n} = 1.$$

Asymptotic equivalence means that f and g grow to be roughly the same size, for large enough n. This is a useful concept for approximating functions that are difficult to calculate: for example, the value of $n!$ ("n *factorial*"), which is the product of the integers from 1 to n inclusive, takes $n - 1$ multiplication operations if it is calculated exactly as it is written. But a formula that is asymptotically equivalent to $n!$, known as *Stirling's approximation*, comes very close to the value of $n!$ for large n:

$$n! = \prod_{i=1}^{n} i \sim \sqrt{2\pi n} \cdot \left(\frac{n}{e}\right)^n. \qquad (21.3)$$

This approximation is useful because it can be computed with just a few standard operations, and is close enough to the exact value of $n!$ for many purposes.

Defining equivalence as "$\sim$" is actually more stringent than we need when talking about algorithmic complexity. As we suggested earlier, we may be able to speed up the execution of any algorithm by a constant factor c just by using c times as many processors, or by buying a new computer that is c times faster. So runtime functions that differ asymptotically by only a constant factor may also be considered roughly equivalent. To achieve this more relaxed concept of equivalence, we will compare functions not by how large they are, but by how quickly they grow.

✳

The growth rate of a function is generally referred to as its *order*, though there are several different ways of specifying the relative orders of functions. The first and most commonly used is big-O notation—the "O" is for "order." As noted earlier, the most important question for us is usually, "How long might I have to wait for the algorithm to give me an answer?" Big-O notation specifies an *upper bound* on the runtime—a function that asymptotically grows as fast as the runtime grows, or faster. To denote the set of functions that, asymptotically, grow no faster than $g(n)$, we write $O(g(n))$. This is pronounced "big O of $g(n)$." (That's the letter "O," not a zero!) An algorithm with runtime function $f(n)$ in $O(g(n))$ is said to have runtime complexity that is $O(g(n))$.

In mathematical terms, $f(n) = O(g(n))$ if and only if

$$\lim_{n \to \infty} \frac{f(n)}{g(n)} < \infty. \tag{21.4}$$

The "$f(n) = O(g(n))$" notation is far from ideal, but it has been used in mathematics since the end of the nineteenth century, so we are stuck with it. To be clear about its meaning: $f(n)$ is a function and $O(g(n))$ is a set of functions, so it would make more sense to write "$f(n) \in O(g(n))$," or perhaps "$f \in O(g)$." Unfortunately, the standard notation uses "$=$," though not to mean equality! It is incorrect, for example, to reverse the expression and write something like "$O(g(n)) = f(n)$" (since this would mean $O(g(n)) \in f(n)$, which makes no sense).

An equivalent way to define big-O notation is to say that $f(n) = O(g(n))$ if and only if there exists a constant c and a minimum value n_0 such that for all $n \geq n_0$,

$$f(n) \leq c \cdot g(n). \tag{21.5}$$

See Figure 21.3. Intuitively, $f(n) = O(g(n))$ suggests a "less than or equal to" relation between the growth rates of f and g—like "$f \leq g$" but ignoring constant factors, for sufficiently large arguments.

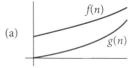

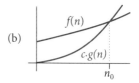

Figure 21.3. (a) Functions f and g. The value of $g(n)$ is less than the value of $f(n)$, at least for all n in the range that is shown. (b) Functions $f(n)$ and $c \cdot g(n)$, where c is some constant greater than 1. Now $c \cdot g(n)$ is still less than $f(n)$ for small values of n; in fact, since $g(0) = 0$, $c \cdot g(0)$ will be less than $f(0)$ no matter how big c is. But for all $n \geq n_0$, $f(n) \leq c \cdot g(n)$, so $f(n) = O(g(n))$.

As an example, we can show that $2^{10} \cdot n^2 + n = O(n^2)$. Using the limit definition (21.4):

$$\lim_{n\to\infty} \frac{2^{10} \cdot n^2 + n}{n^2} = \lim_{n\to\infty} \frac{2^{10} \cdot n^2}{n^2} + \lim_{n\to\infty} \frac{1}{n}$$
$$= 2^{10} + 0$$
$$< \infty.$$

Using the alternate definition (21.5), we can choose, say,

$$c = 2^{10} + 1 \text{ and } n_0 = 0$$

(and there are infinitely many other constants and minimum values that would also work). Then for all $n \geq n_0 = 0$,

$$2^{10} \cdot n^2 + n \leq (2^{10} + 1) \cdot n^2 = c \cdot n^2.$$

Let's examine a few more examples:

- $2^{100} \cdot n^2 + n = O(n^3)$. In fact we shall see that any lower-degree polynomial is big-O of any higher-degree polynomial. Here the left side has degree 2 and the right side has degree 3; the large constant on the left side is irrelevant to the order analysis.
- $n! = O(\sqrt{n} \cdot \left(\frac{n}{e}\right)^n)$. This is just the result of removing the constant from Stirling's approximation (21.3).
- $2^n + n^{100} = O(2^n)$. The term n^{100} is irrelevant in the limit, since it is dwarfed by 2^n as $n \to \infty$:

$$\lim_{n\to\infty} \frac{n^{100}}{2^n} = 0.$$

We shall see that any polynomial is big-O of any exponential.

- $3n + 47 = O(2n)$: Constant factors and additive constants are irrelevant to the rate of growth.
- $135 + \frac{1}{n} = O(1)$. In fact, any constant function is $O(1)$, as is any function of n that tends to 0 as $n \to \infty$.

✳

The class $O(g(n))$ includes all functions that g eventually exceeds, given appropriate scaling by a constant factor and ignoring small arguments. Therefore, a function that is $O(g(n))$ may have values that never approach those of g, and indeed grow farther apart from the values of g as n increases. So any linear function is big-O of any quadratic function, and simply saying that a function is $O(n^2)$, for example, does not suggest that the function is quadratic. (It *does* imply that it can't be cubic, though.)

For the same reason, it is improper to write "$f(n) \neq O(g(n))$" or to say things like "n^3 is more than $O(n^2)$," when trying to explain that the growth rate of the first function exceeds the growth rate of the second. Big-O is intrinsically a class that includes all functions of lower or equal growth rate.

We can get at the idea of the class of functions that have growth rates that are the same as or exceed that of $g(n)$ by introducing a notation symmetrical to big-O for a *lower bound* on a function's rate of growth. The set of functions that asymptotically grow no more slowly than $g(n)$ is $\Omega(g(n))$, pronounced "big omega of $g(n)$." Intuitively, $f(n) = \Omega(g(n))$ corresponds to the idea of "$f \geq g$." Specifically, $f(n) = \Omega(g(n))$ if and only if

$$\lim_{n \to \infty} \frac{f(n)}{g(n)} > 0. \tag{21.6}$$

The relationship between O and Ω is analogous to the relationship between $\leq$ and $\geq$.

Theorem 21.7. $f(n) = O(g(n))$ *if and only if* $g(n) = \Omega(f(n))$.

Proof.

$$f(n) = O(g(n)) \text{ iff } \lim_{n \to \infty} \frac{f(n)}{g(n)} < \infty \quad \text{(by definition of } O\text{)}$$

$$\text{iff } \lim_{n \to \infty} \frac{g(n)}{f(n)} > 0$$

$$\text{iff } g(n) = \Omega(f(n)) \quad \text{(by definition of } \Omega\text{)}.$$

The middle step requires a word of justification. If h is a function such that $\lim_{n \to \infty} h(n) < \infty$, then $h(n)$ might converge either to 0 (in which case its reciprocal increases without bound) or to some nonzero positive number (in which case its reciprocal converges to the reciprocal of that number). In either case, $\lim_{n \to \infty} \frac{1}{h(n)} > 0$. In the proof we are using $h(n) = \frac{f(n)}{g(n)}$. ∎

Big-O and big-Ω notation describe upper and lower bounds, respectively. For a function that can serve as both an asymptotic upper bound and an asymptotic lower bound, we use $\Theta(g(n))$ (pronounced "big theta of $g(n)$") to name the set of functions that grow at the same rate as $g(n)$. See Figure 21.4.

That is, $f(n) = \Theta(g(n))$ just in case

$$\lim_{n \to \infty} \frac{f(n)}{g(n)} = c \tag{21.8}$$

for some constant c, $0 < c < \infty$. Asymptotic equivalence is the special case of (21.8) with $c = 1$.

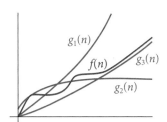

Figure 21.4. Functions f, g_1, g_2, and g_3, where $f(n) = O(g_1(n))$, $f(n) = \Omega(g_2(n))$, and $f(n) = \Theta(g_3(n))$. That is, $g_1(n)$ is an asymptotic upper bound, since it grows more quickly than $f(n)$; $g_2(n)$ is an asymptotic lower bound, since it grows more slowly than $f(n)$; and $g_3(n)$ is both an asymptotic upper bound and an asymptotic lower bound, since it grows at the same rate as $f(n)$.

The Θ relationship plays the role of "=" in our analogy of O to "$\leq$" and Ω to "$\geq$." Like equality, Θ is symmetric.

Theorem 21.9. $f(n) = \Theta(g(n))$ *if and only if* $g(n) = \Theta(f(n))$.

Proof.

$$f(n) = \Theta(g(n)) \text{ iff } \lim_{n \to \infty} \frac{f(n)}{g(n)} = c \quad \text{(where } 0 < c < \infty\text{)}$$

$$\text{iff } \lim_{n \to \infty} \frac{g(n)}{f(n)} = \frac{1}{c} \quad \text{(where } 0 < \frac{1}{c} < \infty\text{)}$$

$$\text{iff } g(n) = \Theta(f(n)).$$ ∎

If $f(n) \sim g(n)$ then $f(n) = \Theta(g(n))$, but the converse is not generally true.

✳

If O and Ω are analogous to "$\leq$" and "$\geq$," then little-o and little-ω notation correspond to the strict inequalities "$<$" and "$>$." ("ω" is a lowercase omega.)

Little-o first. We write $f(n) = o(g(n))$ if and only if the ratio of $f(n)$ to $g(n)$ tends to 0 as n goes to infinity:

$$\lim_{n \to \infty} \frac{f(n)}{g(n)} = 0.$$

For example, $n = o(n^2)$:

$$\lim_{n \to \infty} \frac{n}{n^2} = \lim_{n \to \infty} \frac{1}{n} = 0.$$

Little-omega is the opposite relation. We write $f(n) = \omega(g(n))$ just in case

$$\lim_{n \to \infty} \frac{f(n)}{g(n)} = \infty.$$

In the same way that $a < b$ implies $a \leq b$, $f(n) = o(g(n))$ implies $f(n) = O(g(n))$: if $f(n) = o(g(n))$, then

$$\lim_{n \to \infty} \frac{f(n)}{g(n)} = 0 < \infty,$$

which is the requirement (21.4) for $f(n) = O(g(n))$.

And in the reverse direction, just as $a > b$ implies $a \geq b$, $f(n) = \omega(g(n))$ implies $f(n) = \Omega(g(n))$. If $f(n) = \omega(g(n))$, then

$$\lim_{n \to \infty} \frac{f(n)}{g(n)} = \infty > 0,$$

which is the requirement (21.6) for $f(n) = \Omega(g(n))$.

✳

Now that we have established a language for comparing functions, we'll analyze the relationships between several function classes that appear frequently in runtime analysis.

We'll begin with a lemma that allows us to simplify the analysis of any function that is a sum of multiple terms. It states that a sum of functions has the order of its fastest-growing term.

Lemma 21.10. *Suppose $f(n) = \sum_{i=0}^{k} f_i(n)$, where $f_i(n) = O(f_k(n))$ for $0 \le i < k$. Then $f(n) = \Theta(f_k(n))$.*

Proof. If $\lim_{n \to \infty} f_k(n) = 0$, then $f_i(n)$ must also converge to 0 for each $i < n$ and the result follows immediately. So assume that $f_k(n)$ does not converge to 0. Since $f_i(n) = O(f_k(n))$ for $0 \le i < k$, there are constants c_i, $0 \le c_i < \infty$, such that

$$\lim_{n \to \infty} \frac{f_i(n)}{f_k(n)} = c_i.$$

Then

$$\lim_{n \to \infty} \frac{f(n)}{f_k(n)} = \lim_{n \to \infty} \frac{\sum_{i=0}^{k} f_i(n)}{f_k(n)}$$

$$= \lim_{n \to \infty} \frac{f_k(n)}{f_k(n)} + \sum_{i=0}^{k-1} \lim_{n \to \infty} \frac{f_i(n)}{f_k(n)}$$

$$= 1 + \sum_{i=0}^{k-1} c_i,$$

which is a constant c such that $0 < c < \infty$, so $f(n) = \Theta(f_k(n))$. ∎

The term $f_k(n)$ of Lemma 21.10 is called a "highest-order" term, as opposed to any $f_i(n)$ that are $o(f_k(n))$, which are called "lower-order" terms. Lemma 21.10 simplifies proofs like the following one, by allowing us to drop lower-order terms from the analysis.

Theorem 21.11. *Suppose $f(n)$ is a polynomial of degree a, and $g(n)$ is a polynomial of degree b.*

1. If $a = b$, then $f(n) = \Theta(g(n))$.

2. If $a < b$, then $f(n) = o(g(n))$.

Proof. A function of degree a can be written as

$$f(n) = c_a n^a + \sum_{i \in E} c_i n^i,$$

where $c_a > 0$ and $i < a$ for all $i \in E$. (We write the exponents as $i \in E$, rather than i ranging from 0 to some other value, to indicate that they are not necessarily integers.) Then $c_i n^i = O(c_a n^a)$:

$$\lim_{n \to \infty} \frac{c_i n^i}{c_a n^a} = \frac{c_i}{c_a} \cdot \lim_{n \to \infty} \frac{1}{n^{a-i}} = 0.$$

So $f(n) = \Theta(n^a)$, by applying Lemma 21.10 and then dropping the constant c_a. By similar argument, $g(n) = \Theta(n^b)$. Therefore, there exist constants $0 < k_1, k_2 < \infty$ such that

$$k_1 = \lim_{n \to \infty} \frac{f(n)}{n^a}, \, k_2 = \lim_{n \to \infty} \frac{g(n)}{n^b}.$$

Now we can write

$$\lim_{n \to \infty} \frac{f(n)}{g(n)} = \lim_{n \to \infty} \left(\frac{f(n)}{n^a} \cdot \frac{n^a}{n^b} \cdot \frac{n^b}{g(n)} \right)$$

$$= \lim_{n \to \infty} \left(k_1 \cdot \frac{n^a}{n^b} \cdot \frac{1}{k_2} \right)$$

$$= \frac{k_1}{k_2} \cdot \lim_{n \to \infty} \frac{n^a}{n^b}.$$

If $a = b$, this simplifies to $\frac{k_1}{k_2}$, and $0 < \frac{k_1}{k_2} < \infty$ so $f(n) = \Theta(g(n))$. If $a < b$, this simplifies to 0, so $f(n) = o(g(n))$. ■

So, for example, it is true that

$$5n^3 + 8n + 2 = O(4n^3 + 2n^2 + n + 173)$$

and

$$n^{1.5} = o(n^2).$$

Many practical algorithms have runtimes that are $\Theta(n^k)$, for some (not too large) constant $k \geq 1$. Some examples of such *polynomial-time* algorithms include:

- A linear search algorithm that goes through a list of length n, one element at a time, until the desired element is found. This has runtime of $\Theta(n)$, also called *linear* complexity.

- A sorting algorithm that produces a sorted list from an unsorted list of n elements by starting with an empty list and, one by one, inserting each of the n elements in its proper position in the new list. This algorithm, called *insertion sort*, may compare every element to every element that was inserted before it. That's $\Theta(n^2)$ comparisons in the worst case, so this algorithm has *quadratic* complexity.

What kinds of functions grow more slowly than polynomials? One such class is the *logarithmic functions*: those functions that are $\Theta(\log_c n)$ for some constant $c > 1$, such as $\log_2 n = \lg n$. All logarithmic functions grow at the same rate.

Theorem 21.12. *If $f(n) = \log_c n$ and $g(n) = \log_d n$ for constants $c, d > 1$, then $f(n) = \Theta(g(n))$.*

Proof. We want to find the limit

$$\lim_{n \to \infty} \frac{\log_c n}{\log_d n}.$$

From the standard fact about logarithms that $\log_c n = \log_c d \cdot \log_d n$ for any c, d, and n, it follows that

$$\log_d n = \frac{\log_c n}{\log_c d}.$$

Substituting into the above expression, we get

$$\lim_{n \to \infty} \frac{\log_c n}{\left(\frac{\log_c n}{\log_c d}\right)} = \lim_{n \to \infty} \log_c d.$$

This is just the constant $\log_c d$, independent of n. So $\log_c n = \Theta(\log_d n)$. ∎

Now we can show that these logarithmic functions grow more slowly than any polynomial function.

Theorem 21.13. *For a logarithmic function $f(n) = \log_c n$ and a polynomial function $g(n) = \sum_{i \in E} c_i n^i$, $f(n) = o(g(n))$.*

Proof. By Theorem 21.11, if $g(n)$ has degree d then $g(n) = \Theta(n^d)$, which will simplify the math. Let $0 < k < \infty$ be the constant such that

$$\lim_{n \to \infty} \frac{g(n)}{n^d} = k.$$

We want to find

$$\lim_{n\to\infty} \frac{\log_c n}{g(n)} = \lim_{n\to\infty} \left(\frac{\log_c n}{n^d} \cdot \frac{n^d}{g(n)} \right)$$
$$= \frac{1}{k} \cdot \lim_{n\to\infty} \frac{\log_c n}{n^d}.$$

Both numerator and denominator approach ∞ as $n \to \infty$, so we apply l'Hôpital's rule from calculus—that is, differentiate both the numerator and the denominator of the limit:

$$\frac{1}{k} \cdot \lim_{n\to\infty} \frac{\log_c n}{n^d} = \frac{1}{k} \cdot \lim_{n\to\infty} \frac{\left(\frac{1}{n\ln c}\right)}{d \cdot n^{d-1}}$$
$$= \frac{1}{k} \cdot \lim_{n\to\infty} \frac{1}{\ln c \cdot d \cdot n^d}$$
$$= 0,$$

where the symbol "ln" denotes the natural logarithm, $\log_e$. ∎

Since logarithmic functions grow more slowly than polynomial functions, logarithmic runtimes are preferable to polynomial runtimes, when possible. Many problems require looking at every element of the input, and therefore can't be carried out by any sublinear algorithm. But other algorithms don't actually need to process the entire input.

Consider, for example, the problem of searching for an item in a list that we know to be sorted, assuming we can access any element in constant time if we know its index. A linear search algorithm would take $\Theta(n)$ time, but a *binary search* can do better, taking advantage of the fact that the list is sorted (Figure 21.5): Calculate the middle index (rounding if necessary), and check the element at that index. If the element is bigger than the target, ignore everything to its right and consider the previous index to be the new "end" of the list. If instead the element is smaller than the target, ignore everything to its left and consider the following index to be the new "start" of the list. After adjusting the start or end point as described, check the new middle element of the list. Repeat the process until the target is found, or until the list consists of just one element that is not the target, which indicates that the target was not present.

This algorithm cuts the list in half with each step, retaining only the half that may contain the target. So the maximum number of steps is the number of times we must divide the initial list (which has size n) in half before getting a list of size 1, when we are guaranteed to either find the target or confirm that it is not present. But that is just $\log_2 n$, so binary search has runtime $\Theta(\log n)$.

Are there any classes of functions that grow even more slowly than a logarithmic function? One such class is the *constant* functions, also known

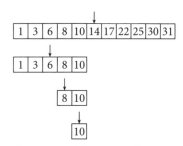

Figure 21.5. A binary search cuts the list of possibilities in half at each step. The above illustrates a search for the target value 10.

as $\Theta(1)$, which take the same amount of time no matter what the size of the input. A few examples of algorithms that can run in constant time include reading the first element of a list; determining whether a list has more than three elements; and calculating $(-1)^n$, which depends only on whether n is even or odd, and so requires knowing only the least significant bit of the binary representation of n.

On the other end of the spectrum, what kind of complexity would render an algorithm so time-inefficient that it would be impractical to use? The most commonly encountered algorithmic complexity class that grows too quickly to be useful is the class of *exponential functions*: those functions that are $\Theta(c^n)$ for some constant $c > 1$. As was the case for polynomial functions, the constant here is important—functions with a smaller constant grow strictly more slowly than those with a larger constant:

Theorem 21.14. *If $c < d$ then $c^n = o(d^n)$.*

Proof. The limit is

$$\lim_{n \to \infty} \frac{c^n}{d^n} = \lim_{n \to \infty} \left(\frac{c}{d}\right)^n = 0$$

since $\frac{c}{d} < 1$. ∎

Finally, let's show that any exponential function grows more quickly than any polynomial.

Theorem 21.15. *Let $f(n) = \sum_{i \in E} c_i n^i$ be a polynomial with degree $d > 0$, and let $g(n) = c^n$ be an exponential function with base $c > 1$. Then $f(n) = o(g(n))$.*

Proof. To simplify the math, it's easiest if we can compare the exponential function $g(n)$ to a single-term polynomial with integer degree. Let $\ell = \lceil d \rceil$; we will compare n^ℓ to $g(n)$. By Theorem 21.11, $f(n) = O(n^\ell)$, since either $f(n) = \Theta(n^\ell)$ if d is itself an integer and therefore $d = \ell$, or $f(n) = o(n^\ell)$ if d is not an integer and therefore $d < \lceil d \rceil = \ell$. Let $0 \leq k < \infty$ be the constant such that

$$\lim_{n \to \infty} \frac{f(n)}{n^\ell} = k.$$

We want the limit

$$\lim_{n \to \infty} \frac{f(n)}{c^n} = \lim_{n \to \infty} \left(\frac{f(n)}{n^\ell} \cdot \frac{n^\ell}{c^n}\right)$$

$$= k \cdot \lim_{n \to \infty} \frac{n^\ell}{c^n}.$$

Again, we use l'Hôpital's rule, since the numerator and denominator both approach ∞. Differentiating both the numerator and denominator gives

$$k \cdot \lim_{n \to \infty} \frac{\ell \cdot n^{\ell-1}}{\ln c \cdot c^n}.$$

But if $\ell > 1$, both the numerator and denominator still approach ∞! So we must apply l'Hôpital's rule ℓ times, after which we obtain

$$k \cdot \lim_{n \to \infty} \frac{\ell!}{(\ln c)^\ell \cdot c^n} = 0. \qquad \blacksquare$$

Exponential runtimes tend to appear when a problem requires an exhaustive search over a space that grows exponentially in the size of the input. We've seen a number of examples already:

- In Chapter 10, we discussed an algorithm that decides whether a propositional-calculus formula with n variables is satisfiable, by testing potentially every one of the 2^n rows of the truth table in search of a satisfying assignment. The runtime of this exhaustive-search algorithm is $\Theta(2^n)$. Satisfiability is NP-complete (page 107) and it is therefore unknown whether there is an algorithm that determines satisfiability in less than exponential time.

- In Chapter 18, we discussed checking a graph for k-colorability. A brute-force search over all the possible k-colorings of a graph with n vertices uses time $\Theta(k^n)$. Faster algorithms are known, but they are still exponential—except in the special case of $k = 2$. A 2-colorable graph is bipartite (page 180), so checking whether a graph can be 2-colored can be done in $\Theta(n)$ time (Figure 21.6). When $k > 2$, determining whether a graph is k-colorable is, like satisfiability, NP-complete; and accordingly, no subexponential-time algorithm is known for this problem.

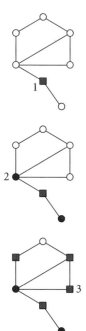

Figure 21.6. An algorithm for checking for 2-colorability in polynomial time, using the fact (page 181) that a bipartite graph is one without odd cycles: Pick any vertex and color it blue. Color each of its neighbors red, and continue alternating colors of neighbors until all vertices are colored or a vertex has the same color as a neighbor. In the above illustration, we color the neighbors of vertex 1 and then vertex 2, but halt upon finding that vertex 3 has a neighbor that is already blue.

✴

Asymptotic notation is sometimes used as convenient shorthand, alone or within an expression, to specify a function that is not fully determined. For example:

- $n^{\Theta(1)}$: n raised to any constant power. Since any polynomial of degree d is $\Theta(n^d)$, $n^{\Theta(1)}$ is the set of all polynomials.

- $\Theta(n \lg n)$: Such functions are sandwiched between linear and quadratic functions—in fact, between the linear functions and $\Theta(n^{1+\epsilon})$ for any constant $\epsilon > 0$:

$$n = o(n \lg n) \text{ and} \qquad (21.16)$$

$$n \lg n = o(n^{1+\epsilon}) \text{ for any } \epsilon > 0. \tag{21.17}$$

- $c^{O(n \lg n)}$: a constant raised to a power $f(n)$ such that $f(n) = O(n \lg n)$.

We have used the base-2 logarithm $\lg n$ in these examples, but the base doesn't matter to the result because of Theorem 21.12.

When asymptotic notation appears on both sides of an equation, remember to interpret the $=$ as $\in$. As always, the left side represents a specific function, so any asymptotic expression on the left stands for some particular function that satisfies the expression. And as always, the right side represents an entire set of functions with a specified rate of growth. For example, it makes sense to write

$$n \log n + O(n) = \Theta(n \log n)$$

since it's true that the function $n \log n$, plus any function that grows no faster than n, grows at the same rate as $n \log n$. But it doesn't make sense to switch the left and right sides (writing "$\Theta(n \log n) = n \log n + O(n)$"), since the specific function $n \log n$ is not part of a definition of a set of functions.

This usage of order notation allows us to write statements like the following:

Example 21.18. *The best known algorithm*[3] *for determining whether two graphs with n vertices are isomorphic (page 165) takes time*

$$2^{O(\log^c(n))} \text{ for some } c > 0. \tag{21.19}$$

We aren't going to prove that; we are just going to try to understand what it says, and compare it to the complexity of the best previously known algorithm.

Solution to example. $2^{O(\log^c(n))}$ is 2 raised to the power of some function that grows no faster than $\log^c(n)$. So as an upper bound, we can suppose this new runtime is $2^{k \log^c(n)}$ for some constant $\underline{k}$. The runtime of the best algorithm that was known previously is $2^{\sqrt{n \log n}}$. How significant was the improvement?

The limit is

$$\lim_{n \to \infty} \frac{2^{k \log^c(n)}}{2^{\sqrt{n \log n}}} = \lim_{n \to \infty} 2^{k \log^c(n) - \sqrt{n \log n}} = 0,$$

since the exponent approaches $-\infty$. So $2^{k \log^c(n)} = o\left(2^{\sqrt{n \log n}}\right)$; the new runtime is asymptotically better. ∎

[3] László Babai, Graph Isomorphism in Quasipolynomial Time, arXiv: 1512.03547 [cs.DS], January 2016.

Now the challenge for computer science researchers is to find an algorithm for which the value of c in (21.19) is 1, because then the algorithm would have runtime $O(2^{\log n})$—which would be a polynomial! (See Problem 21.6.) An algorithm with runtime $2^{O(\log^c(n))}$ is said to run in quasi-polynomial time.

Let's flip the algorithmic complexity question on its side to look at it another way. We have been analyzing how much time an algorithm will take as a function of the size of its input. Another question is this: Suppose we have an algorithm that can solve problems of up to a certain size, given the computing resources and time we can afford to spend on them. Now suppose we are given a bigger budget, and can afford to spend more time, or perhaps to buy a faster computer. By how much will the size of the largest solvable problem increase?

Let's analyze this question just in terms of time budgets (Figure 21.7). Suppose that an algorithm can solve a problem up to size n in time t. How large an input can the algorithm handle if the time budget is increased to $2t$?

Suppose the runtime of the algorithm is $t = f(n)$, where n is the size of the input. The size of a problem that can be solved in twice the amount of time it takes to solve a problem of size n is $f^{-1}(2f(n))$: $2f(n)$ is twice the amount of time it takes to solve a problem of size n, and $f^{-1}(2f(n))$ is the size of the problem that can be solved in that larger amount of time. What is that larger size in relation to n, the size of the original problem, for various possible runtime functions f?

In time t, we can solve a problem of size n.

In time $2t$, how big a problem can we solve?

Figure 21.7. If a computer uses a particular algorithm and can solve a problem of size n in time t, how much bigger a problem can it solve in time $2t$? It depends on the runtime of the algorithm.

- Suppose $f(n) = \lg n$. Then $f^{-1}(t) = 2^t$, and

$$f^{-1}(2f(n)) = 2^{2\lg n} = n^2.$$

If the time complexity of the algorithm grows logarithmically, the size of the largest solvable problem grows quadratically.

- Suppose $f(n) = n^2$. Then $f^{-1}(t) = \sqrt{t}$, and

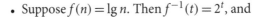

$$f^{-1}(2f(n)) = \sqrt{2n^2} = \sqrt{2} \cdot n.$$

With twice as much time, the algorithm can handle an input that is $\sqrt{2}$ times the size of the original. The size of the maximum feasible input doesn't double, but it does increase by a constant multiplicative factor, about 41%. A similar result would obtain if f was any other polynomial, just with a different constant factor.

- Suppose $f(n) = 2^n$. Then $f^{-1}(t) = \lg t$, and

$$f^{-1}(2f(n)) = \lg(2 \cdot 2^n) = n + 1.$$

With double the time, the algorithm can handle an input that is only 1 larger than the original. Doubling the resource allocation doesn't even result in a multiplicative increase in the size of the problems that can be solved, only an additive increase.

The last result lays bare the futility of exponential-time algorithms. Whatever your budget, you can double it and gain only a small increment in the size of the problems you can solve. And if you aren't satisfied with the result and need to nudge your problem size up again by the same small amount, you will have to double your budget again!

Chapter Summary

- A *runtime* function gives a *worst-case* bound on the number of steps an algorithm may take on an input of a given size.

- Runtime functions are always nonnegative, and usually *monotonic*: the larger the input, the longer the runtime.

- Functions can be compared by taking the *limit* of their ratio. The limit of a sum is the sum of the limits, and similar rules apply for multiplication by a constant, raising to a constant power, subtraction, multiplication, and division, provided that these operations do not require calculating an *indeterminate form*.

- Two functions that are *asymptotically equivalent* (denoted by $\sim$) grow to be roughly equal: $\lim_{n\to\infty} \frac{f(n)}{g(n)} = 1$.

- For runtime analysis, we compare functions according to their growth rates, or *order*, using the following notation:

Notation	Analogy	Name	$\lim_{n\to\infty} \frac{f(n)}{g(n)}$	Interpretation
$f = o(g)$	$<$	little o	0	f grows more slowly than g
$f = O(g)$	$\leq$	big O	$< \infty$	f grows at most as quickly as g
$f = \Omega(g)$	$\geq$	big omega	> 0	f grows at least as quickly as g
$f = \omega(g)$	$>$	little omega	∞	f grows more quickly than g
$f = \Theta(g)$	$=$	big theta	$c, 0 < c < \infty$	f grows at the same rate as g

- A sum of functions has the same order as its highest-order term, so lower-order terms can be dropped from analysis.

- Many runtime functions are either *logarithmic*, *polynomial*, or *exponential*. Functions within these classes can be compared as follows:
  - Any two logarithmic functions are big-Θ of each other.
  - Any logarithmic function is little-o (and big-O) of any polynomial function.
  - Any two polynomial functions of the same *degree* are big-Θ of each other.

- — Any lower-degree polynomial function is little-o (and big-O) of any higher-degree polynomial function.
- — Any polynomial function is little-o (and big-O) of any exponential function.
- — Any exponential function with a lower base is little-o (and big-O) of any exponential function with a greater base.

- ■ Algorithmic complexity can also be framed in terms of how much larger a problem the algorithm can solve if given more time.

Problems

21.1. What is the asymptotic runtime for each of the following algorithms for computing k^n, where k is an integer and $n \geq 1$ is a power of 2?

Algorithm 1	Algorithm 2
1. Set $x \leftarrow k$	**1.** Set $x \leftarrow k$
2. Set $i \leftarrow 1$	**2.** Set $i \leftarrow 1$
3. While $i < n$	**3.** While $i < n$
(a) $x \leftarrow k \cdot x$	(a) $x \leftarrow x^2$
(b) $i \leftarrow i + 1$	(b) $i \leftarrow 2i$
4. Return x	**4.** Return x

21.2. Let $f(n)$ and $g(n)$ be functions from $\mathbb{N} \rightarrow \mathbb{R}$. Prove or disprove the following statements.

(a) $f(n) = O(g(n))$ implies $g(n) = O(f(n))$.

(b) $f(n) = \omega(g(n))$ if and only if $g(n) = o(f(n))$.

(c) $f(n) = \Theta(g(n))$ if and only if $f(n) = O(g(n))$ and $g(n) = O(f(n))$.

(d) $f(f(n)) = \omega(f(n))$ for all strictly increasing f (that is, $f(n+1) > f(n)$ for all n).

(e) $\log_c n = \Theta(\log_d(n^k))$ for any constants c, d, and k.

21.3. Give an example of an algorithm for which the runtime function is not monotonic.

21.4. Prove that asymptotic equivalence of functions is an equivalence relation.

21.5. Prove (21.16) and (21.17).

21.6. Prove that $a^{\log_b n}$ is a polynomial function of n for any $a, b > 0$. What is the degree of the polynomial?

21.7. Prove that

$$n \log n \log \log n = \omega(n \log n).$$

Explain why there is no need to specify a base for any of the four log functions in this formula.

21.8. For an algorithm that has runtime $\Theta(n^3)$, if you double the time budget, by how much does the size of the largest solvable problem increase? More generally, if the budget of an algorithm with runtime $\Theta(n^d)$ is increased by a factor of c, how much larger a problem can be solved?

21.9. Which of the following hold? Prove or disprove in each case.

(a) $64n + 2^5 = o(3n^4)$

(b) $12\log_2(n) = o(2\log_4(n^2))$

(c) $5n = \Omega(6n)$

(d) $2^n n^4 = \Theta(2^n)$

21.10. Rank the following functions by their rates of growth, from slowest to fastest. Specifically, you should rank $f(x)$ before $g(x)$ iff $f(x) = o(g(x))$. There may be some ties; you should indicate this in some way.

(a) 10^x

(b) 2^{4x}

(c) 4^{2x}

(d) $\exp(x)$, or e^x

(e) x^{10}

(f) $x!$

(g) x^x

(h) 5

(i) $\log(x)$

(j) $1/x$

(k) x

(l) $\displaystyle\sum_{n=0}^{10} nx^n$

(m) $\displaystyle\sum_{n=0}^{10} n^n$

Chapter 22

Counting

In mathematics, "counting" means figuring out how many of something there are *without* actually counting them one at a time—by finding a general formula instead.

Example 22.1. *How many ways are there to roll a total of 6 with two standard six-sided dice (Figure 22.1)? If instead n is an arbitrary natural number, how many ways are there to get a total of n by adding together two numbers between 1 and n?*

Solution to example. For the six-sided dice, one way to answer this question is to enumerate the possibilities: $1 + 5, 2 + 4, 3 + 3, 4 + 2, 5 + 1$. So there are 5 ways.

But for the general question, we need to find a formula. The first number could be anywhere from 1 to $n - 1$, you might reason, and the second number would be n minus the first number, so $n - 1$ ways in all. Check it against the case we did by hand: $6 - 1 = 5$, and we found 5 choices in the $n = 6$ case, so the formula seems right. Does it work when n is odd? Well, when $n = 3$, the two possibilities are $1 + 2$ and $2 + 1$, so $n - 1 = 2$ gives the right answer in this case too. What is the smallest n for which this makes sense? In the $n = 2$ case there is just one way to get a total of 2, namely as $1 + 1$. ∎

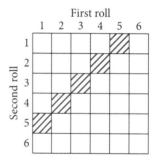

Figure 22.1. In how many ways can two dice be rolled to produce a total of 6?

Counting is integral to creating and analyzing computer algorithms. When designing an algorithm, we need to be able to calculate its runtime as a function of its input, as discussed in Chapter 21, so we know whether it will actually work in a reasonable amount of time for inputs of a reasonable size. For example, when writing a sorting algorithm, we would like to be able to calculate how many comparisons it requires to sort n items. On the other hand, when we create a password or encrypt data, we want to be sure that there *isn't* an algorithm that can guess the password or break the encryption in any reasonable amount of time. For example, how many passwords are there that conform to a specified set of rules? Or, how many different ways are there to jumble the alphabet, using the new ordering to encrypt a message? Are there enough options that trying all of them is infeasible?

These questions boil down to questions about the sizes of sets. We'll start with the basics of set counting: addition and multiplication.

Theorem 22.2. The Sum Rule. *If sets A and B are disjoint, the size of their union is the sum of their sizes:*

$$|A \cup B| = |A| + |B|.$$

For example, if a computer science class consists of 12 sophomores and 9 juniors, the total number of students is $12 + 9 = 21$.

Going the other direction, from the size of a set and a subset, by subtraction you can calculate the number of elements of the set that are *not* in the subset. If there are 100 balls in a jar, each ball is red or blue, and 37 are red, then $100 - 37 = 63$ balls are blue.

Theorem 22.3. The Product Rule. *The size of a Cartesian product of two sets is the product of their sizes:*

$$|A \times B| = |A| \cdot |B|.$$

In the computer science class with 12 sophomores and 9 juniors, if the professor wants to pick a sophomore and a junior to give a presentation together, there are $12 \cdot 9 = 108$ possible pairs of students. And if all the balls in the jar with 37 red balls and 63 blue balls have unique identifiers, there are $37 \cdot 63 = 2331$ ways of picking a set of two balls where one is red and the other is blue.

We can use a similar multiplication rule for combinations of more than two sets:

Example 22.4. *How many different license plates are there, if a license plate must be a sequence of 3 letters followed by 3 digits?*

Solution to example. For each of the three letters, there are 26 possibilities, and for each of the three digits, there are 10 possibilities, for a total of

$$26^3 \cdot 10^3 = 17576000$$

license plates.   ∎

What we used here was actually a more general form of the product rule, which we can use to find the size of the product of any number of sets.

Theorem 22.5. The Generalized Product Rule. *Let S be a set of sequences of length k. Suppose that there are n_1 choices for the first element of a sequence, n_2 for the second, ..., and n_k for the kth element. Then*

$$|S| = n_1 \cdot \ldots \cdot n_k.$$

A special case of this theorem occurs when we have the same number of choices for each element. For example, if we want to count the number of strings of length 8 made up of lowercase letters from the Roman alphabet $\{a, \ldots, z\}$, we have S a set of sequences of length 8, and $n_i = 26$ for all i, so $|S| = 26^8$. Generally, we can say:

Corollary 22.6. *The number of length-n strings from an alphabet of size m is m^n.*

These first few examples were straightforward, but what happens if not all of the sequences we want to count are the same length, or if the available choices for each element can change depending on our choices for other elements?

Example 22.7. *How many 7-digit phone numbers are there, if a phone number can't start with 0, 1, or the sequence 911?*

Solution to example. If we try to just count the choices for each element and multiply, we'll run into trouble. The first digit might be any number from 2 through 9, but then the second and third digits might or might not be allowed to be 1, depending what came beforehand. So instead, let's split this into two problems.

First, consider all the numbers that don't start with 0 or 1—let's call this set P:

$$P = \{d_1 d_2 d_3 d_4 d_5 d_6 d_7 : d_1 \in \{2, \ldots, 9\}, d_i \in \{0, \ldots, 9\} \text{ for } 2 \leq i \leq 7\}$$
$$= \{2, \ldots, 9\} \times \{0, \ldots, 9\}^6.$$

So by the product rule,

$$|P| = 8 \cdot 10^6 = 8000000.$$

Now all the valid numbers are in P, and most of the numbers in P are valid—all except the ones beginning with 911. So we need to subtract from the size of P the size of the set $N \subsetneq P$ of seven-digit numbers starting with 911.

$$N = \{911 d_4 d_5 d_6 d_7 : d_i \in \{0, \ldots, 9\} \text{ for } 4 \leq i \leq 7\}$$

and

$$|N| = 1^3 \cdot 10^4 = 10000.$$

So there are

$$|P| - |N| = 8000000 - 10000 = 7990000$$

valid phone numbers. ∎

Another useful trick is to count a set in a nonobvious order, to avoid the need to split out different cases:

Example 22.8. *How many odd four-digit numbers are there with no leading zeroes and no repeated digits?*

Solution to example. We are looking for numbers n of the form $d_1 d_2 d_3 d_4$. Since n is odd, there are only 5 possible digits for d_4: 1, 3, 5, 7, and 9. Once we've chosen a digit to be d_4, there are only 8 possible digits for d_1 (since it can't equal 0 or d_4). Then we can choose any of the remaining 8 digits for d_2 (not d_1 and not d_4), and then any of the remaining 7 for d_3 (not d_1, d_2, or d_4). So there are $8 \cdot 8 \cdot 7 \cdot 5 = 2240$ possibilities. ∎

What if instead we had started with d_1, and considered each digit in order? We'd have said there are 9 digits for d_1 (anything other than 0), then 9 for d_2 (anything other than d_1), 8 for d_3 (anything other than d_1 and d_2), ...and then we would run into complications. The number of possibilities left over for d_4 depends on whether the numbers we picked for the earlier digits were even or odd! We'd have to split the problem into several cases and consider each separately. Instead, by counting the possibilities "out of order," we saved ourselves some work. When there is more than one way to count a set, some ways may be easier than others.

Certain numerical calculations come up often enough in counting problems to deserve their own notation. Consider the following example:

Example 22.9. *How many ways are there to arrange the 26 letters of the Roman alphabet?*

Solution to example. We can choose any of the 26 letters to go first, any of the remaining 25 to go second, ...all the way down to the 26th choice, when there will only be one option left. So there are

$$26 \cdot 25 \cdot 24 \cdot \ldots \cdot 3 \cdot 2 \cdot 1$$

arrangements of 26 letters. ∎

But this is just an instance of the *factorial* function, defined on page 216. So there are 26! distinct arrangements of the 26 letters of the Roman alphabet.

Because an empty product has the value 1 (page 32), in the $n = 0$ case, the product of the first n positive integers has the value 1. That is, $0! = 1$.

This makes sense, since then $(n+1)! = (n+1) \cdot n!$ for any $n \geq 0$, including $n = 0$.

Factorials arise frequently when considering arrangements of collections, called *permutations*, where each choice leaves one choice fewer for the next element. Formally, a *permutation* of a set S is a bijection $p: \{0, \ldots, |S| - 1\} \to S$, which represents the arrangement of the elements of S in the order

$$p(0), p(1), \ldots, p(|S| - 1).$$

This list includes each element of S exactly once, since p is a bijection. Example 22.9 then generalizes:

Theorem 22.10. *The number of permutations of a set of size n is n!.*

The proof is analogous to the argument in Example 22.9.

What if the relative order of the elements matters, but there is no fixed start or end point? For example, suppose the elements are to be arranged in a circle instead of a line. The number of arrangements can be calculated with a simple modification to the formula for the factorial function.

First, some definitions: If p is a permutation of a set of size n, then a *rotation* of p is any of the permutations p' such that, for some fixed k, $0 \leq k < n$,

$$p'(i) = p(i + k \bmod n) \text{ for } i = 0, \ldots, n - 1.$$

For example, if S is the set $\{1, 2, 3, 4\}$ and p the permutation $3, 2, 1, 4$, then the rotations of p are

$$3, 2, 1, 4 \quad (\text{for } k = 0)$$
$$2, 1, 4, 3 \quad (\text{for } k = 1)$$
$$1, 4, 3, 2 \quad (\text{for } k = 2)$$
$$4, 3, 2, 1 \quad (\text{for } k = 3).$$

Two permutations are said to be *cyclically equivalent* to each other if one is a rotation of the other. (This is in fact an equivalence relation—see Problem 22.9.) For example, Figure 22.2(b) is the same as Figure 22.2(a) after rotation by $k = 4$ positions.

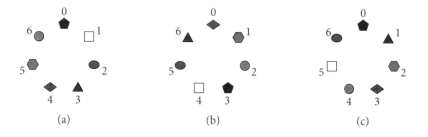

(a)　　　　　　　(b)　　　　　　　(c)

Figure 22.2. Permutations of 7 different colors. The permutations (a) and (b) are the same, after rotation; but (c) is different, because it can't be turned into (a) or (b) by any rotation.

Theorem 22.11. *The number of cyclically inequivalent permutations of a set of size n is $(n-1)!$.*

Proof. The n elements can be permuted in $n!$ ways. But that overcounts, since the n rotations of any permutation (by 0, 1, …, or $n-1$ positions) should all count as identical. So the actual total is

$$\frac{n!}{n} = (n-1)!.$$ ∎

✻

In counting arrangements of the alphabet, we relied on the fact that all of the letters are different, so each choice eliminates that letter from being considered later on. But what if we want to find the number of permutations when some of the elements are the same—for example, the number of anagrams of a word with repeated letters?

Example 22.12. *How many ways can the letters of the word ANAGRAM be arranged?*

Solution to example. Let's give each of those occurrences of A a different color, to help us keep track:
ANAGRAM
As far as we are concerned, all the following arrangements are equivalent to the one just given:
ANAGRAM
ANAGRAM
ANAGRAM
ANAGRAM
ANAGRAM
Similarly, for any arrangement of the 7 letters, we could swap around the 3 occurrences of A without changing the arrangement of the letters.
If all 7 letters were different, there would be $7! = 5040$ arrangements of the letters. But with 3 identical letters A, that figure overcounts by a factor equal to the number of arrangements of the three As, which is $3! = 6$. Dividing 5040 by 6 yields 840 distinct arrangements of the letters of ANAGRAM. ∎

We need some new vocabulary to capture the idea of a collection with multiple occurrences of the same element. A *multiset* is a pair $M = \langle S, \mu \rangle$, where S is a finite set and μ is a function from S to $\mathbb{N}$. For any element $x \in S$, $\mu(x)$ is the *multiplicity* of x in M; or in other words, the number of occurrences of x in M. The *size* of $M = \langle S, \mu \rangle$ is the sum of the multiplicities of the members of S; that is,

$$\sum_{x \in S} \mu(x).$$

For example, the letters of the word ANAGRAM can be regarded as a multiset $\langle S, \mu \rangle$ of size 7, where

$$S = \{A, G, M, N, R\}$$
$$\mu(A) = 3$$
$$\mu(G) = \mu(M) = \mu(N) = \mu(R) = 1.$$

Now let $M = \langle S, \mu \rangle$ be a multiset of size m. Then a *permutation* of M is a mapping p from $\{0, \ldots, m-1\}$ to S such that the number of times an element x occurs as the value of p is equal to its multiplicity in M. That is, for each $x \in S$,

$$|\{i : p(i) = x\}| = \mu(x).$$

The word ANAGRAM is the permutation p of this multiset with the values

$$p(0) = A$$
$$p(1) = N$$
$$p(2) = A$$
$$p(3) = G$$
$$p(4) = R$$
$$p(5) = A$$
$$p(6) = M.$$

There are three values of i for which $p(i) = A$, namely $i = 0$, 2, and 5, which matches the value $\mu(A) = 3$. Each of the other letters has multiplicity 1 and is the value of $p(i)$ for only a single value of i.

Theorem 22.13. *Let $M = \langle S, \mu \rangle$ be a multiset of size m. Then the number of permutations of M is*

$$\frac{m!}{\prod_{x \in S} \mu(x)!}. \tag{22.14}$$

Before proving this theorem, let's work through a few extreme cases to see that it gives the expected answer. If all the multiplicities are 1, then the size of M is just $|S|$, and a permutation of M is just an ordinary permutation of the set S. The product in the denominator of (22.14) is a product of 1s, and hence has the value 1. Then the value of (22.14) is $m!$, which is $|S|!$, matching the statement of Theorem 22.10. At the other extreme, if $|S| = 1$ and the multiplicity of the single element of S is m, then the numerator and denominator of (22.14) are both $m!$ and their ratio is 1—which is correct since there is only a single arrangement of m copies of a single element.

Proof. Instead of permuting the multiset M with multiplicities μ, consider ordinary permutations of the set

$$M' = \bigcup_{x \in S} (\{x\} \times \{1, \ldots, \mu(x)\}).$$

For our example of the multiset of the letters of ANAGRAM, M' includes 7 distinct elements:

$$M' = \{\langle A, 1 \rangle, \langle A, 2 \rangle, \langle A, 3 \rangle, \langle G, 1 \rangle, \langle M, 1 \rangle, \langle N, 1 \rangle, \langle R, 1 \rangle\}.$$

In essence, the second components of these pairs are colors assigned to the various occurrences of the letters so they can be distinguished. The set M' contains m distinct elements, so there are $m!$ permutations of M'. These permutations fall into equivalence classes by ignoring the second components, in exactly the way that the six permutations of ANAGRAM shown above become identical when colors are ignored. The number of equivalence classes is exactly $m!$ divided by the number of permutations of $\{x\} \times \{1, \ldots, \mu(x)\}$ for each $x \in S$. The set $\{x\} \times \{1, \ldots, \mu(x)\}$ contains $\mu(x)$ distinct elements, so it has $\mu(x)!$ permutations. So the denominator is the product of $\mu(x)!$ for each x—which gives exactly (22.14). ∎

Chapter Summary

- The Sum Rule states that if two sets A and B are disjoint, then $|A \cup B| = |A| + |B|$.

- The Product Rule states that for any two sets A and B, $|A \times B| = |A| \cdot |B|$.

- When a choice for one element of a set affects the number of available choices for another element, it may help to split the set into cases, or to count the choices "out of order."

- A *permutation* of a set is an arrangement of its elements; formally, it is a bijection $p : \{0, \ldots, |S| - 1\} \to S$.

- A set of size n has $n!$ ("n factorial") permutations, where $n!$ is the product of the first n positive integers.

- A permutation is a *rotation* of (or *cyclically equivalent* to) another permutation if its elements appear in the same order, but starting with a different element.

- A set of size n has $(n - 1)!$ cyclically inequivalent permutations.

- A *multiset* represents a collection that may include multiple occurrences of each element. Formally, a multiset M is a pair $\langle S, \mu \rangle$, where $\mu(x)$ is the *multiplicity* of x for each element in S.

■ The permutations of a multiset can be counted by first considering each element as distinct, and then dividing by the number of permutations of each set of equivalent elements: if $M = \langle S, \mu \rangle$ has size m, the number of permutations is

$$\frac{m!}{\prod_{x \in S} \mu(x)!}.$$

Problems

22.1. Consider the following ways of encoding a message, using the 26 letters of the Roman alphabet.

(a) A simple encryption scheme is the Caesar cipher (named after Julius Caesar, an early user), which uses rotations of the alphabet: each letter is replaced by the letter that comes k places after it, for some fixed value k. How many different Caesar ciphers are there?

(b) More generally, a substitution cipher is any cipher that substitutes each letter of the alphabet for another. How many different substitution ciphers exist? (Assume that letters are allowed to "substitute" for themselves.[1])

(c) Suppose you want to create a substitution cipher in which the "words" appear more realistic, by always mapping the 5 vowels to vowels, and the 21 consonants to consonants. How many substitution ciphers satisfy this constraint?

22.2. (a) If you have 4 shirts, 3 pairs of pants, and 5 pairs of socks, how many different outfits could you wear?

(b) Suppose 2 of the shirts are striped and 1 of the pairs of pants is polka-dotted, and you don't want to wear stripes and polka dots together. How many possible outfits do you have?

22.3. (a) How many ways can the letters of the word COMPUTER be arranged?

(b) How many ways can the letters of the word MISSISSIPPI be arranged?

22.4. How many ways are there to arrange 5 keys on a keyring (that is, in a circle)? *Hint:* Two permutations correspond to the same arrangement if one can be transformed into the other by rotation and/or reversal.

22.5. Suppose you have a jar with 15 marbles, containing 3 each of 5 different colors. You draw out all 15 marbles, one at a time. How many different sequences of colors are possible?

22.6. How many even four-digit numbers are there with no leading zeroes and no repeated digits?

22.7. Services that require a password often impose constraints to prevent the user from choosing a password that is easy to guess. Consider a service that

[1] If every letter had to be represented by a letter other than itself, this would be the somewhat trickier problem of counting *derangements*: arrangements with no fixed points.

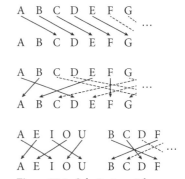

Figure 22.3. Substitution ciphers. Top: a Caesar cipher. Middle: a generic substitution cipher. Bottom: a substitution cipher that maps vowels to vowels and consonants to consonants.

C	D	H	S
2	2	2	2
3	3	3	3
4	4	4	4
5	5	5	5
6	6	6	6
7	7	7	7
8	8	8	8
9	9	9	9
10	10	10	10
J	J	J	J
Q	Q	Q	Q
K	K	K	K
A	A	A	A

Figure 22.4. A standard deck contains 52 different cards, divided into 4 suits: clubs, diamonds, hearts, and spades. Clubs and spades are black suits while diamonds and hearts are red suits. Each suit contains 13 values: the numbers $2, 3, 4, 5, 6, 7, 8, 9, 10$, the face cards J, Q, K (Jack, Queen, and King), and the value A (Ace).

requires an 8-character password, using the 26 Roman letters (both lowercase and uppercase) and the 10 Arabic numerals.

(a) How many possible passwords are there?

(b) How many such passwords use no numerals? No lowercase letters? No uppercase letters?

(c) If a computer can guess one password every billionth of a second, how many hours will it take to guess all the possible passwords from part (a)?

(d) If instead your password can contain only lowercase letters, and you want it to be at least as hard to guess as an 8-character password that may contain lowercase letters, uppercase letters, and numbers, how long must your password be?

22.8. Anisha picks 5 cards at random from a standard card deck (see Figure 22.4). She shows Brandon 4 of the 5 cards, revealing them one at a time. Brandon then correctly guesses the final card. How did they do it? (This requires no tricks or sleight of hand!)

Hint: Anisha can choose which of the 5 cards Brandon will have to guess. She uses the Pigeonhole Principle to choose the suit of the mystery card, and then uses the Pigeonhole Principle again to choose a span of cards within that suit. The first card that Anisha reveals indicates that suit and span, and the next 3 cards distinguish between the remaining possibilities.

22.9. Prove that cyclic equivalence is in fact an equivalence relation on permutations of a set.

Chapter 23

Counting Subsets

Suppose a jar contains five marbles of five different colors. As shown in Chapter 22, the number of permutations of the 5 marbles is 5!. But how many orderings are possible if fewer than 5 marbles are to be selected? For example:

Example 23.1. *How many different permutations of 3 marbles can be chosen from a jar containing 5 different marbles?*

Solution to example. We can think of this as choosing a 5-element permutation, and then ignoring elements after the first 3: any two 5-element permutations that start with the same 3 elements are equivalent. So the total number of permutations must be divided by the number that share the same 3 initial elements.

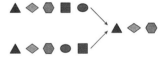

Figure 23.1. Arrangements of marbles. Both permutations of the set of 5 marbles start with the same 3-element permutation.

   If the first 3 marbles are fixed, there are 2! choices for the permutation of the remaining 2 marbles. So there are $\frac{5!}{2!} = 60$ permutations of 3 marbles that can be chosen from the jar. ∎

The term *permutation* is being used in two slightly different senses. There are only 6 permutations of 3 elements once those elements are fixed. But there are 60 permutations of a 3-element subset of a set of size 5 when the choice of the 3 elements is left open. In the previous chapter we counted the number of permutations of a set of size n; we are now counting the number of permutations of a subset of size k drawn from a set of size n.

   The number of permutations of k elements drawn from a set of size n is denoted $_nP_k$. So $_nP_n = n!$; that is the number of ways of picking all n elements from a set of size n (there is only one way to do that) and then permuting them arbitrarily. And $_nP_1 = n$, because there are n possible 1-element subsets of a set of size n, and only one permutation of each.

   Our method of calculating the number of 3-marble permutations from a set of 5 marbles can be generalized:

Theorem 23.2. Permutations of Subsets.

$$_nP_k = \frac{n!}{(n-k)!}. \qquad (23.3)$$

Proof. We map length-n sequences to length-k sequences by taking the first k elements in order. Any two length-n sequences with the same length-k starting sequence will map to the same permutation of k elements, so any permutation of the remaining $n - k$ elements will result in the same length-k permutation. There are $n!$ permutations of the sequence of n elements and $(n-k)!$ permutations of the final $n - k$ elements, so (23.3) results from dividing the first quantity by the second. ∎

When $k = n$, (23.3) becomes $\frac{n!}{0!} = n!$; when $k = 1$, (23.3) becomes $\frac{n!}{(n-1)!} = n$, matching the values already calculated.

✳

In the examples considered so far, the order of the elements has been significant. How should our calculations differ if only the set of elements selected is significant, not their order? In other words, what if the problem is to count the number of ways to make an unordered selection of elements?

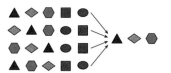

Figure 23.2. Four possible permutations of the set of 5 marbles, all of which correspond to the 3-element combination of red, yellow, blue. In total, there are 12 different permutations that correspond to this combination.

Example 23.4. *A jar contains 5 marbles of 5 different colors. If 3 marbles are selected from the jar, how many different sets of marbles are possible?*

Solution to example. There are $_5P_3$ permutations of 3 marbles chosen from 5. But we care only about the selection of those 3 marbles, not their order, so $_5P_3$ overcounts the number of different *sets* of 3 marbles by a factor of $3!$. The number of sets of 3 marbles chosen from 5 is then

$$\frac{_5P_3}{3!} = \frac{5!}{3!2!} = 10.$$

If the colors are Blue, Green, Purple, Red, and Yellow, the 10 subsets are (abbreviating the names of the colors)

$$\{B, G, P\}, \{B, G, R\}, \{B, G, Y\}, \{B, P, R\}, \{B, P, Y\},$$
$$\{B, R, Y\}, \{G, P, R\}, \{G, P, Y\}, \{G, R, Y\}, \{P, R, Y\}. \qquad ∎$$

A *combination* of k elements from a set S of n elements is a subset of S of cardinality k. The number of combinations of k elements chosen from a set of n distinct elements is denoted $_nC_k$ or

$$\binom{n}{k},$$

and pronounced "n choose k."

Example 23.4 can be generalized.

Theorem 23.5. Combinations.

$$\binom{n}{k} = \frac{n!}{k!(n-k)!}.$$

Proof. Consider how to map permutations of k elements from a set of size n to combinations of k elements. A combination is a set of elements in any order, so any permutation of a given k elements maps to the same combination. There are $k!$ permutations of k distinct elements, so the number of combinations is the result of dividing $_nP_k$ by $k!$:

$$\binom{n}{k} = \frac{_nP_k}{k!} = \frac{n!}{k!(n-k)!}. \qquad \blacksquare$$

Note that this formula is symmetrical: it gives the same result whether k elements or $n - k$ elements are selected:

$$\binom{n}{k} = \frac{n!}{k!(n-k)!} = \frac{n!}{(n-k)!k!} = \binom{n}{n-k}. \tag{23.6}$$

We could have made the argument that $\binom{n}{k}$ and $\binom{n}{n-k}$ are the same even without working through the algebra. The expressions $\binom{n}{k}$ and $\binom{n}{n-k}$ describe two different ways to find the size of the same set: $\binom{n}{k}$ is the number of ways to pick k elements to include from a set of n elements, and $\binom{n}{n-k}$ is the number of ways to pick $n - k$ of those n elements not to include.

This sort of reasoning is called a *counting argument*: it demonstrates that two quantities are equal since they are just two ways of counting the same thing. Counting arguments can be quite useful when proving equalities that might otherwise require tedious algebraic manipulations. In the following example, a counting argument breaks a sum down using an exhaustive case analysis.

Example 23.7. *Prove that*

$$\binom{2n}{n} = \sum_{k=0}^{n} \binom{n}{k}^2.$$

Solution to example. First, since $\binom{n}{k} = \binom{n}{n-k}$, we can rewrite the right side so the identity to be proven is

$$\binom{2n}{n} = \sum_{k=0}^{n} \binom{n}{k}\binom{n}{n-k}.$$

Now we can use a counting argument. Suppose S is a set of size $2n$. Then the left side, $\binom{2n}{n}$, is the number of ways of choosing n elements from S.

The right side describes how to choose n elements from S by splitting S into two disjoint subsets, S_1 and S_2, each of size n, and then choosing k elements from S_1 and $n - k$ elements from S_2, for some value of k, $0 \leq k \leq n$. As these cases are disjoint for different values of k, the total is represented by $\sum_{k=0}^{n} \binom{n}{k}\binom{n}{n-k}$, which therefore also represents the number of ways of choosing n elements from a set of size $2n$. ∎

Theorem 23.5 gives the number of ways to pick out a single subset from a set. But suppose we want to continue, picking out a second subset from the remaining elements, and then perhaps a third subset after that, and so on. The total number of outcomes is the result of taking the product of the number of ways to pick each subset. For example:

Example 23.8. *A jar contains 11 marbles, labeled 1 through 11. In how many ways can they be placed in four boxes, with 3 marbles in the first box, 4 marbles in the second box, 2 marbles in the third box, and the 2 remaining marbles in the fourth box?*

Solution to example. There are $\binom{11}{3}$ ways to pick 3 marbles for the first box. There are 8 marbles left, so there are $\binom{8}{4}$ ways to pick 4 marbles for the second box. Then there are 4 marbles remaining, and $\binom{4}{2}$ ways to pick 2 marbles for the third box. Finally, there are 2 marbles left, both of which go in box 4— there is just 1 way to do this, because $\binom{2}{2} = 1$. The total number of outcomes is the product of these four quantities:

$$\binom{11}{3} \cdot \binom{8}{4} \cdot \binom{4}{2} \cdot \binom{2}{2}.$$

Multiplying this out, many of the terms cancel:

$$\frac{11!}{3!8!} \cdot \frac{8!}{4!4!} \cdot \frac{4!}{2!2!} \cdot \frac{2!}{2!0!} = \frac{11!}{3!4!2!2!}$$
$$= 69300. \qquad ∎$$

The numerator is the factorial of the original set size, and the denominator is the product of the factorials of each of the subset sizes. To put this in general terms:

Theorem 23.9. *Let S be a set of n elements, and let $k_1, k_2, \ldots, k_m$ be integers such that $\sum_{i=1}^{m} k_i = n$. Then the number of ways to partition S into m disjoint labeled subsets, where the ith subset has size k_i, is equal to*

$$\frac{n!}{k_1! \cdot k_2! \cdot \ldots \cdot k_m!}.$$

The proof is a straightforward generalization of the logic of Example 23.8.

For convenience, the number of ways of partitioning a set of size n into disjoint subsets of sizes $k_1, k_2, \ldots, k_m$ is denoted by

$$\binom{n}{k_1, k_2, \ldots, k_m} = \frac{n!}{k_1! \cdot k_2! \cdot \ldots \cdot k_m!}. \tag{23.10}$$

In the $m = 2$ case, this is just the formula for combinations, since the number of ways of partitioning a set of size n into two parts of sizes k_1 and k_2, where $k_1 + k_2 = n$, is just the number of ways of choosing k_1 (or k_2) out of the n elements:

$$\binom{n}{k_1, k_2} = \binom{n}{k_1} = \binom{n}{k_2} = \frac{n!}{k_1! \cdot k_2!}.$$

Except for notation, (23.10) is the same as (22.14) from Theorem 22.13, the number of permutations of a multiset with multiplicities $k_1, \ldots, k_m$. Problem 23.7 explores how these types of questions relate to each other.

Theorem 23.9 gives the number of ways to partition the set into m *labeled* subsets—that is, where the subsets themselves are ordered, though the elements within any subset are unordered. Example 23.8 implicitly specified that the subsets were ordered, by mentioning "the first box," "the second box," and so on. How would the answer be different if instead we wanted to place the 11 marbles into 4 *indistinguishable* boxes?

For instance, in Example 23.8, the following are two different outcomes:

$$\{1, 2, 3\}, \{4, 5, 6, 7\}, \{8, 9\}, \{10, 11\};$$

$$\{1, 2, 3\}, \{4, 5, 6, 7\}, \{10, 11\}, \{8, 9\}.$$

These outcomes are different because the marbles that are placed in the third and fourth boxes are swapped in the second row. Now consider the case where the boxes are indistinguishable, so the above outcomes are regarded as the same.

Example 23.11. *A jar contains 11 marbles, labeled 1 through 11. In how many ways can they be placed in 4 unmarked boxes, with 1 box containing 3 marbles, 1 box containing 4 marbles, and 2 boxes containing 2 marbles each?*

Solution to example. We begin by choosing the marbles for each box in order, as in Example 23.8:

$$\binom{11}{3} \cdot \binom{8}{4} \cdot \binom{4}{2} \cdot \binom{2}{2} = \frac{11!}{3!4!2!2!}.$$

Now, any boxes that contain the same number of marbles can be swapped with each other. So for each box size, we must divide this total by the number of ways to order the boxes of that size. There is 1 box of size 3, 1 of size 4, and 2 of size 2, so we divide by 1!1!2! to get the final result:

$$\frac{11!}{3!4!2!2! \cdot 1!1!2!} = 34650. \qquad \blacksquare$$

The following theorem gives the general formula, which can be proven by similar argument.

Theorem 23.12. *Let S be a set of n elements, to be split up into m disjoint subsets. Let $K \subseteq \mathbb{N}$ be the set of sizes of those subsets (so K is a finite set of positive integers). Let $M = \langle K, \mu \rangle$ be a multiset of size m representing the sizes of those m subsets, so there are $\mu(k)$ subsets of size k:*

$$m = \sum_{k \in K} \mu(k) \qquad \text{(the number of subsets)}$$

$$n = \sum_{k \in K} k \cdot \mu(k) \quad \text{(the number of elements)}.$$

Then the number of ways to split the elements of S into disjoint subsets as specified is

$$\frac{n!}{\prod_{k \in K} \left(\mu(k)!(k!)^{\mu(k)} \right)}.$$

For example, 10 marbles of different colors can be split between 2 unmarked boxes of 5 marbles each in $\frac{10!}{2!(5!)^2} = 126$ ways (the variables in the statement of Theorem 23.12 have the values $n = 10$, $m = 2$, $K = \{5\}$, $\mu(5) = 2$).

✳

A permutation is sometimes called an *ordering without replacement*, since a permutation of a set S is the result of removing elements from S one at a time and putting them in the order in which they were drawn. The "without replacement" suggests that an element, once removed, cannot be used in the permutation a second time. In the same way, a permutation of a multiset is referred to as an ordering without replacement. (When an element appears multiple times in a multiset permutation, it is because it has multiplicity greater than 1—not because it was replaced.) Analogously, a combination could be termed a *selection without replacement*: an element once chosen can't be chosen again, though the order of the chosen elements is not significant.

A selection where the same item can be chosen multiple times is called a *combination with replacement*. There is no special notation for combinations with replacement—because any combination with replacement can be enumerated in terms of a combination without replacement.

Example 23.13. *As in Example 23.4, suppose that a jar contains 5 marbles, one each of the colors blue, green, purple, red, and yellow. Suppose now marbles are drawn from the jar one at a time and immediately replaced. As marbles are drawn, their colors are tallied—we are just keeping count of how many times each color is drawn. This process is repeated 15 times, and the resulting counts are reported. How many different results are possible?*

Solution to example. A result is essentially a 5-tuple of nonnegative integers that sum to 15, each component representing the number of marbles of a particular color—the first component being the count of blue marbles, the second the count of green marbles, and so on. Because the total is 15, such a 5-tuple can be depicted as a sequence of 15 tally marks (we'll use "$*$" to represent one marble) separated by 4 dividers (we'll use a vertical bar). For example, the result 3 blue, 4 green, 1 purple, 5 red, and 2 yellow would be represented as follows:

$$* * * | * * * * | * | * * * * * | * *$$

Every result can be written in this manner, as an arrangement of 19 symbols: 15 asterisks to represent the 15 drawn marbles, and 4 vertical separators to separate the 5 "buckets" in which marbles of different colors are tallied. And any such arrangement corresponds to exactly one result, denoting how many times each color is drawn, without regard for order. So there is a bijection between 5-tuples of nonnegative integers summing to 15 and such arrangements of 15 asterisks and 4 vertical bars. This representation is called a *stars and bars diagram*.

Now this has become a problem of counting ordinary combinations without replacement. There are 19 places in which to put symbols, and 15 must be selected as places to put the asterisks, with the 4 separators filling in the rest. So the answer is

$$\binom{19}{15} = \frac{19!}{15!4!}.$$

Of course, if instead 4 of the 19 are chosen as places to put the divider bars, with the asterisks filling the leftover spots, the same result is derived, since $\binom{19}{4} = \binom{19}{15}$, by (23.6). ∎

Example 23.13 describes a scenario with n distinct elements that are replaced after each choosing, but the reasoning would be the same for a situation with n types of elements and (practically speaking) an unlimited supply of elements of each type. To reframe Example 23.13 in those terms, we could perform the same experiment without using marbles: instead we could choose a color from the 5 options (so the colors are the "types"), and repeat 15 times. This alternative framing is often used when the things to be chosen are not physical objects—for example, letters from an alphabet. Whether we are choosing with replacement from n distinct elements, or simply choosing from n types of elements, the problem is a combination with replacement. And the same approach works for both: use a stars and bars diagram to transform the question into an equivalent question about combinations with replacement. To state the general result:

Theorem 23.14. *The number of ways of choosing k elements from n types of elements is*

$$\binom{k+n-1}{k}.$$

Proof. Label the types $t_1, t_2, \ldots, t_n$. Any set of k elements can be represented as a series of k asterisks and $n-1$ separators, where the asterisks before the first separator represent elements of type t_1, those between the ith and $(i+1)^{\text{st}}$ separators represent elements of type t_i, and those after the $(n-1)^{\text{st}}$ separator represent elements of type t_n. The number of such arrangements is the number of ways to select the positions of k asterisks in a row of $k+n-1$ possible locations; that is, $\binom{k+n-1}{k}$. ■

This value is sometimes called "*n multichoose k*," since it is the number of multisets of size k that can be constructed from an underlying set of size n.

*

So far in this chapter, we have considered variations on the problem of selecting k elements from a set of n distinct elements. With a bit of imagination, we can apply some of the same techniques to questions where the n elements are identical, rather than distinct. In this situation, selecting a single group of elements is not very interesting—any group of k elements is the same, since all the elements are identical—but we might instead ask how many ways there are to divide the n identical elements into k groups.

First, we consider the case where the k groups are ordered; that is, distinguishable from each other. Suppose the groups are labeled 1 through k; we want to know the number of ways to assign the n identical elements to the k labeled groups. For example, perhaps we are placing n identical marbles into k boxes, where each box is a different color.

But this is just Theorem 23.14 with the roles of k and n reversed: now there are k distinct types, where each type is the label of a group; and we need to choose n labels, one for each of the n items to be placed in a group. We apply Theorem 23.14 with k and n swapped:

Corollary 23.15. *The number of ways of grouping n identical elements into k distinguishable groups, where $k \geq 1$, is*

$$\binom{n+k-1}{n}.$$

As before, we can represent any such grouping with a stars and bars diagram, containing n asterisks to represent the n identical elements, and $k-1$ separators to divide the elements into k groups. The quantity $\binom{n+k-1}{n}$ is just the number of ways of choosing n positions for the asterisks out of the $n+k-1$ possible locations. For instance, we can repurpose the same diagram from Example 23.13 to represent this new scenario: Suppose we have $n = 15$ identical marbles that we would like to place in $k = 5$ colored boxes: blue, green, purple, red, and yellow. Then the selection

$$* * * \,|\, * * * * \,|\, * \,|\, * * * * * \,|\, * *$$

represents 3 blue, 4 green, 1 purple, 5 red, and 2 yellow, and there are $\binom{15+5-1}{15} = \binom{19}{15}$ possible selections, just as before.

When grouping n elements into k groups, we may want to add an additional constraint: that each group contains at least one element. If not, then supposing exactly one of the k groups is empty (the case where two separators are adjacent), it could be argued that in fact we have $k-1$ groups, rather than k. This constraint can be modeled by allocating 1 element to each of the k groups at the start, and then dividing the remaining $n-k$ elements between the k groups.

Theorem 23.16. *The number of ways of grouping n identical elements into k distinguishable groups, where $1 \leq k \leq n$ and each group contains at least one element, is*

$$\binom{n-1}{n-k}.$$

Proof. Let G_1 be the set of groupings of n elements into k groups, where each group contains at least one element, and G_2 the set of groupings of $n-k$ elements into k groups, with no restrictions on the number of elements per group. Then the function $f: G_1 \to G_2$ that removes one element from each of the k groups is a bijection: each member of G_2 is the value of f for exactly one element of G_1. So G_1 and G_2 are of the same size. By Corollary 23.15, the number of groupings in G_2 is just

$$\binom{(n-k)+k-1}{(n-k)} = \binom{n-1}{n-k}.$$ ∎

✳

Now we turn to the case where the n elements are identical, and the k groups are indistinguishable; for example, placing those same $n = 15$ white marbles into $k = 5$ boxes, where now the boxes are indistinguishable from each other unless they contain different numbers of marbles. In this scenario, the stars and bars diagram

$$* * * | * * * * | * | * * * * * * | * *$$

describes the same outcome as the diagram

$$* * * * | * * * | * * | * * * * * * | *$$

which we might as well show by listing the boxes from smallest to largest:

$$* | * * | * * * | * * * * | * * * * *$$

There are 5! such diagrams that all correspond to the same outcome: the set of numbers $\{1, 2, 3, 4, 5\}$. But not every outcome has 5! corresponding diagrams. For instance, the outcome represented by

$$* * * | * * * | * * * | * * * | * *$$

has just this one diagram.

Counting the ways of putting indistinguishable elements into indistinguishable boxes is essentially the problem of expressing a positive integer n as the sum of k smaller positive integers, called a *partition* of n, where the k terms being added up are its *parts*.[1] If we write $p_k(n)$ for the number of ways of partitioning n into k parts, then, for example,

$$
\begin{array}{lll}
p_1(2) = 1 & & (2) \\
p_2(2) = 1 & & (1+1) \\
p_1(3) = 1 & & (3) \\
p_2(3) = 1 & & (1+2) \\
p_3(3) = 1 & & (1+1+1) \\
p_1(4) = 1 & & (4) \\
p_2(4) = 2 & & (1+3, 2+2)
\end{array}
$$

[1] Problem 15.8, regarding Bulgarian solitaire, was actually about such partitions.

$$p_3(4) = 1 \qquad (1+1+2)$$
$$p_4(4) = 1 \qquad (1+1+1+1).$$

There is no simple formula for $p_k(n)$. However, these numbers can be calculated recursively by breaking the problem down into smaller, similar problems until we reach base cases that are easy to solve.

Theorem 23.17. Recursive Calculation of Partitions of an Integer.

$$p_k(n) = p_{k-1}(n-1) + p_k(n-k),$$

with base cases

$$p_1(n) = 1 \text{ for all } n,$$
$$p_n(n) = 1 \text{ for all } n,$$
$$p_k(n) = 0 \text{ for } k > n.$$

Proof. Let $P_{k,n}$ be the set of partitions of n into k parts. (For example, $P_{2,4} = \{1+3, 2+2\}$.) Let $A \subseteq P_{k,n}$ be the set of such partitions for which at least one part is equal to 1. Any member of A can be written as $\{1\} \cup A'$, where A' is a partition of $n-1$ into $k-1$ parts; that is, A' is a member of $P_{k-1,n-1}$. The function that removes one part of size 1 from such a partition is a bijection between A and $P_{k-1,n-1}$.

Let $B \subseteq P_{k,n}$ be the set of such partitions for which no part is equal to 1—that is, each of the k parts is greater than 1. For any member of B, we can subtract 1 from each part to get a partition of $n-k$ into k parts. The function that subtracts 1 from each part of such a partition is a bijection between B and $P_{k,n-k}$.

Then A and B are disjoint, and $P_{k,n}$ is their union. So

$$|P_{k,n}| = |A| + |B|$$
$$= |P_{k-1,n-1}| + |P_{k,n-k}|,$$

or in other words,

$$p_k(n) = p_{k-1}(n-1) + p_k(n-k).$$

As for the base cases, the only partition of n into 1 part is n itself, and the only partition of n into n parts is

$$n = \overbrace{1 + \ldots + 1}^{n \text{ times}}.$$

There are no partitions of n into more than n parts, since each part must be at least 1. ∎

For example, consider the partitions of 7 into 3 parts. There are three such partitions that contain one or more 1s, and each can be written as 1 plus a partition of 6 into 2 parts ($n - 1 = 6$ and $k - 1 = 2$):

$$7 = 5 + 1 + 1 = (5 + 1) + 1,$$

$$7 = 4 + 2 + 1 = (4 + 2) + 1,$$

$$7 = 3 + 3 + 1 = (3 + 3) + 1.$$

There is one such partition that does not contain any 1s:

$$7 = 3 + 2 + 2 = (2 + 1 + 1) + 3.$$

After subtracting 1 from each part, we are left with $2 + 1 + 1$, which is a partition of 4 into 3 parts ($n - k = 4$ and $k = 3$).

We can use Theorem 23.17 to check that the four partitions listed above are the only partitions of 7 into 3 parts:

$$p_3(7) = p_2(6) + p_3(4)$$

$$= p_1(5) + p_2(4) + p_2(3) + p_3(1)$$

$$= 1 + p_1(3) + p_2(2) + p_1(2) + p_2(1) + 0$$

$$= 1 + 1 + 1 + 1 + 0 + 0$$

$$= 4.$$

This calculation is not too difficult when k and n are small, but with each step, each term splits into two, unless it is a base case. For larger values of k and n, the arithmetic becomes cumbersome—and may even call for redundant work. Problem 23.12 explores more efficient strategies for carrying out such calculations.

A lovely symmetry exists between partitioning an integer into k parts and partitioning that integer in such a way that the largest part is k. Remarkably, the number of partitions of these two kinds is the same.

Theorem 23.18. *The number of partitions of n into k parts is equal to the number of partitions of n for which the largest part is equal to k.*

Proof. Consider a diagram like those in Figure 23.3, called *Young diagrams*, consisting of n boxes arranged in rows, where the rows are arranged in nonincreasing order from top to bottom, and each row is left-justified.

Every partition of n into k parts corresponds to exactly one Young diagram of n boxes in k rows, and vice versa. To map a partition to a diagram, order the parts from largest to smallest, then construct a diagram for which

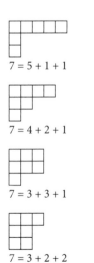

$7 = 5 + 1 + 1$

$7 = 4 + 2 + 1$

$7 = 3 + 3 + 1$

$7 = 3 + 2 + 2$

Figure 23.3. Young diagrams representing each of the 4 partitions of 7 into 3 parts.

the number of boxes in the ith row is equal to the ith part. To map from diagram to partition, include one part equal to the length of each row.

But every partition of n for which the largest part is equal to k also corresponds to exactly one such diagram, with n boxes and k rows, and vice versa. To map a partition to a diagram, order the parts from largest to smallest, and then construct a diagram for which the number of boxes in the ith *column* is equal to the ith part, with the columns top-justified. This results in the desired diagram of n boxes, with its rows in nonincreasing order from top to bottom, and with k rows (since the leftmost column represents the largest part, which is k). To map from diagram to partition, include one part equal to the height of each column.

So the number of Young diagrams containing n boxes in k rows counts both the number of partitions of n into k parts and the number of partitions of n for which the largest part is equal to k. Therefore, these counts are equal.   ∎

✳

This chapter has covered several variations on the problem of selecting subgroups from a collection of elements. In the case of distinct elements, we specified the size of the subgroup (k) and the size of the original collection (n), and the problems varied mainly in two ways: either the outcome is ordered, or not; and either replacement is allowed, or not. Figure 23.4 shows the four cases, with $n = 5$ and $k = 3$. The selection is ordered in the first two cases and unordered in the last two. Replacement is made in the first and third cases, so some of the chosen marbles can be the same color, but not in the second and fourth cases, so the colors of the chosen marbles must be different.

For some of these variations it made sense to also analyze the case where multiple subgroups are chosen. In the case of identical elements, we specified the number of subgroups (k) and the size of the original collection (n), and the problems varied according to whether the subgroups are ordered or not. We have consistently used the metaphor of picking marbles from a jar, but the same techniques are applicable to a wide range of concrete situations:

- Distinct elements, ordered, with replacement—though we didn't use this description at the time, this scenario is discussed in Chapter 22: how many 10-character passwords are there, using an alphabet of 26 characters?

- Permutations—distinct elements, ordered, without replacement: In a tournament with 8 teams, how many different rankings of the teams are possible (on the assumption that ties cannot occur)? How many possibilities are there for just first, second, and third place? (The first question is about permutations of the entire set, and the second question about subset permutations.)

Ordered, with replacement

1   2   3

Ordered, without replacement

1   2   3

Unordered, with replacement

Unordered, without replacement

Figure 23.4. Various kinds of selections from a jar containing 5 different marbles.

- Combinations—distinct elements, unordered, without replacement: if 10 students run for student council and 3 can be elected, how many outcomes are possible?

- Combinations with replacement—distinct elements, unordered, with replacement: how many different orders of 12 pizzas can be purchased from a shop that sells 3 types of pizza?

- Identical elements, ordered—simply another perspective on distinct elements, unordered, with replacement: if a shop has 12 identical plain pizzas and each pizza must be topped with exactly 1 of 3 possible toppings, how many ways are there to choose how many pizzas receive each topping?

- Partitions—identical elements, unordered: How many ways are there to write 25 as a sum of 9 positive integers? How many ways are there to divide a $1000 prize among first-, second-, and third-place winners, with first place getting at least as much as second place and second place at least as much as third place?

Chapter Summary

■ The number of ways of ordering k elements chosen from a set of n elements with *replacement*, or from n types of elements, is n^k.

■ The number of ways of ordering k elements chosen from a set of n elements without replacement is $_nP_k = \frac{n!}{(n-k)!}$. Such an ordering is called a *permutation*.

■ The number of ways of choosing k elements from a set of n elements without replacement is $_nC_k = \binom{n}{k} = \frac{n!}{k!(n-k)!}$. Such a selection is called a *combination*.

■ A *counting argument* demonstrates that two quantities are the same by showing that they are different ways of counting the same set.

■ Suppose a set of size n is partitioned into m disjoint subsets of sizes $k_1, k_2,$ $\ldots, k_m$. If the order of the subsets matters, the number of ways to do so is

$$\binom{n}{k_1, k_2, \ldots, k_m} = \frac{n!}{k_1! \cdot k_2! \cdot \ldots \cdot k_m!}.$$

If the order of the subsets does not matter, divide the above by the number of ways of ordering the subsets of each size. If each size k occurs with multiplicity $\mu(k)$, this is

$$\frac{n!}{\prod_{k \in K} \left(\mu(k)! (k!)^{\mu(k)} \right)}.$$

■ The number of ways of choosing k elements from n elements with replacement, or from n types of elements, is equal to $\binom{k+n-1}{k}$. Such a selection is called a *combination with replacement*. The number of such selections is also called *n multichoose k*.

- A *stars and bars diagram* suggests a correspondence between a combination with replacement of k elements chosen from n types of elements, and a combination without replacement of k positions selected as stars (or $n - 1$ as bars) out of $k + n - 1$ positions.

- The number of ways of grouping n identical elements into k distinguishable groups is $\binom{n+k-1}{n}$, or if each group must contain at least one element, $\binom{n-1}{n-k}$.

- A grouping of n identical elements into k indistinguishable groups is called a *partition* of n into k parts. The number of such partitions can be calculated recursively, using the equality $p_k(n) = p_{k-1}(n-1) + p_k(n-k)$.

- The value $p_k(n)$ also counts the number of partitions of n with largest part equal to k.

Problems

23.1. Suppose a jar contains r red balls and b blue balls, each with a unique identifier on it. How many ways are there to choose a set of two balls of the same color? Of different colors? Show that the sum of these two numbers is the number of ways of choosing two balls from the total, ignoring color.

23.2. Consider a distributed computer network, comprising n computers.
  (a) If one computer is designated as the central computer, and every other computer is connected to the central computer, how many connections are needed?
  (b) If instead, each computer is connected to every other computer, how many connections are needed?

23.3. What proportion of all 5-character strings, composed from the 26 letters of the English alphabet, use 5 distinct letters?

23.4. Suppose a computer program takes a sequence of letters without spaces (for example, a URL), and finds where the spaces could go, by testing all possible ways of breaking the sequence into segments and then checking the segments against a dictionary to determine whether each is a word.
  (a) Given a sequence that is 20 characters long, how many ways are there to break the sequence into 3 nonempty segments?
  (b) How many ways are there to break a sequence of n characters into any number of segments?

23.5. How many different undirected graphs can be made from a set of n labeled vertices? (Consider two graphs to be different even if they are isomorphic, but have different vertex labels.)

23.6. Consider a grid that is 7 units wide and 9 units tall (Figure 23.5). A *monotonic path* is one that begins at point $(0, 0)$ (the bottom left corner) and

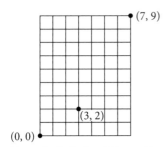

Figure 23.5. Paths within a grid.

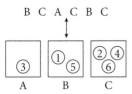

Figure 23.6. We can create a bijection between permutations of the letters and combinations of the marbles. The above image suggests a natural way to do so.

traverses to $(7, 9)$ (the top right corner), using only upward and rightward moves.
(a) How many different monotonic paths are possible?
(b) How many such paths go through the point $(3, 2)$?

23.7. In this problem we'll explore the connection between permutations of multisets, and combinations in which a set is split into several subsets.
(a) How many permutations are there of the letters ABBCCC?
(b) Suppose there are 6 marbles, labeled with the integers 1 through 6; and 3 boxes, labeled with the letters A, B, and C. In how many ways is it possible to arrange the marbles with 1 marble in box A, 2 marbles in box B, and 3 marbles in box C?
(c) Describe a bijection between the permutations of ABBCCC from part (a) and the arrangements of the marbles 1 through 6 in boxes A, B, and C from part (b). *Hint:* See Figure 23.6.

23.8. Use counting arguments to prove the following:
(a)
$$\sum_{k=0}^{n} \binom{n}{k} = 2^n.$$
(b)
$$\sum_{k=1}^{n} \binom{n}{k} \cdot k = n2^{n-1}.$$

Hint: Consider the number of ways to choose a committee of at least 1 person from a group of size n, with 1 committee member designated as its leader.

23.9. In the word game Scrabble, players end their turns by picking letter tiles to replace the ones they have just used. Suppose a player picks 4 letter tiles from the 7 tiles that remain: A, A, B, C, D, E, and F.
(a) How many different selections of letters are possible?
(b) How many different "words" (sequences of letters, not necessarily valid English words) can be made from 4 of these 7 letters?

23.10. Suppose that 27 students get together to play soccer.
(a) There are three different fields for the students to practice on. How many different ways are there to assign the 27 players to the 3 fields in teams of 9?
(b) How many ways are there to assign the 27 players to 3 teams of 9, without regard for which team is on which field?
(c) How many ways are there to assign the 27 players to 3 teams of 9, and for each team to choose 1 of its players as captain?
(d) One of the teams plays 10 games against teams from other schools, ending the season with a 7-3 record. How many different sequences of wins and losses could have led to this outcome?

23.11. Consider a standard deck of playing cards (as described on page 242).

(a) How many hands (sets) of 5 cards can be dealt from such a deck (which contains 52 distinct cards)?

(b) In poker, such a 5-card hand is called a *full house* if it contains two cards of one value and three of another (for example, two 5s and three 9s). How many 5-card hands are a full house?

(c) A much more common poker hand is *two pair*: two cards of one value, two cards of another value, and one card of a third value (for example two 5s, two 9s, and one K). How many 5-card hands are a two pair?

23.12. On page 254 we noted that calculating a single value $p_k(n)$ by the recursive definition may require many steps, since each term splits into two terms that must also be calculated recursively, and may call for redundant calculations.[2]

(a) If we want to find not just one $p_k(n)$ but all $p_k(n)$ up to some maximum values of k and n, there is a more efficient way, which requires just a single addition to calculate each one. This is achieved by starting with the base cases and then working up from smaller cases to larger ones, tabulating the results along the way, such that the values $p_{k-1}(n-1)$ and $p_k(n-k)$ are already present in the table when we need to calculate $p_k(n)$.

Complete the following table with the values of $p_k(n)$ for $1 \le k, n \le 10$, proceeding from top to bottom and within each row from left to right. The base cases are already filled in.

[2] Problem 23.12 describes two forms of *dynamic programming*, an algorithmic strategy that is useful for problems that can be defined in terms of overlapping subproblems. The first method, which proceeds from the bottom up and computes all values smaller than the target, is called *tabulation*. The second method, which proceeds from the top down and computes only the necessary smaller values, is called *memoization*.

n \ k	1	2	3	4	5	6	7	8	9	10
1	1	1	1	1	1	1	1	1	1	1
2	0	1								
3	0	0	1							
4	0	0	0	1						
5	0	0	0	0	1					
6	0	0	0	0	0	1				
7	0	0	0	0	0	0	1			
8	0	0	0	0	0	0	0	1		
9	0	0	0	0	0	0	0	0	1	
10	0	0	0	0	0	0	0	0	0	1

(b) To find just a single value $p_k(n)$, it is still useful to record intermediate results in a table, though it is not necessary to fill in every cell. Find $p_5(10)$ as follows: Start by circling its corresponding cell. Then circle the two cells whose values are used to compute $p_5(10)$. For each newly circled cell, circle the two cells that it depends on, and repeat until reaching the base cases. Finally, proceed from top to bottom and left to right, filling in only the circled cells, until reaching $p_5(10)$.

$\diagdown$ n k	1	2	3	4	5	6	7	8	9	10
1	1	1	1	1	1	1	1	1	1	1
2	0	1								
3	0	0	1							
4	0	0	0	1						
5	0	0	0	0	1					

23.13. How many partitions of n have all their parts equal to 1 or 2?

23.14. Three children are given 10 identical marbles, which they must divide among themselves.

    (a) How many ways can the 10 marbles be divided among the 3 children?

    (b) How many ways can the 10 marbles be divided among the 3 children, if each child must receive at least 1 marble?

    (c) How many ways can the 10 marbles be divided, if each child must receive at least 1 marble, and the oldest child takes the largest share (or one of the two largest, if the two largest are equal), the second-oldest child takes the next-largest share (or one of the two next-largest, if the two are equal), and the youngest child takes the last remaining share? *Hint:* Use one of the strategies described in Problem 23.12.

    (d) How many ways can the 10 marbles be divided, if a child may receive no marbles, and the oldest child takes the largest share (or one of the two largest, if the two largest are equal), the second-oldest child takes the next-largest share (or one of the two next-largest, if the two are equal), and the youngest child takes the last remaining share? *Hint:* View this as a division of the 10 pieces into 3, 2, or 1 piles, and use one of the strategies described in Problem 23.12.

23.15. Are there more ways to split 12 people up into 4 groups of 3 each, or into 3 groups of 4 each?

Chapter 24

Series

A series is a sum of similar terms, or sometimes a product of similar terms. Finite and infinite series come up frequently in computer science. For example, suppose a short piece of code is executed over and over as the body of a loop, with the values of certain variables changing on each iteration. Then important cumulative parameters of the computation—running time or memory usage, for example—might be expressed as the sum of a series of terms, each of which has a particular algebraic resemblance to the others in the series. As another example, if some quantity increases or decreases over time at a predictable rate, a series may express the total amount after a given period of time.

To start with a concrete example, let's consider the legend of the king and the inventor of chess.

Example 24.1. *The inventor proposes that the king put a single grain of wheat on the first square of the chessboard, and two on the second, four on the third, and so on, putting on each square twice as much as on the previous square. In exchange for the game, the king agrees to give the inventor all the grain on the chessboard when the 64 squares of the board have all been covered. How much wheat does the king owe the inventor?*

Solution to example. This is an example of exponential growth:[1] the number of grains of wheat on the nth square is 2^n, where the first square is numbered 0 and therefore contains $2^0 = 1$ grain. The number of grains on the last square would be 2^{63}, which turns out to be hundreds of times the annual production of wheat for the entire world.

But what the inventor is due is not just the amount of wheat on the last square, but all the wheat on the entire chessboard. If we let

$$S_n = \sum_{i=0}^{n} 2^i,$$

then the inventor is due S_{63} grains. Certainly $S_n = \Omega(2^n)$, since 2^n is the last term in the sum. But what is S_n exactly?

[1] The phrase "exponential growth" has entered common parlance as synonymous with "growing really fast." There are examples of growth that continues at an exponential rate for a few dozen iterations (epidemics and Moore's law, for example), but exponential growth of anything can't continue for a hundred generations since the result would exhaust the physical universe. In any case, what looks like exponential growth may really be "just" growth at a rate properly described by some low-degree polynomial.

We have already solved this problem; the answer is given by (3.6) on page 27. Let's review how we got the answer by working out a few examples, making a conjecture, and then proving it by induction. The first few values are

$$S_0 = 1$$
$$S_1 = 3$$
$$S_2 = 7$$
$$S_3 = 15,$$

which led to the conjecture that

$$\sum_{i=0}^{n} 2^i = 2^{n+1} - 1. \tag{24.2}$$

Let's check that. When $n = 0$,

$$2^{n+1} - 1 = 2^1 - 1 = 2 - 1 = 1 = S_0,$$

so the formula (24.2) holds in the base case. And if $S_n = 2^{n+1} - 1$, then

$$
\begin{aligned}
S_{n+1} &= S_n + 2^{n+1} && \text{(by the definition of } S_n) \\
&= 2^{n+1} - 1 + 2^{n+1} && \text{(by the induction hypothesis)} \\
&= 2^{n+2} - 1 && \text{(since } 2^{n+1} + 2^{n+1} = 2^{n+2})
\end{aligned}
$$

so (24.2) works also for the next larger value of n, and hence for all n.

There is nothing wrong with this approach, but there is a trick for deriving the formula (24.2) much more directly, without working out examples and making a conjecture.

Note that

$$1 + 2 + 4 + \cdots + 2^{n+1} = 1 + 2 \cdot (1 + 2 + \cdots + 2^n),$$

so we can write

$$
\begin{aligned}
S_{n+1} &= \sum_{i=0}^{n+1} 2^i \\
&= 1 + 2 \sum_{i=0}^{n} 2^i \\
&= 1 + 2 S_n.
\end{aligned}
$$

On the other hand, we can break off the largest term in the sum S_{n+1} and see that $S_{n+1} = 2^{n+1} + S_n$. Putting these two ways of writing S_{n+1} together,

$$1 + 2S_n = 2^{n+1} + S_n,$$

and solving this simple linear equation for S_n yields $S_n = 2^{n+1} - 1$ directly. ∎

Once equipped with the idea of multiplying an entire series by some factor and distributing that factor across the terms of the series, we can derive a number of other useful results from (24.2). For example, suppose we wanted to find the sum of *negative* powers of 2:

$$1 + \frac{1}{2} + \frac{1}{4} + \frac{1}{8} + \ldots + \frac{1}{2^n} = \sum_{i=0}^{n} 2^{-i}. \qquad (24.3)$$

No problem. Just divide (24.2) by 2^n.

$$\sum_{i=0}^{n} 2^i = 2^{n+1} - 1 \qquad \text{(so, dividing by } 2^n\text{)}$$

$$\sum_{i=0}^{n} 2^{i-n} = 2 - 2^{-n}.$$

The sum on the left is simply $\sum_{i=0}^{n} 2^{-i}$ with the terms added in the opposite order, so

$$\sum_{i=0}^{n} 2^{-i} = 2 - 2^{-n}. \qquad (24.4)$$

That is, adding more terms brings the sum $\sum_{i=0}^{n} 2^{-i}$ ever closer to 2, and in the limit, the sum[2] has the value exactly 2:

$$\sum_{i=0}^{\infty} 2^{-i} = 2. \qquad (24.5)$$

The methods used to derive (24.2), (24.4), and (24.5) can be generalized to handle similar sums in which the ratio of successive terms is any fixed number other than 1. (If the ratio is 1 then all terms are equal, say with value t, and the sum of n terms is just nt.) Such a series is said to be a *geometric series*; that is, it has the form

[2] This equation too has a history in legend. Zeno imagined someone repeatedly splitting the distance to be traveled, say two feet, by a factor of 2 at each step, and argued that the journey could never be completed, since the person would have to take an infinite number of steps. The supposed paradox is resolved using convergent sums: if the person is moving at a steady rate, say one foot per second, then each step takes half as long as the previous step, and the time to travel two feet is exactly two seconds.

$$\sum_{i=0}^{n} q^i \text{ or } \sum_{i=0}^{\infty} q^i, \tag{24.6}$$

so q is the ratio between terms. Let $q \in \mathbb{R}$ be any real number except 1. To calculate $S_{q,n} = \sum_{i=0}^{n} q^i$, note that

$$qS_{q,n} = q\sum_{i=0}^{n} q^i$$

$$= \sum_{i=1}^{n+1} q^i$$

$$= \left(\sum_{i=0}^{n} q^i\right) + q^{n+1} - 1$$

$$= S_{q,n} + q^{n+1} - 1$$

and solving for $S_{q,n}$ yields

$$\sum_{i=0}^{n} q^i = \frac{q^{n+1} - 1}{q - 1}. \tag{24.7}$$

If $|q| < 1$, then $q^{n+1} \to 0$ as $n \to \infty$, so

$$\sum_{i=0}^{\infty} q^i = \lim_{n\to\infty} \frac{q^{n+1} - 1}{q - 1} = \frac{1}{1 - q}. \tag{24.8}$$

(24.5) matches (24.8) with $q = \frac{1}{2}$.

A formula that expresses an infinite sum using a finite number of operations, like the right side of (24.8), is called a *closed form* of the series.

A series that converges to a finite value as $n \to \infty$, such as (24.7) with $|q| < 1$, is called a *convergent series*; while a series that exceeds all fixed finite values as $n \to \infty$, such as (24.7) with $|q| > 1$, is said to be a *divergent series*.

Sums such as (24.8) are generally derived in analysis courses as examples of *Taylor series*, which are series expansions of a great variety of functions. With a little bit of calculus, we can get useful closed-form expressions for a variety of similar sums.

Let's start with (24.8), now expressed as a function of the variable x:

$$\frac{1}{1-x} = \sum_{i=0}^{\infty} x^i. \qquad (24.9)$$

We've seen that if we plug in $x = \frac{1}{2}$, we get the value 2. But suppose what we wanted was not the sum (24.5), but

$$1 + \frac{1}{2^2} + \frac{1}{2^4} + \cdots = \sum_{i=0}^{\infty} 2^{-2i}. \qquad (24.10)$$

We could try to do some shifting, multiplying, and rearranging as we did in the last sections, but (24.9) invites another idea. If we just substitute x^2 for x in (24.9), we get

$$\frac{1}{1-x^2} = \sum_{i=0}^{\infty} x^{2i}. \qquad (24.11)$$

Setting $x = \frac{1}{2}$ in (24.11) immediately yields that the value of (24.10) is $\frac{4}{3}$.

Let's be even more daring! Suppose we need to evaluate

$$\frac{0}{2^0} + \frac{1}{2^1} + \frac{2}{2^2} + \frac{3}{2^3} + \ldots = \sum_{i=0}^{\infty} i \cdot 2^{-i}. \qquad (24.12)$$

It's not at all obvious that this series converges, but let's forge ahead.

We'll start by replacing the negative powers of 2 by powers of a variable x, and giving the value of the series a name:

$$F(x) = \sum_{i=0}^{\infty} i \cdot x^i. \qquad (24.13)$$

Now there's a trick: if we add (24.9), the series for $\frac{1}{1-x}$, to (24.13), we get something that looks a lot like (24.13) again; the exponents are just off by 1:

$$F(x) + \frac{1}{1-x} = \sum_{i=0}^{\infty} (i+1) \cdot x^i$$

$$= \frac{1}{x} \cdot \sum_{i=0}^{\infty} (i+1) \cdot x^{i+1} \quad (\text{since } x^i = \frac{x^{i+1}}{x})$$

$$= \frac{1}{x} \cdot \sum_{i=1}^{\infty} i \cdot x^i \quad \text{(shifting the index } i \text{ up by 1)}$$

$$= \frac{1}{x} \cdot \sum_{i=0}^{\infty} i \cdot x^i \quad \text{(since } 0 \cdot x^0 = 0)$$

$$= \frac{1}{x} \cdot F(x).$$

Then solving for the expression $F(x)$, we get

$$F(x) = \frac{x}{(1-x)^2} \tag{24.14}$$

and so $F(\frac{1}{2})$, the quantity we wanted to find, is equal to 2.

A function like (24.13) is called a *generating function* or a *formal power series*. The idea is to avoid all the questions about the values of x for which it is actually true, and to manipulate it algebraically until the end, only then plugging in a particular value of x. For students of calculus who have been taught to be strict about the range of values for which a Taylor series converges, manipulating generating functions feels like living in a state of sin. Nonetheless, we will use them with pleasure, as they produce very useful results.

Convergent geometric series are easily recognized by the fact that the denominators of the coefficients are successive powers of a fixed parameter. Another series that converges is the *exponential series*:

$$e^x = \sum_{i=0}^{\infty} \frac{x^i}{i!} = 1 + \frac{x^1}{1!} + \frac{x^2}{2!} + \frac{x^3}{3!} + \dots . \tag{24.15}$$

Let's look at the individual terms in this series, thinking of x as fixed and i as variable. The absolute value of the ratio of two consecutive terms is

$$\left| \frac{x^{i+1}}{(i+1)!} \middle/ \frac{x^i}{i!} \right| = \left| \frac{x}{i+1} \right|,$$

which tends toward 0 as i increases. As a result, (24.15) provides a practical way to compute the value of e^x to any desired precision, and therefore to approximate c^x for any constant c, since $c^x = e^{x \ln c}$: just add up as many terms of the series as are needed to achieve the desired precision. (Exactly how best to do such calculations is an important question of computational applied mathematics, beyond the scope of this book.)

Regarded as a generating function, the series (24.15) allows the evaluation of sums such as

$$\sum_{i=0}^{\infty} \frac{1}{i!} = e^1 = e,$$

$$\sum_{i=0}^{\infty} \frac{2^i}{i!} = e^2, \text{ and}$$

$$\sum_{i=0}^{\infty} \frac{1}{2^i \cdot i!} = e^{1/2} = \sqrt{e}.$$

*

Generating functions can be used to find closed forms equal to the sums of convergent series, as in the previous examples. Sometimes it is useful to go in the other direction: given a sequence a_0, a_1, ..., to find a generating function for the infinite sum

$$\sum_{i=0}^{\infty} a_i x^i.$$

We earlier on referred to such a sum as a "formal power series" because the goal is not to plug in a particular value of x to evaluate an infinite sum, but to manipulate the generating function itself to help calculate the coefficients.

For example, consider the number of partitions of an integer, a question we introduced in Chapter 23. We there defined $p_k(n)$ to be the number of partitions of n into k parts. Now let

$$p(n) = \text{ the number of partitions of } n = \sum_{k=1}^{n} p_k(n).$$

The first few values of $p(n)$ are shown in Figure 24.1. While there is no simple formula for $p(n)$, there *is* a simple generating function producing the values of $p(n)$ as the coefficients.

Theorem 24.16. Generating Function for Partitions of an Integer.

$$\sum_{n=1}^{\infty} p(n)x^n = \prod_{i=1}^{\infty} \frac{1}{1 - x^i}.$$

Proof. To get $p(n)$ as the coefficient of the x^n term, we will express x^n as a product of x^{ij} terms, where i is one part of a partition of n and j is the number of times it occurs in the partition. The coefficient of x^n in the series will then be the number of different ways that n can be partitioned.

n	$p(n)$
1	1
2	2
3	3
4	5
5	7
6	11
7	15
8	22
9	30
10	42

Figure 24.1. Values of $p(n)$, the number of partitions of n, for $1 \leq n \leq 10$.

Consider this infinite product of infinite sums:

$$(x^0 + x^1 + x^2 + x^3 + \ldots) \cdot (x^0 + x^2 + x^4 + x^6 + \ldots)$$

$$\cdot (x^0 + x^3 + x^6 + x^9 + \ldots) \cdot \ldots$$

$$= \prod_{i=1}^{\infty} \left(\sum_{j=0}^{\infty} x^{ij} \right)$$

$$= \prod_{i=1}^{\infty} \frac{1}{1 - x^i}. \tag{24.17}$$

The last equality is a version of (24.9), which we can rewrite as $\sum_{j=0}^{\infty} z^j = \frac{1}{1-z}$. Substituting x^i for z, we get $\sum_{j=0}^{\infty} x^{ij} = \frac{1}{1-x^i}$.

If the product $\prod_{i=1}^{\infty} \frac{1}{1-x^i}$ is multiplied out, the coefficient of the x^n term is the number of ways to pick one term from each factor

$$(x^0 + x^i + x^{2i} + x^{3i} + \ldots)$$

so that the exponents of all the chosen terms add up to n. In order for a product of terms to have exponent n, all but finitely many of those terms must be x^0. So any choice of terms whose product is x^n can be written as a finite product, by omitting all the x^0 terms following the last term with a nonzero exponent. For example, one way to choose terms so that the exponents add up to 34 is

$$x^{34} = x^{1 \cdot 0} \cdot x^{2 \cdot 2} \cdot x^{3 \cdot 1} \cdot x^{4 \cdot 2} \cdot x^{5 \cdot 1} \cdot x^{6 \cdot 0} \cdot x^{7 \cdot 2} \cdot x^{8 \cdot 0} \cdot x^{9 \cdot 0} \cdot \ldots$$

$$= x^0 \cdot x^4 \cdot x^3 \cdot x^8 \cdot x^5 \cdot x^0 \cdot x^{14}. \tag{24.18}$$

Each such choice of terms corresponds to a partition of n: if the jth term is chosen from the ith factor, then there are j occurrences of the integer i in the corresponding partition. (24.18) corresponds to the partition $34 = 2 + 2 + 3 + 4 + 4 + 5 + 7 + 7$. So (24.17) is a generating function for the number of partitions of n, into any number of parts. ∎

Of course, we can't compute anything by multiplying out an infinite product of infinite sums! Problem 24.13 explores how to turn (24.17) into a finite process for calculating $p(n)$.

We can modify (24.17) to get a generating function for the number of partitions of n into k parts, using the fact (Theorem 23.18) that the number of partitions of n into k parts is equal to the number of partitions of n for which the largest part is equal to k.

Theorem 24.19. *For any fixed k,*

$$\sum_{n=1}^{\infty} p_k(n)x^n = x^k \prod_{i=1}^{k} \frac{1}{1-x^i}.$$

Proof. Since the term chosen from the ith factor of (24.17) is the number of copies of i in the corresponding partition, taking the product up to only the kth factor,

$$\prod_{i=1}^{k} \frac{1}{1-x^i} \tag{24.20}$$

is a generating function for the number of partitions of n in which the largest part is at most k. But this is not quite what we want, since that count includes partitions in which there are 0 parts equal to k.

To guarantee that one part is actually equal to k, we count instead the number of partitions of $n - k$ in which the largest part is at most k. Each such partition of $n - k$ corresponds to one and only one partition of n in which the largest part is equal to k. The coefficient of x^{n-k} from (24.20) is exactly the desired value. So for the generating function that gives the number of partitions of n with largest part equal to k, we want a series for which the coefficient of x^n is equal to the coefficient of x^{n-k} from (24.20). Multiplying the entire function (24.20) by x^k has the effect of moving each coefficient to the right by k places. So

$$x^k \prod_{i=1}^{k} \frac{1}{1-x^i} \tag{24.21}$$

is the generating function for the number of partitions of n with largest part equal to k, and therefore (by Theorem 23.18) the number of partitions of n into exactly k parts. ∎

❈

A series commonly encountered in the analysis of algorithms is

$$F_k(n) = \sum_{i=1}^{n} i^k. \tag{24.22}$$

Here k is a fixed integer. For example, $F_1(n)$ is the sum of the first n positive integers. It might measure, for example, the time required to execute a doubly nested loop, such as

1. for $i \leftarrow 1, \ldots, n$

   (a) for $j \leftarrow 1, \ldots, i$

      (i) (code taking constant time t)

Such a loop might occur in a sorting algorithm, like the insertion sort described on page 223. The outer loop gets executed n times; on the ith iteration, the inner loop gets executed i times; and the code executed each time takes time t. So the total time for the program is

$$\sum_{i=0}^{n}(i \cdot t) = \left(\sum_{i=1}^{n} i\right) \cdot t.$$

Since t is a constant that does not vary with n, the time taken by this nested loop is proportional to $F_1(n)$; t disappears into the proportionality constant.

$F_1(n)$ has a familiar expression:

$$F_1(n) = \sum_{i=1}^{n} i = \frac{n \cdot (n+1)}{2}.$$

It is possible to reconstruct this formula if you forget it, by adding the series to itself, with one copy in increasing order and the other in decreasing order:

$$2F_1(n) = \sum_{i=1}^{n} i + \sum_{i=1}^{n}(n+1-i)$$

$$= \sum_{i=1}^{n}(i+n+1-i)$$

$$= \sum_{i=1}^{n}(n+1)$$

$$= n \cdot (n+1),$$

so $F_1(n)$ is half this quantity.

Similarly, a triply nested loop with a constant innermost body would have running time proportional to $F_2(n)$, and so on. There are also exact expressions for $F_k(n)$ when $k > 1$, but for the purposes of algorithm analysis, it generally suffices to know that if $k \geq 0$,

$$F_k(n) = \Theta(n^{k+1}). \tag{24.23}$$

The running time of an algorithm is affected by many other constant and additive factors, in addition to the exact value of the multiplicative constant and lower-order terms implicit in (24.23), so we focus on just showing that the exponent is correct. We can prove (24.23) by establishing the upper and lower bounds separately:

- $F_k(n) = O(n^{k+1})$, because $i^k \leq n^k$ for each i, $1 \leq i \leq n$, and therefore

$$\sum_{i=1}^{n} i^k \leq n \cdot n^k = n^{k+1}.$$

- To see that $F_k(n) = \Omega(n^{k+1})$, note that the terms are strictly increasing: $i^k < (i+1)^k$ for every i, $1 \leq i < n$. Now consider the halfway term, $\left(\frac{n}{2}\right)^k$ if n is even, or $\left(\frac{n+1}{2}\right)^k$ if n is odd. The value of this term is at least $2^{-k} \cdot n^k$, and there are at least $\frac{n}{2}$ terms of equal or greater value, so the sum of all terms is $\Omega(n \cdot n^k)$; that is, $\Omega(n^{k+1})$.

✳

What can we say about the sum (24.22) when k is negative? For example, what is

$$F_{-2}(n) = \sum_{i=1}^{n} i^{-2} = 1 + \frac{1}{4} + \frac{1}{9} + \ldots + \frac{1}{n^2}?$$

Such sums have an illustrious history in mathematics. For us it suffices to show that for each $k \leq -2$, this sum converges to a value between 1 and 2 as $n \to \infty$.[3] Plainly each sum is at least 1, and $F_{-k}(n) > F_{-\ell}(n)$ if $k < \ell$, so it suffices to show that $F_{-2}(n)$ converges to a value that is at most 2. To see this, note that

$$\frac{1}{i^2} < \frac{1}{i \cdot (i-1)} = \frac{1}{i-1} - \frac{1}{i}$$

if $i > 1$. But then

$$F_{-2}(n) = \sum_{i=1}^{n} \frac{1}{i^2}$$

$$= 1 + \sum_{i=2}^{n} \frac{1}{i^2}$$

$$< 1 + \sum_{i=2}^{n} \left(\frac{1}{i-1} - \frac{1}{i}\right)$$

$$= 2 - \frac{1}{n},$$

since each negative term but the last is canceled out by the next positive term.

[3] The value that we call $\lim_{n\to\infty} F_{-k}(n)$ is known elsewhere in math as $\zeta(k)$, the Riemann zeta function. In 1734, Euler found a closed form of $\zeta(2)$, a question that mathematicians had been trying to answer for nearly a century. It is, remarkably enough, $\frac{\pi^2}{6}$, or about 1.644934.

✳

We've established that $F_k(n)$ diverges for every $k \geq 0$ ($F_0(n)$ is just n, of course). And for every $k \leq -2$, $F_k(n)$ converges to a number between 1 and 2, with the exact value depending on k. That leaves open the question of $F_{-1}(n)$, more commonly known as H_n, the nth *harmonic number*:

$$H_n = \sum_{i=1}^{n} \frac{1}{i} = 1 + \frac{1}{2} + \frac{1}{3} + \frac{1}{4} + \ldots + \frac{1}{n}. \qquad (24.24)$$

This is known as the *harmonic series*. Does it diverge?

It does diverge; $H_n \to \infty$ as $n \to \infty$. One way to see this is to consider the disjoint sets created by grouping the terms from

$$\frac{1}{2^\ell + 1} \quad \text{to} \quad \frac{1}{2^{\ell+1}}$$

for each integer $\ell \geq 0$. Such a set of terms contains 2^ℓ elements, and each term in the group is greater than or equal to $\frac{1}{2^{\ell+1}}$, so when all the terms are added together, their sum is at least $\frac{2^\ell}{2^{\ell+1}} = \frac{1}{2}$.

$$1 + \overbrace{\frac{1}{2}}^{=\frac{1}{2}} + \overbrace{\frac{1}{3} + \frac{1}{4}}^{>\frac{1}{2}} + \overbrace{\frac{1}{5} + \frac{1}{6} + \frac{1}{7} + \frac{1}{8}}^{>\frac{1}{2}} + \ldots. \qquad (24.25)$$

It follows that the terms up to and including $\frac{1}{2^\ell}$ sum to a number that is at least $1 + \frac{\ell}{2}$. That is,

$$H_{2^\ell} \geq 1 + \frac{\ell}{2}.$$

So the value of H_n exceeds any bound as $n \to \infty$.

The partial sums of the harmonic series grow very slowly, however. Since

$$\frac{d}{dx} \ln x = \frac{1}{x},$$

the sum H_n can be approximated by the natural logarithm. In fact,

$$\ln n \leq H_n \leq \ln(n+1). \qquad (24.26)$$

Figure 24.2. A book 2 units long, resting with its center of gravity right at the edge of the table.

One striking application of the divergence of the harmonic series is to show that a stack of identical books can, in principle, extend any distance over the edge of a table on which the bottom book is resting. Let's say the books are 2 units long, so a single book can just balance if it is a hair less than 1 unit over the edge of the table (Figure 24.2).

A second book can be slipped under the first, and the first book balanced with half its length (1 unit) over the edge of the second book, just as it had previously balanced on the table. The stack of 2 books can then balance on the table with the bottom book half a unit over the edge. The mass of 1 book is over the table, and the mass of 1 book is over the void (Figure 24.3).

So with $n = 1$ or $n = 2$ books, they are perfectly balanced with maximum overhang just in case the rightmost edge of the top book is exactly H_n from the right edge of the table. Let's show that the pattern continues.[4]

If we put an $(n+1)^{\text{st}}$ book underneath the first n with its right edge at the center of gravity of the first n, how far from the right end of the ensemble is the center of gravity of all $n+1$ books? Let's say that the mass of a single book is 1. The first n books act as a single mass of size n situated at distance H_n from the right end of the stack. The new book has mass 1 and its center of gravity is at distance $1 + H_n$ from the right edge of the stack, since it is 2 units long. So the center of gravity of the ensemble is at distance

$$\frac{1}{n+1}(nH_n + 1 + H_n) = H_n + \frac{1}{n+1}$$
$$= H_{n+1}$$

from the right end of the stack, as was to be shown. For example, the third book can jut out by only $\frac{1}{3}$ of a unit to make the stack balance perfectly (Figure 24.4).

Because the harmonic series diverges, it is possible (in principle, without taking into account breezes, earth tremors, or irregularities in the books) to build a stack of books that extends arbitrarily far over the edge of the table. It takes 31 books to get 2 book lengths (that is, 4 units) away from the edge of the table (Figure 24.5).

In Chapter 23, $\binom{n}{k}$ was defined as the number of ways of choosing a k-element subset of a set of size n, namely

$$\binom{n}{k} = \frac{n!}{k! \cdot (n-k)!}.$$

The numbers $\binom{n}{k}$ are also known as the *binomial coefficients*, because of the following theorem.

Theorem 24.27. The Binomial Theorem. *For any integer $n \geq 0$ and any $x, y \in \mathbb{R}$,*

$$(x+y)^n = \sum_{i=0}^{n} \binom{n}{i} x^i y^{n-i}. \tag{24.28}$$

Figure 24.3. A second book is slipped under the first. The first book rests on the second as it used to rest on the table. The two books together will just balance if the second book sticks out from the table by $\frac{1}{2}$ a unit.

[4]This assumes that the books must be stacked one on top of another. If we are allowed to have two books at the same level with a gap between them, sandwiched between books above and below, this limit can be exceeded.

Figure 24.4. The third book can extend $\frac{1}{3}$ off the table.

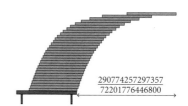

Figure 24.5. With 31 books, the overhang is more than 2 book lengths.

To see why (24.28) is true, note that multiplying $x + y$ by itself n times and fully distributing all the multiplications across the additions creates 2^n terms, since each of the n factors offers 2 choices: x or y. The number of those 2^n terms that are equal to $x^i y^{n-i}$—that is, the coefficient of $x^i y^{n-i}$ in the summation—is the number of ways of choosing x from i of the factors and y from the remaining $n - i$ of the factors, namely $\binom{n}{i}$.

Equation (24.28) provides a gateway to summing series involving binomial coefficients. For example, simply plugging in $x = y = 1$ yields

$$\sum_{i=0}^{n} \binom{n}{i} = 2^n. \tag{24.29}$$

As another example, what is

$$\sum_{i=0}^{n} 2^i \binom{n}{i}?$$

To evaluate this sum, just set $x = 2$ and $y = 1$ in (24.28):

$$\sum_{i=0}^{n} \binom{n}{i} 2^i 1^{n-i} = (2 + 1)^n = 3^n.$$

Chapter Summary

- A *series* is a sum (or product) of algebraically similar terms (or factors).

- A *geometric series* is a series in which the ratio of any two consecutive terms is a fixed number other than 1.

- The finite geometric series $\sum_{i=0}^{n} q^i$ is equal to $\frac{q^{n+1}-1}{q-1}$ for any $q \neq 1$.

- The infinite geometric series $\sum_{i=0}^{\infty} q^i$ converges and is equal to $\frac{1}{1-q}$ for any $|q| < 1$.

- If a series converges to a finite value as $n \to \infty$, it is called a *convergent series*. If not, it is called a *divergent series*.

- A *generating function* or *formal power series* expresses a sequence of numbers as the coefficients of a polynomial. Such a function can be manipulated algebraically to find a *closed form* for the sequence.

- There are simple generating functions for $p(n)$, the number of partitions of an integer n, and $p_k(n)$, the number of partitions of an integer n into k parts.

- The *exponential series* $\sum_{i=0}^{\infty} \frac{x^i}{i!}$ converges, and is equal to e^x.

- The series $F_k(n) = \sum_{i=1}^{n} i^k$ appears often in analysis of algorithms—for example, it is the number of iterations of a loop with k levels of nesting—and is $\Theta(n^{k+1})$ when $k \geq 0$.

- The *harmonic series* $H_n = \sum_{i=1}^{n} \frac{1}{i}$ diverges, and can be approximated by the natural logarithm $\ln n$.

- The values $\binom{n}{k}$, representing the number of ways of choosing k elements from a set of size n, appear as the coefficients in the series that results from expanding the binomial $(x + y)^n$. Due to this application, they are also known as the *binomial coefficients*.

Problems

24.1. (a) Paul offers to let you play a game: he'll flip a fair coin until he flips tails, then pay you k dollars, where k is the number of heads he flipped. For instance, the sequence *HHHT* would earn you $3, while a tails on the first flip would pay you nothing. What is the expected value of your winnings if you play this game once; or in other words, how much should you be willing to pay Paul to play this game one time?

   (b) What would your answer be if Paul instead paid you 2^k dollars if you first got a tails after k flips?

24.2. Simplify $\frac{1}{1 \cdot 2} + \frac{1}{2 \cdot 3} + \frac{1}{3 \cdot 4} + \ldots + \frac{1}{(n-1) \cdot n}$.

24.3. Simplify $\left(1 + \frac{1}{a}\right)\left(1 + \frac{1}{a^2}\right)\left(1 + \frac{1}{a^4}\right)\ldots\left(1 + \frac{1}{a^{2^{100}}}\right)$.
   Hint: The formula for the "difference of two squares" may be helpful.

24.4. (Uses calculus) Derive (24.14) another way, by taking the first derivative of both sides of (24.9).

24.5. What is $\sum_{i=0}^{n} 3^{-3i}$?

24.6. (Uses calculus) Prove that $\frac{d}{dx} e^x = e^x$ by differentiating the series (24.15) term by term.

24.7. What is $\sum_{i=0}^{\infty} \frac{x^{2i}}{i!}$?

24.8. Use (24.26) to estimate $\sum_{i=1}^{n} \frac{1}{2i}$ and $\sum_{i=0}^{n} \frac{1}{2i+1}$.

24.9. (a) Generalize (24.29) to prove that for any $n \geq 0$ and $k \geq 0$,

$$\sum_{i=0}^{n} k^i \binom{n}{i} = (k+1)^n.$$

   (b) What is the value of

$$\sum_{i=0}^{n} 2^{-i} \binom{n}{i}?$$

24.10. A version of the Binomial Theorem (24.28) is true even when n is not an integer. If $|x| < 1$, then for any real α,

$$(1+x)^\alpha = \sum_{i=0}^{\infty} \binom{\alpha}{i} x^i, \tag{24.30}$$

where

$$\binom{\alpha}{i} = \frac{\alpha \cdot (\alpha - 1) \cdot \ldots \cdot (\alpha - i + 1)}{i!}.$$

When α is a nonnegative integer and the sum is truncated after the $i = n$ term, (24.28) is (24.30). The formula (24.30) is called a *Maclaurin series*.

Expand the series with $\alpha = \frac{-1}{2}$ and $x = \frac{-1}{2}$ to determine the value of

$$\sum_{n=0}^{\infty} \left(\frac{1}{4^n n!} \prod_{i=0}^{n-1} (2i+1) \right).$$

24.11. Let S be the set $\{1, 2, 3, \ldots, 9, 11, \ldots, 19, 21, \ldots\}$, consisting of all natural numbers that do not contain the digit zero when written using decimal notation. Does the sum of the reciprocals of the elements of S converge or diverge? Justify your answer.

24.12. Let $q(n)$ be the number of partitions of n into odd parts. For example, $q(4) = 2$, since of the 5 partitions of 4 (4, $1+3$, $1+1+2$, $1+1+1+1$, and $2+2$), only $1+3$ and $1+1+1+1$ consist solely of odd parts. Write a generating function for $q(n)$.

24.13. (a) Suppose n is fixed. Show how to find $p(n)$ by evaluating a version of (24.17) that entails multiplying out a finite product of finite sums.

(b) Use this method to compute $p(6)$.

Chapter 25

Recurrence Relations

Let $a_0, a_1, \ldots$ be any infinite sequence (perhaps of numbers, but perhaps of strings or some other kind of mathematical objects). A *recurrence relation* is an equation or set of equations that makes it possible to find the value of a_n in terms of the values of the a_i for $i < n$. Recurrence relations are accompanied by one or more base cases, giving the value of a_i for one or more fixed small values of i.

Proofs by induction such as the ones in Chapters 3 and 4 include recurrence relations. For example, the proof of (3.3) on page 27 relies on the proof that

$$a_n = \sum_{i=0}^{n} 2^i$$
$$= 2^{n+1} - 1$$

by means of the recurrence relation and base case

$$a_{n+1} = a_n + 2^{n+1}$$
$$a_0 = 1.$$

We also used a recurrence relation in Theorem 23.17 on page 253 to help calculate the number of partitions of an integer n into k parts.

Recurrence relations arise naturally in the analysis of algorithms. Let's start by revisiting (3.12), the formula for the sum of the first n nonnegative integers:

$$\sum_{i=0}^{n} i = \frac{n \cdot (n+1)}{2}. \tag{25.1}$$

This formula arises in the analysis of the simple sorting algorithm that follows. This selection sort puts a list of numbers in increasing order by repeatedly finding the greatest number in the unprocessed part of the list and moving it to the beginning of the list. (The name "selection sort" refers to any sorting algorithm that operates by repeatedly selecting the next item and processing it.) We assume that lists are maintained as linked data structures, so that items can be removed and prepended in constant time. The notation $A \cdot B$ will mean a list that contains the elements of list A followed by the elements of list B.

Algorithm $S(L)$, to sort a list L:

1. Set $M \leftarrow \lambda$

2. While $L \neq \lambda$

    (a) Find the greatest element of L, call it x

    (b) Let A and B be sublists such that $L = A \cdot x \cdot B$

    (c) $L \leftarrow A \cdot B$; that is, the result of removing x

    (d) $M \leftarrow x \cdot M$

3. Return M

For example, if L starts out as the list $4 \cdot 2 \cdot 7 \cdot 3$, then on successive iterations of the loop, M and L are as follows:

$$
\begin{array}{cc}
\overbrace{\lambda}^{M} & \overbrace{4 \quad 2 \quad 7 \quad 3}^{L} \\[2ex]
\overbrace{7}^{M} & \overbrace{4 \quad 2 \quad 3}^{L} \\[2ex]
\overbrace{4 \quad 7}^{M} & \overbrace{2 \quad 3}^{L} \\[2ex]
\overbrace{3 \quad 4 \quad 7}^{M} & \overbrace{2}^{L} \\[2ex]
\overbrace{2 \quad 3 \quad 4 \quad 7}^{M} & \overbrace{\lambda}^{L}
\end{array}
$$

If L starts off having length n and M having length 0, then the main loop is executed n times, with L having length $n, n - 1, \ldots, 1$ at the beginning of each iteration, and M having length $0, 1, \ldots, n - 1$. After the last iteration, L has length 0 and M has length n. Finding the greatest item in a list of length k by a simple search takes time proportional to k, and those search operations dominate the running time, so the running time of the entire algorithm is proportional to the sum of the first n integers; that is, to (25.1), which is $\Theta(n^2)$.

A *recursive* algorithm is an algorithm that calls itself, but with arguments that are different (usually smaller or simpler) than the original argument.

Let's consider a recursive version of the sorting algorithm, which finds and removes the smallest element of L, recursively sorts the rest of L, and prepends that smallest element to the result.

Algorithm $S_R(L)$, to sort a list L:

1. If $L = \lambda$ then return λ

2. Else

    (a) Find the smallest element of L, call it x

    (b) Let A and B be sublists such that $L = A \cdot x \cdot B$

    (c) Return $x \cdot S_R(A \cdot B)$

The successive calls can be depicted as follows on the example above.

$$S_R(4 \cdot 2 \cdot 7 \cdot 3)$$
$$2 \cdot S_R(4 \cdot 7 \cdot 3)$$
$$2 \cdot 3 \cdot S_R(4 \cdot 7)$$
$$2 \cdot 3 \cdot 4 \cdot S_R(7)$$
$$2 \cdot 3 \cdot 4 \cdot 7 \cdot S_R(\lambda)$$
$$2 \cdot 3 \cdot 4 \cdot 7$$

Each call on S_R results in prepending a single element to the result of calling S_R with an argument list one shorter. So if S_R is initially called with an argument of length n, the running time is proportional to the solution to the recurrence relation

$$a_n = a_{n-1} + n \qquad (25.2)$$

with the base case $a_0 = 0$.

The "$+n$" term in (25.2) is the time to find the smallest element. The solution of this recurrence relation is quadratic, exactly like the solution to (25.1).

These two examples are essentially the same algorithm, one with an iterative control structure and one recursive. Accordingly, they have the same time complexity, and nothing is really gained by changing to a recursive formulation. But recursive algorithms can be startlingly more efficient than iterative algorithms, when the recursion divides the argument evenly—unlike the unbalanced division of S_R, which chops one element off a size-n argument before calling itself recursively with an argument of size $n - 1$—and the recurrence relations are more interesting. The classic example is *merge sort*.

Algorithm MergeSort(L), to sort a list L:

1. If $|L| \leq 1$, then return L

2. Else

   (a) Divide $L = A \cdot B$ into two sublists A and B of nearly equal length

   (b) $A' \leftarrow$ MergeSort(A)

   (c) $B' \leftarrow$ MergeSort(B)

   (d) Return Merge(A', B')

Here, Merge(L_1, L_2) is a linear-time algorithm that takes two sorted lists, L_1 and L_2, and produces a single sorted list of the same elements. Merge simply interleaves its arguments together into a single ordered list.

Figure 25.1 depicts the execution of MergeSort, using the same example input as above.

Figure 25.1. Execution of MergeSort on the input list $4 \cdot 2 \cdot 7 \cdot 3$. The recursive calls are executed in the order designated by the arrows, traveling down and up the tree.

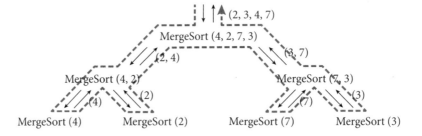

Let $t(n)$ be the number of steps MergeSort uses to sort a list of length n. To simplify the math in the analysis, let's assume that n is a power of 2, so that every time the list is divided in half, it is divided into two sublists of equal length. Then the steps take time as follows:

1. $\Theta(1)$

2. (a) $\Theta(n)$

   (b) $t(\frac{n}{2})$

   (c) $t(\frac{n}{2})$

   (d) $\Theta(n)$

So the time complexity of MergeSort is governed by the recurrence relation

$$t(n) = 2t\left(\frac{n}{2}\right) + \Theta(n) \tag{25.3}$$

with the base case $t(1) = \Theta(1)$. $\tag{25.4}$

Note that in (25.3) we are using $\Theta(n)$ in its sense as a particular function that is not fully specified, but is known to be a member of $\Theta(n)$.

The recurrence (25.3)–(25.4) fits a pattern that comes up repeatedly. We'll first work through the runtime of this particular example, and then state a more general theorem from which this special case can be derived.

Since we assumed that n is a power of 2, let's say that $n = 2^m$. Then $m = \lg n$. Since the length of the list is halved at each recursion level, the recursion is to depth $m = \lg n$. If we call the top level 1 and the bottom level m, then there are 2^i calls at level i, each on a list of length 2^{m-i}. The time spent within each of these calls (that is, excluding the recursive calls to the next level) is proportional to the length of argument: $\Theta(2^{m-i})$. Therefore, the total time expended in all the recursive calls at level i is $\Theta(2^i \cdot 2^{m-i}) = \Theta(2^m) = \Theta(n)$, independent of i. So including all m recursion levels, the total time $t(n)$ is given by

$$t(n) = \Theta(m \cdot n)$$

$$= \Theta(n \lg n). \tag{25.5}$$

This is a big improvement over the $\Theta(n^2)$ time required by selection sort, but our concern here is less about the efficiency of the algorithm than the way we analyzed its complexity. Let's state a more general form of the recurrence (25.3)–(25.4) and its solution.

Theorem 25.6. The Master Theorem. *Consider the recurrence*

$$T(n) = aT\left(\frac{n}{b}\right) + \Theta(n^d) \tag{25.7}$$
$$T(1) = \Theta(1).$$

Assume that $b > 1$, so the recurrence gives a script for determining the value of T on argument n in terms of its value on a smaller argument. Let $e = \log_b a$. Then

1. $T(n) = \Theta(n^d)$, *if $e < d$;*

2. $T(n) = \Theta(n^d \log n)$, *if $e = d$;*

3. $T(n) = \Theta(n^e)$, *if $e > d$.*

That is,

$$T(n) = \Theta(n^{\max(d,e)}) \text{ if } d \neq e,$$

$$T(n) = \Theta(n^d \log n) = \Theta(n^e \log n) \text{ if } d = e.$$

The analysis of MergeSort follows the pattern in the statement of Theorem 25.6. In terms of the parameters of the statement of the Master Theorem, (25.3) has $a = 2$, $b = 2$, and $d = 1$. Since then $e = \log_2 2 = 1$, $e = d$ and the second case applies. According to the Master Theorem, $T(n) = \Theta(n^1 \log n) = \Theta(n \log n)$, exactly as derived in (25.5).

Proof. To simplify the math, let's assume that n is a power of b—say $n = b^m$—so repeatedly dividing by b always gives a whole number value, and after $m - 1$ recursions reduces that value to 1. Also, we'll replace $\Theta(n^d)$ by a term $f \cdot n^d$, fixing the implicit multiplicative constant. (Proving the theorem in full generality is technically messier but not fundamentally different.) Then

$$
\begin{aligned}
T(n) &= aT\left(\frac{n}{b}\right) + f \cdot n^d \\
&= a^2 T\left(\frac{n}{b^2}\right) + a \cdot f \cdot \left(\frac{n}{b}\right)^d + f \cdot n^d \\
&= a^3 T\left(\frac{n}{b^3}\right) + a^2 \cdot f \cdot \left(\frac{n}{b^2}\right)^d + a \cdot f \cdot \left(\frac{n}{b}\right)^d + f \cdot n^d \\
&= \ldots \\
&= a^m T(1) + f \cdot n^d \cdot \sum_{i=0}^{m-1} a^i b^{-id}.
\end{aligned}
\tag{25.8}
$$

Now $m = \log_b n$ since $n = b^m$, and therefore

$$
\begin{aligned}
a^m &= a^{\log_b n} \\
&= a^{\log_b a \cdot \log_a n} \\
&= n^{\log_b a} \\
&= n^e.
\end{aligned}
$$

Therefore (25.8) reduces to

$$
T(n) = \Theta(n^e) + \Theta\left(n^d \cdot \sum_{i=0}^{m-1} a^i b^{-id}\right).
\tag{25.9}
$$

In case (2), $e = d$; that is, $\log_b a = d$. Then $a = b^d$, so $a^i b^{-id} = 1$ for any i, and (25.9) simplifies to

$$
\begin{aligned}
&\Theta(n^e) + \Theta(n^d \cdot m) \\
&= \Theta(n^e) + \Theta(n^d \log n) \\
&= \Theta(n^d \log n) \qquad \text{(since } e = d\text{)}.
\end{aligned}
$$

In case (1), $e < d$, so $a < b^d$ and $a^i b^{-id} < 1$ for any i. Then the summation in (25.9) is an initial segment of a convergent geometric series, so is bounded by a constant, and (25.9) reduces to $\Theta(n^e) + \Theta(n^d)$, which is $\Theta(n^d)$ since $e < d$.

In case (3), $e > d$, so $a > b^d$. Then each of the terms in the summation is dominated by the highest-degree term, so

$$n^d \cdot \sum_{i=0}^{m-1} a^i b^{-id} = \Theta(n^d \cdot a^{m-1} b^{-(m-1)d})$$

$$= \Theta(n^d \cdot (ab^{-d})^{m-1})$$

$$= \Theta(n^d \cdot (ab^{-d})^m) \qquad \text{(since } ab^{-d} \text{ is a constant)}$$

$$= \Theta(n^d \cdot (ab^{-d})^{\log_b n})$$

$$= \Theta(n^d \cdot n^{\log_b (ab^{-d})})$$

$$= \Theta(n^d \cdot n^{\log_b a - \log_b (b^d)})$$

$$= \Theta(n^d \cdot n^{e-d})$$

$$= \Theta(n^e).$$

So (25.9) is the sum of two $\Theta(n^e)$ terms, and is therefore $\Theta(n^e)$.

∎

✳

Merge sort is an example of a *divide and conquer* algorithm, in which the argument is split into successively smaller pieces, with the algorithm operating recursively on each. We have actually seen an example of a divide and conquer algorithm already, on page 224: binary search. In terms of the Master Theorem, the recurrence relation for the analysis of binary search has

$$T(n) = T\left(\frac{n}{2}\right) + \Theta(1)$$
$$T(1) = \Theta(1),$$

so $a = 1$, $b = 2$, and $d = 0$. Since then $\log_b a = 0 = d$, case (2) applies, and $T(n) = \Theta(n^0 \log n) = \Theta(\log n)$, exactly as we discovered.

We'll now go through several examples in which divide and conquer yields surprising improvements to the obvious algorithms, and the Master Theorem gives us almost instantaneous complexity analyses.

Example 25.10. *Integer arithmetic.*

Consider the problem of multiplying two integers represented in positional notation (such as binary or decimal). Multiplying two n-digit numbers in general produces a $2n$-digit result. The *grade-school algorithm* is quadratic, since it produces n partial products, each of length $\Theta(n)$, which must be added together at the end. The $\Theta(n)$ additions each require $\Theta(n)$

```
      10110
  ×   11011
  ─────────
      10110
     10110
    00000
   10110
  10110
  ──────────
  1001010010
```

Figure 25.2. Grade-school multiplication of two 5-bit numbers to produce a 10-bit result requires writing down 25 bits in the intermediate calculation.

time, and indeed it takes $\Theta(n^2)$ time just to write down all the digits of the partial products. For example, Figure 25.2 shows a grade-school multiplication of two 5-bit numbers.

There is a recursive, divide and conquer version of this algorithm. Let's assume that n is a power of 2, say $n = 2^m$, and that the task is to multiply two n-bit numbers a and b. Then a and b can be written as

$$a = a_1 \cdot 2^{n/2} + a_0$$
$$b = b_1 \cdot 2^{n/2} + b_0,$$

where a_0, a_1, b_0, and b_1 are $n/2$-bit numbers, so

$$a \cdot b = a_1 b_1 \cdot 2^n + (a_1 b_0 + a_0 b_1) \cdot 2^{n/2} + a_0 b_0. \tag{25.11}$$

To take a concrete example, let $a = 1001$ and $b = 1110$. Then, following the above formulas, we can write

$$1001 = 10 \cdot 100 + 01$$
$$1110 = 11 \cdot 100 + 10$$

and then

$$1001 \cdot 1110 = (10 \cdot 11) \cdot 10000 + (10 \cdot 10 + 01 \cdot 11) \cdot 100 + 01 \cdot 10.$$

Multiplying by a power of 2 is really shifting, not multiplication—it just means appending 0s to the number. So (25.11) shows how to multiply two n-bit numbers by four multiplications of $n/2$-bit numbers (that is, $a_1 b_1$, $a_1 b_0$, $a_0 b_1$, and $a_0 b_0$), plus some linear-time operations (shifting and adding). So the time complexity of this recursive integer multiplication algorithm is given by the recurrence relation

$$T(n) = 4T\left(\frac{n}{2}\right) + \Theta(n). \tag{25.12}$$

Applying the Master Theorem with $a = 4$, $b = 2$, and $d = 1$, we find that $e = \log_b a = 2 > d = 1$, so case (3) applies and $T(n) = \Theta(n^e) = \Theta(n^2)$, the same as the grade-school algorithm. The effort to divide and conquer recursively seems not to have been worth the bother.

Except that (25.11) can be rewritten another way, which involves only *three* $n/2$-bit multiplications, plus some shifts and additions:

$$a \cdot b = a_1 b_1 \cdot 2^n + (a_1 b_0 + a_0 b_1) \cdot 2^{n/2} + a_0 b_0$$
$$= a_1 b_1 \cdot (2^n + 2^{n/2}) + (a_1 - a_0)(b_0 - b_1) \cdot 2^{n/2} + a_0 b_0 \cdot (2^{n/2} + 1). \tag{25.13}$$

First, check to see how terms cancel out to yield the right result, and then persuade yourself that there are only three "real" multiplications—multiplying an n-bit number by $2^{n/2} + 1$, for example, just involves copying the bits and shifting them into the right positions, which takes linear time. But now the recurrence relation for the time complexity is

$$T(n) = 3T\left(\frac{n}{2}\right) + \Theta(n),$$

which fits the Master Theorem with $a = 3$ (rather than 4), $b = 2$, and $d = 1$. Now $e = \log_2 3 \approx 1.58$. That number is greater than $d = 1$, so case (3) still applies, but the net result is that $T(n) = \Theta(n^{1.58\ldots})$—a substantial improvement over the $\Theta(n^2)$ grade-school algorithm!

This method is known as *Karatsuba's algorithm*, after its discoverer Anatoly Karatsuba, then a 23-year-old student at Moscow State University. Karatsuba was challenged to find a better algorithm when his professor, the distinguished mathematician Andrey Kolmogorov, announced his hypothesis that the grade-school algorithm was asymptotically optimal. It took Karatsuba only a few days to discover his improvement. Further research has established that there is an algorithm for multiplying n-bit numbers that runs in time $\Theta(n \cdot \log n \cdot \log \log n)$—that is, not quite linear, but better than $n^{1+\varepsilon}$ for any $\varepsilon > 0$.

Example 25.14. *Matrix multiplication.*

Multiplying two $n \times n$ matrices to produce the resulting $n \times n$ matrix product requires $\Theta(n^3)$ operations using the standard technique, which calculates each entry in the result matrix as the dot (inner) product of two n-bit vectors. So n^2 entries, each of which requires $\Theta(n)$ time to calculate (n products, which are then combined via $n - 1$ additions), adds up to $\Theta(n^3)$ operations in all.[1]

Let's again assume that n is a power of 2. We can divide the two matrices to be multiplied into quarters, and recursively block-multiply those quadrants as shown in Figure 25.3.

Unfortunately, while this produces the right result, it doesn't improve the time complexity of the iterative algorithm. To multiply two $n \times n$ matrices by carrying out eight $\frac{n}{2} \times \frac{n}{2}$ matrix multiplications leads to the recurrence

$$T(n) = 8T\left(\frac{n}{2}\right) + \Theta(n^2),$$

since linear-time operations on $\frac{n}{2} \times \frac{n}{2}$ matrices use $\Theta(n^2)$ operations. In terms of the Master Theorem, $a = 8$, $b = 2$, and $d = 2$, so $e = \log_2 8 = 3 > d = 2$. Accordingly, case (3) applies and $T(n) = \Theta(n^3)$, exactly like the standard method.

In 1970, the German mathematician Volker Strassen stunned the computational world by noticing that the four quantities of the quadrants in Figure 25.3 can be calculated using only seven multiplications of $\frac{n}{2} \times \frac{n}{2}$

[1] In this example we are treating the entries as having constant size, so that integer multiplication and addition are both $\Theta(1)$. As we saw in Example 25.10, multiplication and addition grow with the size of their inputs, but here we are considering the running time as a function of the size of the matrix itself, not a function of the size of its entries.

$$\begin{pmatrix} A & B \\ C & D \end{pmatrix} \times \begin{pmatrix} E & F \\ G & H \end{pmatrix}$$

$$= \begin{pmatrix} AE + BG & AF + BH \\ CE + DG & CF + DH \end{pmatrix}$$

Figure 25.3. Multiplication of two $n \times n$ matrices by splitting each into four $\frac{n}{2} \times \frac{n}{2}$ matrices and recursively constructing the four $\frac{n}{2} \times \frac{n}{2}$ quadrants of the product.

matrices, with the aid of a few extra additions and subtractions before and after the multiplications. Let

$$P_1 = A(F - H)$$
$$P_2 = (A + B)H$$
$$P_3 = (C + D)E$$
$$P_4 = D(G - E)$$
$$P_5 = (A + D)(E + H)$$
$$P_6 = (B - D)(G + H)$$
$$P_7 = (A - C)(E + F).$$

Now, almost magically,

$$AE + BG = P_5 + P_4 - P_2 + P_6$$
$$AF + BH = P_1 + P_2$$
$$CE + DG = P_3 + P_4$$
$$CF + DH = P_5 + P_1 - P_3 - P_7.$$

Seven multiplications! For *Strassen's algorithm*, the recurrence relation describing the complexity is

$$T(n) = 7T\left(\frac{n}{2}\right) + \Theta(n^2),$$

the solution of which is $T(n) = \Theta(n^{\log_2 7}) = \Theta(n^{2.81\cdots})$.

It is hard to overstate how amazing Strassen's result was and still seems. Matrix multiplication, and equivalent operations such as Gaussian elimination, had been studied for centuries, perhaps millennia, by the world's greatest mathematical minds. But it was not until 1970 that anyone noticed a way to perform this basic operation in less than cubic time. And any high school student motivated to push sums and products around on scratch paper might have noticed exactly the same thing!

Strassen's discovery set off a furious competition to find similar algorithms with exponents smaller than $\log_2 7$. As of this writing, the asymptotically fastest known algorithm has time complexity $\Theta(n^{2.3728639\cdots})$, though the constant factor implicit in the $\Theta(\cdot)$-notation is so huge as to render that algorithm useless in practice.[2] Nobody thinks that strange exponent is the last word on this problem!

[2] François Le Gall, "Powers of Tensors and Fast Matrix Multiplication," Proceedings of the 39th International Symposium on Symbolic and Algebraic Computation (ISSAC 2014).

Example 25.15. *Euclid's algorithm.*

How long does it take for Euclid's algorithm (page 154) to find the greatest common divisor of m and n? To estimate the number of iterations of the "while" loop, recall that after the first iteration, p is guaranteed to be greater than q. And after two iterations, p has been replaced by $p \bmod q$. Now if $p > q$, then $p \bmod q < p/2$. (If $q \leq p/2$, then $p \bmod q < p/2$ because $p \bmod q$ is always a number less than q. And if $q > p/2$, then $\lfloor p/q \rfloor = 1$ and $p \bmod q = p - q < p/2$.) So except for perhaps the first and last iterations, the value of p is at least halved every two iterations; and the value of q is always less than the value of p.

Without loss of generality, let's assume that $n \geq m$. Then an upper bound on one-half the number of loop iterations—which is asymptotically equivalent to the overall running time of the algorithm—satisfies the following recurrence relation:

$$T(n) = T(n/2) + \Theta(1),$$

which by the Master Theorem has a solution of the form $T(n) = \Theta(\log n)$. That is, Euclid's algorithm runs in time proportional to the *length* of its arguments—the number of bits it takes to write down m and n.

Example 25.16. *Gray codes.*

As a final example of a divide and conquer algorithm, consider the following practical problem: find a way to associate the numbers from 0 to $2^n - 1$ with bit strings of length n (which we will refer to as *codewords*), in such a way that the codewords representing successive integers differ in only one bit position. For example, the ordinary representation of numbers as binary numerals fails this condition rather dramatically. If $n = 3$, then the codeword for 3 is 011, but the codeword for 4 is 100, which differs from the codeword for 3 in all three bit positions. We'll use the word *code* to refer to the mapping from integers to the codewords that represent those integers. A code that achieves the desired effect is

$$G_3 = \langle 000, 001, 011, 010, 110, 111, 101, 100 \rangle. \tag{25.17}$$

In fact, G_3 delivers a little more than was asked: the last codeword differs from the first in only one bit position. This means that the codewords can be arranged in a circle, and successive integers could be assigned to successive codewords by proceeding around the circle in either direction, starting anywhere (Figure 25.4).

This is a combinatorial puzzle, but also a problem of real practical importance. For example, imagine a computer with a grid-like input device, with a stylus that moves over the grid. As the stylus slides from one position to the next, the coded representation of the position should change smoothly. In a physical device, however, if the representations differ by more than one bit, those bits may not change at exactly the same time, causing jumps back and forth to other values as the bits change one by one. Using a Gray code solves this problem, as shown in Figure 25.5. In addition, any uncertainty about the exact position of the stylus between two grid lines would affect the reported position only minimally.

Figure 25.4. The 3-bit Gray code G_3, arranged in a circle. The numbers from 0 through 7 can be assigned successive codewords by going around the circle, starting anywhere and proceeding either clockwise or counterclockwise.

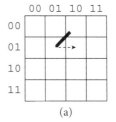

(a)

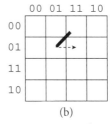

(b)

Figure 25.5. A stylus moving over a grid. In (a), the grid cells are labeled in the usual numerical order. As the stylus moves from $\langle 01, 01 \rangle$ to the neighboring $\langle 10, 01 \rangle$, two bits of the representation change. If the bits do not change at exactly the same time—say 01 changes to 11 and then to 10—it will appear as if the stylus jumped from $\langle 01, 01 \rangle$ to $\langle 11, 01 \rangle$ before arriving at $\langle 10, 01 \rangle$. In (b), the grid cells are labeled according to G_2, so the same movement is represented as a change from $\langle 01, 01 \rangle$ to $\langle 11, 01 \rangle$. Only one bit of the representation changes, eliminating the possibility of an apparent jump.

The *n*-bit *Gray code*[3] is the result of concatenating two copies of the $(n-1)$-bit Gray code: the first with 0 prepended to all the codewords, and the second in reverse order and with 1 prepended to all the codewords. So to construct the 4-bit Gray code, start with G_3 (25.17) and prepend a 0 to each codeword—we'll write this as $0 \cdot G_3$:

$$0 \cdot G_3 = \langle 0000, 0001, 0011, 0010, 0110, 0111, 0101, 0100 \rangle.$$
(25.18)

Next, reverse G_3 (that is, reverse the sequence, not the bit strings within it) and prepend a 1 to each codeword:

$$1 \cdot G_3^R = \langle 1100, 1101, 1111, 1110, 1010, 1011, 1001, 1000 \rangle.$$
(25.19)

Finally, G_4 is the result of concatenating these two sequences. We'll use "∘" to denote the concatenation of two sequences of bit strings:

$$(0 \cdot G_3) \circ (1 \cdot G_3^R)$$
$$= \langle 0000, 0001, 0011, 0010, 0110, 0111, 0101, 0100,$$
$$1100, 1101, 1111, 1110, 1010, 1011, 1001, 1000 \rangle$$
$$= G_4.$$

In general, Gray codes are defined by the recurrence

$$G_0 = \langle \lambda \rangle$$

and for each $n \geq 0$,

$$G_{n+1} = (0 \cdot G_n) \circ (1 \cdot G_n^R).$$
(25.20)

Theorem 25.21. *The n-bit Gray code includes each string of n bits exactly once, and successive codewords differ in only one bit position. Also, the last and first codewords differ in only one bit position.*

Proof. The proof is by induction on n. When $n = 0$, there is only one string of n bits, namely λ, so the theorem is true trivially.

Now suppose that G_n has the stated properties and consider $G_{n+1} = (0 \cdot G_n) \circ (1 \cdot G_n^R)$. G_n includes all n-bit sequences, so G_n is of length 2^n; and G_{n+1} is twice as long as G_n, so G_{n+1} has length 2^{n+1}. The first 2^n bit strings in G_{n+1} are all different from each other, since all the strings in G_n are different from each other. Similarly, the last 2^n strings in G_{n+1} are different from each other. And no string is in both the first and second halves of the sequence, because all the strings in the first half begin with 0 and all the strings in the second half begin with 1. So each string of $n + 1$ bits appears exactly once.

In the same way, successive strings in the first half of G_{n+1} differ in only one bit position, because that is true of G_n, and they all have the same first bit, namely 0. The same is true for the strings in the second half of G_{n+1}. And the last string in the first half is identical to the first string in the second half, except that the former begins with 0 and the latter begins with 1. So throughout the sequence, successive codewords differ in only one bit position.

Finally, the last string in G_{n+1} is 1 followed by the first string in G_n, while the first string in G_{n+1} is 0 followed by that same string, the first string in G_n. So the first and last strings in G_{n+1} differ only in their first bit. ∎

How much time does it take to generate the n-bit Gray code?[4] Let $m = 2^n$, the number of elements in the n-bit Gray code. If we mechanically follow the formula (25.20), then we get the recurrence

$$T(m) = 2T\left(\frac{m}{2}\right) + \Theta(m)$$
$$T(1) = \Theta(1),$$

which fits the Master Theorem with $a = b = 2$, so $\log_b a = 1 = d$. So case (2) applies and $T(m) = \Theta(m \log m) = \Theta(n2^n)$.

Can we get rid of the log factor? Indeed. There is no need to generate G_n twice while generating G_{n+1}; it can be generated once, stored, and copied later on when it is needed the second time. At the end, there will be one stored copy of G_i for every i. The recurrence for the time complexity would then be

$$T(m) = T\left(\frac{m}{2}\right) + \Theta(m)$$
$$T(1) = \Theta(1).$$

Now $a = 1$, $b = 2$, and $\log_b a = 0 < d = 1$, so case (1) applies. The result is $T(m) = \Theta(m) = \Theta(2^n)$; the Gray code can be generated in time proportional to the length of the code itself.[5]

✳

The Master Theorem is extremely useful, but it won't solve every recurrence relation problem. The *Fibonacci numbers* provide an example of a different approach: using generating functions to solve recurrence relations.

Example 25.22. *You need to climb a flight of n stairs, and on each step you can go up either one stair or two. In how many different ways can you climb the whole flight of n stairs?*

Solution to example. For consistency with the standard indexing of Fibonacci numbers, we'll define f_{n+1} to be the number of ways of climbing n stairs if $n \geq 0$, and set $f_0 = 0$. Then $f_1 = 1$ (the only way to climb zero stairs is not to step at all), and $f_2 = 1$ (there is only one way to step up one stair). When

[4] The following analysis assumes that operations on a single codeword are done in $\Theta(1)$ time. An individual codeword has length n; so naively, it might seem that copying it or prepending a symbol to it would take $\Theta(n)$ time. However, rather than literally copying the codeword, the computer can simply assign a new reference to the existing codeword (a $\Theta(1)$ operation) and refer back to that same object whenever it is needed, rather than writing it out multiple times.

[5] Andrew T. Phillips and Michael R. Wick, "A Dynamic Programming Approach to Generating a Binary Reflected Gray Code Sequence" 2005.

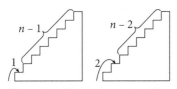

Figure 25.6. To climb n stairs, either climb 1 and then $n-1$, or climb 2 and then $n-2$.

n	f_n
0	0
1	1
2	1
3	2
4	3
5	5
6	8
7	13
8	21
9	34
10	55

Figure 25.7. Values of the Fibonacci numbers.

climbing more than two stairs, you have two choices for the first step, to climb up one stair and leave one fewer stair to climb, or to climb up two stairs and leave two fewer stairs to climb (Figure 25.6). So the recurrence relation is

$$f_n = f_{n-1} + f_{n-2} \tag{25.23}$$

when $n \geq 2$, with the base cases

$$f_0 = 0 \tag{25.24}$$
$$f_1 = 1. \tag{25.25}$$

The $f_2 = 1$ result is then the $n=2$ case of the recurrence relation (25.23). The first few Fibonacci numbers are shown in Figure 25.7.

But what is the solution of the recurrence relation (25.23) for general n? The Master Theorem doesn't help, but generating functions (page 266) do.

Let

$$F(x) = \sum_{n=0}^{\infty} f_n x^n.$$

Then, multiplying by x and by x^2 and shifting the indices to make it easier to collect terms, we get the following equations:

$$F(x) = \sum_{n=0}^{\infty} f_n x^n$$

$$xF(x) = \sum_{n=1}^{\infty} f_{n-1} x^n$$

$$x^2 F(x) = \sum_{n=2}^{\infty} f_{n-2} x^n.$$

Subtracting the second and third equations from the first,

$$(1 - x - x^2)F(x) = f_0 + (f_1 - f_0)x + \sum_{n=2}^{\infty} (f_n - f_{n-1} - f_{n-2})x^n$$

$$= 0 + (1 - 0)x + \sum_{n=2}^{\infty} 0x^n$$

$$= x,$$

by (25.23), (25.24), and (25.25). So

$$F(x) = \frac{x}{1 - x - x^2}.$$ (25.26)

The quadratic polynomial $1 - x - x^2$ can be factored:

$$1 - x - x^2 = -(\phi + x) \cdot (\psi + x).$$ (25.27)

Here ϕ and ψ are the roots of the equation $x^2 - x - 1 = 0$, and it is easy to check that $-\phi$ and $-\psi$ are roots of the equation

$$1 - x - x^2 = 0.$$ (25.28)

The quadratic formula gives the values of ϕ and ψ,

$$\phi = \frac{1 + \sqrt{5}}{2} \approx 1.618$$

$$\psi = \frac{1 - \sqrt{5}}{2} \approx -0.618.$$ (25.29)

So substituting (25.27) in the denominator of (25.26),

$$F(x) = \frac{-x}{(x + \phi)(x + \psi)}.$$ (25.30)

The number ϕ has been known since antiquity as the *golden ratio*; it is the proportion (comparing the long dimension to the short dimension) of a rectangle with the property that chopping off a square leaves a rectangle with the same proportion (Figure 25.8).

If we can express (25.30) as the sum of fractions of the form $\frac{1}{1-cx}$, where c is a constant, we will be able to use (24.9) (page 265) to find the value of f_n as the coefficient of x^n.

We can find values of a and b such that[6]

$$F(x) = \frac{-x}{(x + \phi)(x + \psi)} = \frac{a}{x + \phi} + \frac{b}{x + \psi}$$ (25.31)

by multiplying through by $(x + \phi)(x + \psi)$ and plugging in $x = -\phi$ and $x = -\psi$:

$$-x = a(x + \psi) + b(x + \phi)$$

$$\phi = a(\psi - \phi)$$

$$\psi = b(\phi - \psi).$$

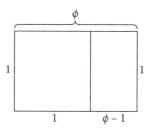

Figure 25.8. The golden ratio ϕ is the length of the longer dimension of a rectangle that is of length 1 in its shorter dimension and has the property that chopping off a unit square leaves a rectangle of the same proportions. That is, $\frac{\phi}{1} = \frac{1}{\phi - 1}$, so $\phi^2 - \phi - 1 = 0$.

[6] This is known as a *partial fraction decomposition*. The general method is used to express a fraction of the form $P(x)/Q(x)$, where P and Q are polynomials in x, as a sum of terms of the form $P'(x)/Q'(x)$, where P' and Q' are polynomials and the degrees of the Q' polynomials are lower than the degree of Q.

So

$$a = \frac{\phi}{\psi - \phi} = \frac{-\phi}{\sqrt{5}}$$

$$b = \frac{\psi}{\phi - \psi} = \frac{\psi}{\sqrt{5}}$$

$$F(x) = \frac{1}{\sqrt{5}} \left(\frac{-\phi}{x + \phi} + \frac{\psi}{x + \psi} \right). \tag{25.32}$$

To proceed further, we need a simple identity involving ϕ and ψ.

$$\phi \cdot \psi = \frac{1 + \sqrt{5}}{2} \cdot \frac{1 - \sqrt{5}}{2} = \frac{1^2 - \sqrt{5}^2}{2^2}$$

$$= -1,$$

$$\text{so } \psi = \frac{-1}{\phi}$$

$$\phi = \frac{-1}{\psi}.$$

Therefore

$$F(x) = \frac{1}{\sqrt{5}} \left(\frac{-\phi}{x + \phi} + \frac{\psi}{x + \psi} \right)$$

$$= \frac{1}{\sqrt{5}} \left(\frac{-1}{(x/\phi) + 1} + \frac{1}{(x/\psi) + 1} \right)$$

$$= \frac{1}{\sqrt{5}} \left(\frac{1}{1 - \phi x} - \frac{1}{1 - \psi x} \right)$$

$$= \frac{1}{\sqrt{5}} \left(\sum_{n=0}^{\infty} \phi^n x^n - \sum_{n=0}^{\infty} \psi^n x^n \right) \qquad \text{(using (24.9))}.$$

The nth Fibonacci number f_n is the coefficient of x^n, namely

$$f_n = \frac{1}{\sqrt{5}} (\phi^n - \psi^n). \tag{25.33}$$

Remarkably, both terms in (25.33) are irrational, and yet f_n is an integer! Since $\phi > 1$ while $|\psi| < 1$, it follows that

$$f_n \sim \frac{1}{\sqrt{5}} \phi^n.$$

So the numbers in the Fibonacci series grow exponentially, with the base of the exponent being about 1.6. ∎

*

We close with one final example of a recurrence based on an example we have already seen: balanced parentheses (page 84).

Example 25.34. Catalan Numbers. *How many properly balanced strings of parentheses are there, with n left parentheses and n matching right parentheses?*

Solution to example. There are two ways to arrange one left and one right parenthesis, "( )" and ") (", but only one of them is balanced. The number of properly balanced strings of n left and n right parentheses is called the nth Catalan number[7] C_n. (The Catalan numbers turn out to count a great many things besides balanced strings of parentheses; some further applications are explored in Problem 25.6.) The first few Catalan numbers, starting with C_1, are $1, 2, 5, 14, 42, 132, 429, \ldots$; they seem to increase fairly rapidly, but there is no obvious pattern. It makes sense to also set $C_0 = 1$, since the empty string is the one and only balanced string of 0 pairs of parentheses.

A recurrence relation for C_n is easier to write down than to solve. To express C_n in terms of C_i for $i < n$, consider any balanced string w of $n \geq 1$ pairs of parentheses. The string w begins with a left parenthesis, and there is a matching right parenthesis somewhere later in the string. So

$$w = (u)v,$$

where u and v are balanced strings of parentheses, u consists of some number $i < n$ of pairs of parentheses, and v then has $n - i - 1$ pairs of parentheses. Multiplying the number of possibilities for u and v and summing over all possible i yields

$$C_n = \sum_{i=0}^{n-1} C_i C_{n-i-1} \tag{25.35}$$

$$C_0 = 1.$$

For example, $C_2 = C_0 C_1 + C_1 C_0 = 1 \cdot 1 + 1 \cdot 1 = 2$, corresponding to the two possible strings "(())" for which $u = ()$ and $v = \lambda$, and "()()" for which $u = \lambda$ and $v = ()$.

The Master Theorem doesn't help with (25.35). There exist more general methods that can be applied to the problem, but we'll solve it by looking at the original problem in a different way.

Instead of a balanced string of n pairs of parentheses, consider *any circular arrangement* of $n + 1$ left parentheses and n right parentheses (Figure 25.9). One of the left parentheses turns out to be special—it is the only one that can't be matched with a subsequent right parenthesis, where "subsequent" means going around the circle clockwise. To see this, start with any left parenthesis. There is at least one, since $n + 1 \geq 1$. If $n + 1 = 1$ then you have found the unmatched left parenthesis. If $n + 1 > 1$, then $n \geq 1$ and there is at least one right parenthesis, so keep going clockwise until you encounter

[7]Eugène Charles Catalan (1814–94) was a Belgian mathematician.

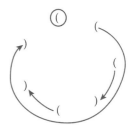

Figure 25.9. $n + 1 = 4$ left and $n = 3$ right parentheses arranged in a circle. Each left parenthesis except one is matched to a unique right parenthesis further around the circle in the clockwise direction and shown in the figure at the other end of a purple arrow. The unmatched left parenthesis is uniquely determined and is circled in red. This circular arrangement corresponds to the balanced string $(( )( ))$.

a right parenthesis. That right parenthesis and the immediately preceding left parenthesis are a matching pair; remove them and repeat the process, now with n left parentheses and $n-1$ right parentheses. Eventually you will get down to the $n=1$ case and identify the unmatched left parenthesis. Now starting at the unmatched left parenthesis and reading clockwise yields a balanced string of parentheses, and any balanced string of parentheses corresponds to a unique circular arrangement of $n+1$ left and n right parentheses, with the unmatched left parenthesis at the beginning of the string.

We have succeeded in changing a question about *balanced* strings of parentheses into a question about *arbitrary circular arrangements* of $n+1$ left and n right parentheses. How many such circular arrangements are there? As many as there are ways of choosing which $n+1$ of the $2n+1$ positions in the circle should have left parentheses. That is,

$$C_n = \frac{1}{2n+1}\binom{2n+1}{n+1}, \tag{25.36}$$

where the division by $2n+1$ is because all $2n+1$ ways of rotating the circle by one position are equivalent (see page 238). Equation (25.36) can be simplified by writing out the formula for the "choose" operator:

$$\begin{aligned}
C_n &= \frac{1}{2n+1}\binom{2n+1}{n+1} \\
&= \frac{1}{2n+1} \cdot \frac{(2n+1)!}{n!(n+1)!} \\
&= \frac{1}{n+1} \cdot \frac{(2n)!}{n!n!} \\
&= \frac{1}{n+1}\binom{2n}{n}.
\end{aligned}$$ ∎

Chapter Summary

- A *recurrence relation* defines the values of an infinite sequence in terms of smaller values of the sequence, starting with base cases.

- Recurrence relations arise in the analysis of recursive algorithms.

- Merge sort and binary search are *divide and conquer* algorithms, a common type of recursive algorithm. Such an algorithm splits the problem into successively smaller subproblems, operating recursively on each, and then combines the answers to the subproblems into an answer to the original problem.

- The runtime of a divide and conquer algorithm can often be described by a recurrence of the form

$$T(n) = aT\left(\frac{n}{b}\right) + \Theta(n^d),$$

where a is the number of subproblems, b is the size of the original problem relative to the subproblems, and the algorithm to combine the answers to the subproblems has polynomial runtime with degree d. The *Master Theorem* gives the closed form for recurrences that take this form, based on the relationship between a, b, and d.

■ Divide and conquer algorithms may yield surprising improvements on the obvious algorithms; for example, in integer arithmetic (*Karatsuba's algorithm*), matrix multiplication (*Strassen's algorithm*), and finding the greatest common divisor (*Euclid's algorithm*).

■ A *Gray code* of length n arranges $\{0, 1\}^n$ in such a way that each codeword differs from the next in only one bit position.

■ The *Fibonacci numbers* satisfy the recurrence $f_n = f_{n-1} + f_{n-2}$, with $f_0 = 0$ and $f_1 = 1$. The values of f_n grow exponentially with n; the base of the exponential is about 1.6.

■ The *Catalan numbers* are the numbers of balanced strings of n left and n right parentheses.

Problems

25.1. The *Towers of Hanoi* puzzle (Figure 25.10) is to move the stack of disks on peg 1 to peg 2, while obeying the following constraints: you can move only one disk at a time from peg to peg; and you can never put a bigger disk on top of a smaller disk.

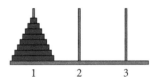

Figure 25.10. Towers of Hanoi. How can you move the stack of disks to the second peg, moving one disk at a time between pegs and never putting a bigger disk on top of a smaller one?

  (a) Describe a recursive algorithm for solving the puzzle. How many steps does it take, if there are n disks?
  (b) Show that an n-bit Gray code also solves the problem. Simply run through the codes in order, at each step moving the disk corresponding to the bit position that changes (where disk 0 is the smallest disk, and bit position 0 is the rightmost). If the disk to be moved is any but the smallest, there is only one peg to which it can legally be moved. The smallest disk moves cyclically, either $2 \to 3 \to 1 \to 2 \to \ldots$ if n is odd, or $3 \to 2 \to 1 \to 3 \to \ldots$ if n is even.

25.2. A *Hamiltonian circuit* is a circuit that visits each vertex of a graph exactly once. Explain the sense in which the n-bit Gray code describes a Hamiltonian circuit of an n-dimensional hypercube.

25.3. How many binary strings of length n have no consecutive occurrences of 1? Write a recurrence relation with two cases, based on whether the string begins with 0 or 1, and solve it.

25.4. Find a recurrence relation for the number of n-digit ternary sequences without any occurrence of the subsequence 012. (A ternary sequence is a sequence composed of 0s, 1s, and 2s.)

25.5. Find a recurrence relation for the number of different ways to make change for n cents when you have an unlimited number of coins of 5, 10, and 25 cents. Find the values for $n = 25$, 30, and 35.

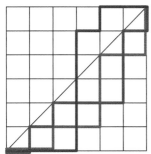

Figure 25.11. Three different monotonic paths on a 6×6 grid. The blue and green paths do not cross the diagonal, while the red path does.

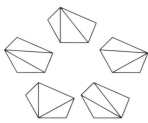

Figure 25.12. There are five ways to divide a convex pentagon into triangles with nonintersecting diagonals.

25.6. More applications of the Catalan numbers.

(a) A *monotonic path* (also mentioned in Problem 23.6) on an $n \times n$ grid is a path that starts at the bottom left corner, ends at the top right corner, and includes only upward and rightward moves (see Figure 25.11). Show that the number of such paths that do not cross the diagonal (the one extending from bottom left to top right) is C_n.

(b) Show that the number of ways of drawing nonintersecting diagonals in a convex n-gon so as to divide it into triangles is C_{n-2}, the $(n-2)^{\text{nd}}$ Catalan number. For example (Figure 25.12), this can be done five ways in a pentagon, and $C_{5-2} = 5$. *Hint:* Show that the number of such triangulations satisfies the same recurrence relation as do the Catalan numbers.

25.7. Using generating functions, solve these recurrence relations.

(a)
$$a_0 = 0$$
$$a_1 = 1$$
$$a_n = a_{n-1} + 2a_{n-2} \text{ for } n \geq 2.$$

(b)
$$a_0 = 1$$
$$a_1 = 1$$
$$a_n = a_{n-1} + 6a_{n-2} \text{ for } n \geq 2.$$

25.8. A bank pays 2% annual interest. Find and solve recurrence relations for the amount of money you would have after n years on the following assumptions.

(a) You put $1000 in the bank and just let it compound.

(b) You start by depositing $100 and at the end of each year, after the interest has been paid, you add another $100.

25.9. Prove that

$$\psi = -(\phi - 1),$$

which explains the similarity between the digits of ϕ and ψ in (25.29).

Chapter 26

Probability

Counting is a means to an end. Chapters 22 and 23 develop mathematical machinery that is included in computer science courses for practical reasons: it helps scientists predict how systems will behave.

How frequently should we expect one outcome or another to occur if we repeat some procedure many times? The procedure might be drawing marbles from a jar, or anything for which marbles and jars serve as a suitable metaphor. It might be rolling a pair of standard dice and getting a total between 2 and 12, or any other random phenomenon for which the possible results are discrete and specific. This frequency with which an outcome is expected to occur is the event's *probability*.

Let's start by defining our terms. A *trial* is a repeatable procedure, which results in one of a well-defined set of *outcomes*. A trial could be drawing a marble from a jar, for example, in which case the possible outcomes are the marbles; or flipping a coin, in which case the possible outcomes are heads and tails; or rolling a pair of dice, in which case the outcomes are pairs of integers, with each component between 1 and 6 inclusive.

For our purposes, the set of possible outcomes will always be finite. (A different approach is required if the set of outcomes is infinite.) This means, for example, that the physical coordinates of the dice on the tabletop—a pair of real numbers—could not be construed as an outcome.

An *experiment* is a sequence of individual trials. An experiment can be repeated, and it is somewhat arbitrary whether, for example, a sequence of 4 coin flips is considered a single trial with 16 possible outcomes or an experiment consisting of 4 trials, each with 2 possible outcomes.

The *sample space* of a trial is the set of all possible outcomes. So the sample space of rolling a standard die is $\{1, 2, 3, 4, 5, 6\}$, and of rolling a pair of dice is $\{1, 2, 3, 4, 5, 6\}^2$. An *event* is any set of outcomes; that is, a subset of the sample space. For example, if the sample space is the possible results of rolling a pair of dice, then rolling numbers that sum to 4 would be the event $\{\langle 1, 3 \rangle, \langle 2, 2 \rangle, \langle 3, 1 \rangle\}$.

The set of possible events of a trial is the power set of the sample space—the set of all subsets of the set of outcomes. So (by Problem 5.3) there are 2^n possible events in a sample space of size n.

Consider a trial with a finite sample space S. A *probability function* is a function $\Pr : \mathcal{P}(S) \to \mathbb{R}$ that assigns probabilities (which are real numbers) to events. Any probability function must satisfy the following conditions, called the *axioms of finite probability*:

Axiom 1. $0 \le \Pr(A) \le 1$ for any event A; that is, $\Pr(A) \in [0, 1]$, the set of real numbers between 0 and 1 inclusive.

Axiom 2. $\Pr(\emptyset) = 0$ and $\Pr(S) = 1$. Together, these conditions state that in any trial, one of the outcomes in S must occur. For finite sample spaces, probability 0 can be taken to mean "impossible" and probability 1 can be taken to mean "certain."[1]

Axiom 3. If $A, B \subseteq S$ and $A \cap B = \emptyset$, then

$$\Pr(A \cup B) = \Pr(A) + \Pr(B).$$

That is, the probability of the union of two disjoint events is the sum of their probabilities (Figure 26.1).

From these axioms, several useful basic properties can be inferred.

Theorem 26.1. *The probability of an event* not *happening is* $\Pr(\overline{A}) = 1 - \Pr(A)$.

Proof. Since S exhausts all possible outcomes, if the event A does not happen then the event $S - A = \overline{A}$ does happen. The events A and $\overline{A}$ are disjoint (Figure 26.2). So by Axioms 2 and 3,

$$1 = \Pr(S) = \Pr(A \cup \overline{A}) = \Pr(A) + \Pr(\overline{A}).$$

Therefore $\Pr(\overline{A}) = 1 - \Pr(A)$ by rearranging terms. ∎

The event $\overline{A}$ is referred to as the *complement* or *negation* of event A.

Theorem 26.2. *If one event is a subset of another, the first has no greater probability than the second. That is, if* $A, B \subseteq S$ *and* $A \subseteq B$, *then* $\Pr(A) \le \Pr(B)$.

Proof. Consider the sets A and $B - A$. These are disjoint, so since $A \subseteq B$,

$$\Pr(B) = \Pr(A \cup (B - A)) = \Pr(A) + \Pr(B - A).$$

Rearranging,

$$\Pr(A) = \Pr(B) - \Pr(B - A).$$

But $\Pr(B - A) \ge 0$ by Axiom 1, so $\Pr(A) \le \Pr(B)$. ∎

[1] This is not true in the infinite case. Flipping a fair coin over and over forever and never getting anything but heads has probability 0, but is not technically impossible. In the infinite case, an event that is a subset of the sample space but has probability 1 is said to happen "almost surely."

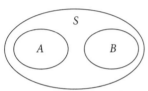

Figure 26.1. *A* and *B* are disjoint, so their probabilities can be summed to find the probability of their union.

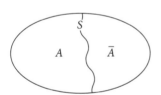

Figure 26.2. Together, A and $\overline{A}$ comprise the entire sample space.

Theorem 26.3. *The probability that one of a set of disjoint events occurs is the sum of their individual probabilities. That is, suppose* $A_1, A_2, \ldots, A_n \subseteq S$ *are all disjoint (that is,* $A_i \cap A_j = \emptyset$ *whenever* $i \neq j$*); then*

$$\Pr\left(\bigcup_{i=1}^n A_i\right) = \sum_{i=1}^n \Pr(A_i).$$

Proof. The proof is by induction on n. The base case, when $n = 2$, follows immediately from Axiom 3:

$$\Pr(A_1 \cup A_2) = \Pr(A_1) + \Pr(A_2).$$

Now suppose that $n \geq 2$ and

$$\Pr\left(\bigcup_{i=1}^n A_i\right) = \sum_{i=1}^n \Pr(A_i).$$

The union $\bigcup_{i=1}^n A_i$ is disjoint from A_{n+1}, since each of its component sets is disjoint from A_{n+1}, so again by Axiom 3,

$$\Pr\left(\left(\bigcup_{i=1}^n A_i\right) \cup A_{n+1}\right) = \Pr\left(\bigcup_{i=1}^n A_i\right) + \Pr(A_{n+1}),$$

which, by the induction hypothesis, is equal to $\sum_{i=1}^{n+1} \Pr(A_i)$. ∎

Theorem 26.4. Inclusion-Exclusion Principle. *Add If A and B are any two events (not necessarily disjoint), the probability that one or the other will happen is the sum of their probabilities minus the probability of their intersection:*

$$\Pr(A \cup B) = \Pr(A) + \Pr(B) - \Pr(A \cap B).$$

Proof. Consider the three disjoint events $A - B$, $B - A$, and $A \cap B$. The union of these three events is $A \cup B$ (Figure 26.3). By Theorem 26.3,

$$\Pr(A \cup B) = \Pr(A - B) + \Pr(B - A) + \Pr(A \cap B). \qquad (26.5)$$

Since $A = (A - B) \cup (A \cap B)$ and $B = (B - A) \cup (A \cap B)$, by Axiom 3,

$$\Pr(A - B) = \Pr(A) - \Pr(A \cap B) \text{ and}$$

$$\Pr(B - A) = \Pr(B) - \Pr(A \cap B).$$

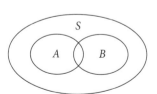

Figure 26.3. When A and B overlap, in calculating the probability of their union, their intersection must not be double-counted.

Substituting into (26.5),

$$\Pr(A \cup B) = \Pr(A) - \Pr(A \cap B) + \Pr(B) - \Pr(A \cap B) + \Pr(A \cap B)$$
$$= \Pr(A) + \Pr(B) - \Pr(A \cap B). \qquad \blacksquare$$

✴

For a trial with n possible outcomes, one simple probability function assigns equal probability $\frac{1}{n}$ to each outcome. Equally likely outcomes are a good model for many phenomena—for example, a fair coin (two outcomes), a fair cubic die (six outcomes), or a jar of marbles of equal size and weight.

Theorem 26.6. Probability Function for Equally Likely Outcomes. *Consider an experiment with finite sample space S, where each outcome in S is equally likely. Define, for any event $E \subseteq S$,*

$$\Pr(E) = \frac{|E|}{|S|}. \tag{26.7}$$

Then Pr *is a probability function.*

Proof. We need to check that (26.7) satisfies the axioms of finite probability:

1. $\Pr(A) = \frac{|A|}{|S|}$. Both numerator and denominator are nonnegative and $0 \le |A| \le |S|$ since $A \subseteq S$, so $0 \le \frac{|A|}{|S|} \le 1$.

2. $\Pr(\emptyset) = \frac{|\emptyset|}{|S|} = 0$, and $\Pr(S) = \frac{|S|}{|S|} = 1$.

3. If A and B are disjoint, then $|A \cup B| = |A| + |B|$. So

$$\Pr(A \cup B) = \frac{|A \cup B|}{|S|} = \frac{|A| + |B|}{|S|} = \frac{|A|}{|S|} + \frac{|B|}{|S|} = \Pr(A) + \Pr(B). \qquad \blacksquare$$

Example 26.8. *If a fair coin is flipped 4 times, what is the probability that exactly 2 of the flips land heads up?*

Solution to example. Rather than taking a trial to be a single coin flip, let's consider a trial to be flipping the coin 4 times.

An outcome is any length-4 sequence of heads and tails. Let event E be the set of sequences that contain exactly 2 heads. Then the sample space and the event of interest are

$$S = \{H, T\}^4$$

$$E = \{x_1 x_2 x_3 x_4 \in \{H, T\}^4 : |\{i : x_i = H\}| = 2\}.$$

E comprises $\binom{4}{2} = 6$ of the 16 possible outcomes (Figure 26.4). So the probability of getting exactly 2 heads is

$$\Pr(E) = \frac{|E|}{|S|} = \frac{\binom{4}{2}}{2^4} = \frac{6}{16} = \frac{3}{8}. \qquad \blacksquare$$

Here is a slightly more complicated example of calculating probabilities. In a group of 366 people (none born in a leap year), some two must have the same birthday, by the Pigeonhole Principle. In a group of 365 or fewer people, duplicate birthdays are possible but not inevitable, and the smaller the group the less likely they are. What is the mathematical relation between the size of the group and the likelihood that at least one birthday is doubled up?

Example 26.9. *In a randomly selected set of $n \leq 365$ people, what is the likelihood that some two have the same birthday? (Assume that all 365 birthdays in a nonleap year are equally likely, that no one is born on February 29, and that the births are independent of each other—no twins.)*

Solution to example. It is easier to calculate the probability of the complement (that all the birthdays are unique), and then subtract that probability from 1.

To start, imagine a group of only two people. The first person could have been born on any day, and the second person could have been born on any day except the birthday of the first—364 possible days. So the probability that the second person's birthday is different from the first person's is $\frac{364}{365}$.

Now assume that the first two birthdays are different, and add a third person. The probability that his or her birthday is different from the first two is $\frac{363}{365}$, so the probability that all three are different is

$$\frac{364}{365} \cdot \frac{363}{365}.$$

In general, suppose that after i people have been added, all their birthdays are distinct—so i birthdays are already claimed, and $365 - i$ birthdays are unclaimed. The probability that the next person's birthday is different from all the previous birthdays is $\frac{365-i}{365}$. To obtain the probability that all n birthdays are different, multiply these probabilities together for all values of i from 0 to $n - 1$:

$$\prod_{i=0}^{n-1} \frac{365 - i}{365} = \frac{365!}{(365 - n)! \cdot 365^n}. \qquad (26.10)$$

The probability that some two people do share a birthday is 1 minus this number. $\qquad \blacksquare$

HHHH
HHHT
HHTH
HHTT
HTHH
HTHT
HTTH
HTTT
THHH
THHT
THTH
THTT
TTHH
TTHT
TTTH
TTTT

Figure 26.4. The 16 elements of S, with the 6 elements of E in red.

When $n = 23$, the value of (26.10) is approximately 0.49; that is the smallest value of n for which (26.10) is less than $\frac{1}{2}$. That is, in a randomly chosen group of 23 or more people (none born on February 29), it is more likely than not that some two will have the same birthday.

✳

Both Examples 26.8 and 26.9 assume that all outcomes (coin flips or birthdays) are equally likely. It is worth focusing on Example 26.8, the coin-flipping example, because it illustrates the important fact that *if success and failure are equally likely, then a long run of successes, or a long run of failures, is no more unlikely than any other sequence of events of the same length.* Specifically, the assumption that all 16 sequences of four flips are equally likely depends on two facts about coins and probabilities. First, that the coin is fair; it is physically uniform and not weighted in some way that makes it more likely to come to rest on one side than the other. And second, that the outcome of one flip does not affect the outcome of any other flip—so for example, *HHHH* and *HHHT* are equally likely. Another way of saying this is that *the coin has no memory.* This is a matter of objective fact, and yet people may think that a certain outcome is "due" if it hasn't happened recently.

Example 26.11. The Gambler's Fallacy. *Let n be a large number. If a coin is flipped n times and comes up heads the first $n - 1$ times, what is the probability that the nth flip comes up heads?*

Solution to example. The probability is $\frac{1}{2}$, notwithstanding any intuition that the coin is "due" to come up tails. This false intuition is named for its potential to mislead gamblers into placing bets that are not mathematically sound.   ∎

When the knowledge of whether an event A occurred gives no information about whether another event B occurred, A and B are said to be independent. Mathematically, events A and B are *independent* if and only if

$$\Pr(A \cap B) = \Pr(A) \cdot \Pr(B). \qquad (26.12)$$

We can use the definition of independence to check Example 26.11 numerically in the $n = 4$ case. Define the sample space S and events A and B as follows.

$$S = \{H, T\}^4, \qquad\qquad |S| = 2^4 = 16,$$

$$A = H^3\{H, T\}, \qquad\qquad |A| = 2,$$

$$B = \{H, T\}^3 H, \qquad\qquad |B| = 2^3 = 8.$$

First let's calculate the probability of each event:

$$\Pr(A) = \frac{|A|}{|S|} = \frac{2}{16} = \frac{1}{8}$$

$$\Pr(B) = \frac{|B|}{|S|} = \frac{8}{16} = \frac{1}{2}.$$

Now since $A \cap B = \{HHHH\}$, $|A \cap B| = 1$. But then

$$\Pr(A \cap B) = \frac{|A \cap B|}{|S|}$$
$$= \frac{1}{16}$$
$$= \frac{1}{8} \cdot \frac{1}{2}$$
$$= \Pr(A) \cdot \Pr(B),$$

so (26.12) holds and A and B are independent. It doesn't matter that the first three flips came up heads; the probability of the fourth flip being heads remains $\frac{1}{2}$. See Figure 26.5 for a visual representation of this example.

The feeling that tails is "due" is an illusion caused by the unlikelihood of the earlier string of heads. But *any other string of flips is equally unlikely—any* particular sequence of n flips has probability exactly 2^{-n}. If anything, a long string of heads should make one suspicious that, fairness claims about the coin notwithstanding, the coin is biased and another heads is *more* likely, not less!

What do nonindependent events look like? (Events are said to be *dependent* if they are not independent.) We have already seen one instance. In Example 26.9, we stipulated that there were no twin births, and indeed such births would fail the independence test. For example, if A represents the event that Amir was born on January 1 and B represents the event that Barbara was born on January 1, then $\Pr(A) = \Pr(B) = \frac{1}{365}$, and $\Pr(A) \cdot \Pr(B) = \frac{1}{365^2}$. But if Amir and Barbara are twins, then $\Pr(A \cap B) = \frac{1}{365}$, since we know both were born on the same day,[2] and this quantity is very different from $\Pr(A) \cdot \Pr(B)$.

Example 26.13. *Suppose a coin is flipped 4 times, and consider the events "at least 2 flips land on heads" (event A) and "at least 2 flips land on tails" (event B). Are A and B independent?*

Solution to example. No. Intuitively, if we know that A occurred, we may suspect that B is less likely since the possibility of 3 or 4 tails has been ruled out. That intuition is correct, as can be demonstrated mathematically.

```
HHHH HHHT
HHTH HHTT
HTHH HTHT
HTTH HTTT
THHH THHT
THTH THTT
TTHH TTHT
TTTH TTTT
```

Figure 26.5. The outcomes in A are in a box, and the outcomes in B are printed in red. If we represent probabilities spatially, when two events are independent, each event occupies the same percentage of space within the other event as it does within the entire space. Here A is $\frac{1}{8}$ of B and $\frac{1}{8}$ of the entire space, and B is $\frac{1}{2}$ of A and $\frac{1}{2}$ of the entire space.

[2] Ignoring the possibility that twins might be born a few minutes apart on successive days.

Event A occurs if 2, 3, or 4 heads are flipped. Let's add up the number of ways for each to happen:

$$|A| = \binom{4}{2} + \binom{4}{3} + \binom{4}{4} = 6 + 4 + 1 = 11.$$

Event B occurs if 2, 3, or 4 tails are flipped, so $|B| = 11$ as well, by symmetry. The event $A \cap B$ represents the scenario where exactly 2 flips land on heads and exactly 2 land on tails:

$$|A \cap B| = \binom{4}{2} = 6.$$

As before, $|S| = 16$. So

$$\Pr(A \cap B) = \frac{6}{16} = \frac{3}{8}, \text{ while}$$
$$\Pr(A) \cdot \Pr(B) = \frac{11}{16} \cdot \frac{11}{16} = \frac{121}{256}.$$

These probabilities are not equal (in fact $\frac{3}{8} = \frac{96}{256} < \frac{121}{256}$), so events A and B are dependent. ∎

Sometimes it is hard to spot the dependency between two events, and the only way to be sure whether they are independent is to do the calculation.

Example 26.14. *Suppose three marbles are drawn (without replacement) from a jar containing six red marbles and four blue marbles. Consider these events:*

A: *The three marbles are not all the same color.*

B: *The first marble drawn is red.*

Are A and B independent events?

Solution to example. No.

The number of ways of drawing 3 marbles of the *same* color is $\binom{6}{3} + \binom{4}{3}$; that is, the number of ways of choosing 3 red marbles or 3 blue marbles. That number is 24, so the number of ways of choosing 3 marbles that are not all the same color is $\binom{10}{3} - 24 = 120 - 24 = 96$. The probability of choosing 3 marbles not all of the same color is $p_A = \frac{96}{120} = \frac{4}{5}$.

The probability that the first marble drawn was red is just the proportion of red marbles; that is, $p_B = \frac{6}{10} = \frac{3}{5}$.

Let's denote by $p_{A \cap B}$ the probability of $A \cap B$; that is, drawing a red marble first and then having at least 1 blue marble in the other 2 draws. After the first draw there are 5 red marbles and 4 blue marbles left. There are $\binom{5}{2}$ ways of getting 2 more red marbles, and therefore $\binom{9}{2} - \binom{5}{2} = 36 - 10 = 26$ ways of drawing at least 1 blue marble. So

$$
\begin{aligned}
p_{A \cap B} &= \frac{6}{10} \cdot \frac{26}{36} \\
&= \frac{13}{30} \\
&\neq \frac{12}{25} \\
&= \frac{4}{5} \cdot \frac{3}{5} \\
&= p_A \cdot p_B.
\end{aligned}
$$

So A and B are not independent events.   ∎

The concept of independence can be applied to more than two events. Several events are said to be *pairwise independent* if every pair of two of the events are independent of each other: that is, if X is a set of events, then for every $A, B \in X$, A and B are independent.

A stronger condition on a set of events is *mutual independence*: several events are mutually independent if each event is independent of the intersection of any number of the remaining events. That is, if X is a set of events, then for every $A \in X$ and every $Y \subseteq X - \{A\}$, A and $\bigcap_{B \in Y} B$ are independent:

$$
\Pr\left(A \cap \bigcap_{B \in Y} B\right) = \Pr(A) \cdot \Pr\left(\bigcap_{B \in Y} B\right). \tag{26.15}
$$

Intuitively, this means that no subset of the events gives any information about any of the other events.

A set of events may be pairwise independent but not mutually independent, as the following example shows.

Example 26.16. *A fair coin is flipped twice. Define the events*

$$A = \textit{the first flip lands on } H,$$

$$B = \textit{the second flip lands on } H,$$

$$C = \textit{the two flips have the same outcome.}$$

Are A, B, and C pairwise independent? Are they mutually independent?

Solution to example. Let's write out the sample space S, the events, and each of their pairwise intersections:

$$
\begin{aligned}
S &= \{HH, HT, TH, TT\}, \\
A &= \{HH, HT\}, \\
B &= \{HH, TH\}, \\
C &= \{HH, TT\}, \\
A \cap B &= \{HH\}, \\
A \cap C &= \{HH\}, \\
B \cap C &= \{HH\}.
\end{aligned}
$$

Their probabilities are

$$
\Pr(A) = \Pr(B) = \Pr(C) = \frac{1}{2},
$$

$$
\Pr(A \cap B) = \Pr(A \cap C) = \Pr(B \cap C) = \frac{1}{4}.
$$

Then

$$
\Pr(A \cap B) = \frac{1}{4} = \frac{1}{2} \cdot \frac{1}{2} = \Pr(A) \cap \Pr(B),
$$

$$
\Pr(A \cap C) = \frac{1}{4} = \frac{1}{2} \cdot \frac{1}{2} = \Pr(A) \cap \Pr(C), \text{ and}
$$

$$
\Pr(B \cap C) = \frac{1}{4} = \frac{1}{2} \cdot \frac{1}{2} = \Pr(B) \cap \Pr(C),
$$

so the events are pairwise independent. The occurrence of any one event does not impact the probability that any other event occurred.

But the three events are not mutually independent! Intuitively, if we know that both B and C occurred, then we know that A also occurred, since B and C together imply that the outcome was HH. So A is not independent of the combination of B and C. Such reasoning helps in this specific case, but in general the only way to be sure is to do the math. So let Y be the set of events $\{B, C\}$. Then $A \cap (B \cap C) = \{HH\}$ and $\Pr(A \cap (B \cap C)) = \frac{1}{4}$, so

$$
\begin{aligned}
\Pr(A \cap (B \cap C)) &= \frac{1}{4} \\
&\neq \frac{1}{2} \cdot \frac{1}{4} \\
&= \Pr(A) \cdot \Pr(B \cap C).
\end{aligned}
$$

Knowing whether each of B and C occurred gives information about whether A occurred—it actually tells us with certainty—so the three events are not mutually independent. ■

Note that we need to find just one event $A \in X$ and a corresponding $Y \subseteq X - \{A\}$ for which the equality (26.15) does not hold, in order to prove that the events in X are not mutually independent. To show that a set of events is mutually independent, one must show that the equality holds for every such A and Y.

Chapter Summary

- If a procedure is repeated many times, the *probability* of any *outcome* is the frequency with which the outcome is expected to occur.

- A *trial* is one run of a repeatable procedure, resulting in one of the finitely many outcomes in the *sample space*. An *experiment* is a sequence of trials. An *event* is a set of outcomes.

- A *probability function* assigns each event a probability in accordance with the *axioms of finite probability*: every event has probability between 0 and 1, exactly one outcome in the sample space occurs, and the probabilities of two disjoint events sum to the probability of their union.

- The *complement* of an event A is written $\overline{A}$, and its probability is $\Pr(\overline{A}) = 1 - \Pr(A)$.

- The probability of any subset of an event is no greater than the probability of the event itself: $\Pr(A) \leq \Pr(B)$ for $A \subseteq B$.

- The probability of a union of disjoint events is the sum of the events' individual probabilities:

$$\Pr\left(\bigcup_{i=1}^{n} A_i\right) = \sum_{i=1}^{n} \Pr(A_i)$$

  for disjoint events A_i.

- The probability of a union of two nondisjoint events can be calculated using the *Inclusion-Exclusion Principle*—include the probability of each event, and then exclude the probability of their overlap: $\Pr(A \cup B) = \Pr(A) + \Pr(B) - \Pr(A \cap B)$.

- In a trial where all outcomes are equally likely, the probability of any event is just the number of outcomes it contains divided by the size of the sample space: $\Pr(E) = \frac{|E|}{|S|}$.

- Two events are *independent* if knowledge of one gives no information about the other. That is, for events A and B, $P(A \cap B) = P(A) \cdot P(B)$. If events are not independent, they are *dependent*.

- Several events are *pairwise independent* if no individual event gives information about any other event: for every $A, B \in X$, A and B are independent.

■ Several events are *mutually independent* (a stronger condition than pairwise independence) if no subset of the events gives information about any other event: for every $A \in X$ and every $Y \subseteq X - \{A\}$, A and $\bigcap_{B \in Y} B$ are independent.

Problems

26.1. A jar contains 5 marbles: blue, green, purple, red, and yellow.

(a) If you draw 3 marbles from the jar, with replacement, what is the probability of drawing the purple marble twice and the red marble once?

(b) If you draw 3 marbles from the jar, without replacement, what is the probability that the first marble is blue and the third is green?

26.2. A card is drawn from a standard deck (described on page 242). Which is more likely: (1) that it is either a black card or a face card (value J, Q, or K); or (2) that it is neither a heart nor a face card?

26.3. In this problem we will derive fair coin flips from biased coins: that is, given coins for which the probabilities of landing heads and tails are not equal, the problem is to construct an experiment for which there are two disjoint events, with equal probabilities, that we can call "heads" and "tails."

(a) Given c_1 and c_2, where c_1 lands heads up with probability $\frac{2}{3}$ and c_2 lands heads up with probability $\frac{1}{4}$, construct a "fair coin flip" experiment.

(b) Given just one coin, with unknown probability p of landing heads up, where $0 < p < 1$, construct a "fair coin flip" experiment. (Note: the sample space may contain outcomes that are not a member of either event; if such an outcome occurs we ignore the result and perform a new trial of the experiment.)

(c) In your solution to part (b), what is the chance that a single trial of your experiment results in an outcome of either heads or tails?

26.4. A professor wants to estimate how many students skipped the reading assignment for today's class. She does not ask directly, knowing that the students would be embarrassed to answer that they did not do the reading. Instead the professor instructs the students to each flip a coin without showing anyone else the result, and then raise a hand if either (1) the student's coin landed heads up or (2) the student didn't do the reading assignment. She figures students will answer honestly since any individual student might have raised a hand because of the result of the coin flip. If $\frac{3}{4}$ of the students raise their hands, what fraction did not do the reading?

26.5. Under the assumptions of Example 26.9, what is the probability (in terms of n) that there is a single date on which exactly two people were born, with all other birthdays in the group unique? What is this value for $n = 23$?

26.6. The Binomial Distribution. Suppose a message is n bits long, and for each bit, the probability that it has been corrupted is p. What is the probability that exactly k bits are corrupted, for $0 \le k \le n$?

26.7. An *error-correcting code* is a technique for signaling and correcting errors when sending a message over an unreliable communication channel. One simple example is a repetition code, in which the message is repeated a fixed number of times. If any bit differs between repetitions, it is known to be corrupted (and the true value is assumed to be the value that appears most frequently).

Suppose that for a given communication channel, the probability that any given bit is corrupted is 0.05, and the outcome for each bit is independent. As an error-correcting code, messages are repeated 4 times.

(a) Suppose a message of length 20 is sent (for a total of 80 bits after repetition). What is the probability that at least one bit is corrupted?

(b) What is the maximum-length message that can be sent such that the probability of at least one bit being corrupted is less than 0.5?

(c) The weakness of a repetition code is that if a given bit is corrupted in every repetition, there is no signal of the error. Consider again a message of length 20. What is the probability that there exists a bit that is corrupted in all 4 repetitions of the message (that is, for some value $0 \le k < 20$, the bits at positions $k, 20 + k, 40 + k, 60 + k$ are all corrupted)?

26.8. A *Bloom filter*[3] is a data structure for representing a set, with some probability of false positives but with no false negatives. That is, the data structure is used to answer questions of the form "is $x \in S$?," and there is some likelihood that the answer will be that $x \in S$ when in fact $x \notin S$, but there is no possibility that the answer will be that $x \notin S$ when in fact $x \in S$. As a tradeoff against the possibility of false positives, a Bloom filter can be much smaller than the set it represents.

A Bloom filter consists of a bit array of length m, together with k *hash functions* $h_1, \ldots, h_k$ ("hash functions" were introduced in Problem 1.13). Each hash function h_j maps a set element (a string, perhaps) to a position in the array, called its hash value. The functions h_j assign elements to positions in such a way that (1) each element has equal probability of being mapped to any of the m possible positions; (2) the hash values of any hash function for different elements are independent of each other; and (3) the hash functions are themselves independent of each other. Initially, all m positions in the bit array are 0. To insert an element e, compute the hash values $h_1(e), \ldots, h_k(e)$ and set those bit positions to 1. Then to check whether an element e is in the set, compute $h_1(e), \ldots, h_k(e)$; if at least one position $h_j(e)$ has value 0, e is absent; if all of the positions $h_j(e)$ have value 1, e is probably present. The memory efficiency of a Bloom filter comes because an element is represented by the distribution of positions that are the hash values of the hash functions. The element itself is not stored at all.

Suppose n elements have been inserted into a Bloom filter, and we wish to check for the presence of an element e' that has not been inserted. Calculate the probability of a false positive, in terms of m, k, and n. What is this probability if $m = 1000$, $k = 10$, and $n = 100$?

[3] Devised by computer scientist Burton Bloom in 1970.

26.9. Prove that if events A and B are disjoint and both have nonzero probability, then they are dependent.

26.10. Many email providers have a spam filter, which classifies emails as likely to be unwanted and sends them directly to a spam folder. One possible indicator of spam is whether the sender's email address is in the recipient's list of contacts.

Suppose that 30% of all emails are spam (event S_P); that 20% of emails are from an unknown sender, not in the recipient's contact list (event U); and that 15% of emails are spam from unknown senders. Are U and S_P independent?

26.11. Two dice are rolled. Define the events

$$A = \text{the first roll is a } 1, 2, \text{ or } 3;$$
$$B = \text{the second roll is a } 4, 5, \text{ or } 6;$$
$$C = \text{the two rolls sum to } 7.$$

(a) Are A, B, and C pairwise independent?

(b) Are A, B, and C mutually independent?

Chapter 27

Conditional Probability

More information makes for better predictions. For example, the odds of rolling two 1s with two dice are $\frac{1}{36}$, because rolling 1 with either die has probability $\frac{1}{6}$ and the rolls of the two dice are independent events. But if we know for some reason that the roll of at least one of the dice is an odd number, then the likelihood of two 1s increases somewhat, since it would have been impossible if one of the dice had come up with 2, 4, or 6 (Figure 27.1). The odds of two 1s drop to 0 if we know that one of the dice has landed with an even number!

A *conditional probability* is the probability that an event will occur, on the condition that some other event occurs—such as rolling a total of 2 with two dice, on the condition that the roll of one of the dice is odd. This intuition is formalized in a straightforward way: Suppose A and B are events in a sample space S, and $\Pr(B) > 0$. Then the *conditional probability* of A given B is

$$\Pr(A|B) = \frac{\Pr(A \cap B)}{\Pr(B)}. \tag{27.1}$$

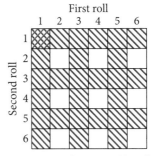

Figure 27.1. The 36 possible rolls of a pair of dice comprise the set $\{1, 2, 3, 4, 5, 6\}^2$. The outcome of two 1s is indicated in blue, and the set of outcomes where at least one roll is an odd number is indicated in red. Knowing that the outcome is in the red region increases the probability that the outcome is two 1s.

In the case $\Pr(B) = 1$, then knowing that B occurred gives no information, and $\Pr(A|B)$ is just $\Pr(A)$. If $\Pr(B) < 1$, then knowing that B happened increases the probability that both A and B happened by exactly the reciprocal of $\Pr(B)$.

In the special case where all outcomes are equally likely, (27.1) can be simplified:

$$\begin{aligned}
\Pr(A|B) &= \frac{\Pr(A \cap B)}{\Pr(B)} \\
&= \frac{|A \cap B|/|S|}{|B|/|S|} \\
&= \frac{|A \cap B|}{|B|}. \tag{27.2}
\end{aligned}$$

In the example of rolling two 1s, S is the set $\{1, \ldots, 6\} \times \{1, \ldots, 6\}$, and $A = \{\langle 1, 1 \rangle\}$. If B represents the set of pairs of rolls for which one of the dice has come up with an odd value, then

$$B = \{1, 3, 5\} \times \{1, 2, 3, 4, 5, 6\} \cup \{1, 2, 3, 4, 5, 6\} \times \{1, 3, 5\}.$$

Using the Inclusion-Exclusion Principle to avoid double-counting, $|B| = 18 + 18 - 9 = 27$. The event $A \cap B$ is the same as A, so (27.2) says that

$$\Pr(A|B) = \frac{1}{27} > \frac{1}{36}.$$

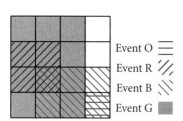

Event O =
Event R ///
Event B \\\
Event G ▨

Figure 27.2. A sample space and four events.

Figure 27.2 illustrates several events and their conditional probabilities. The sample space includes 16 equally likely outcomes (the squares of the grid), and the probability of each of the four events is just its area divided by 16:

$$\Pr(O) = \frac{1}{16}$$

$$\Pr(R) = \frac{4}{16}$$

$$\Pr(B) = \frac{6}{16}$$

$$\Pr(G) = \frac{12}{16}.$$

Now $\Pr(R|G) = \frac{4}{12}$, but $\Pr(G|R) = 1$ since every red square is also green. $\Pr(B|G) = \frac{4}{12}$, but $\Pr(G|B) = \frac{4}{6}$. And $\Pr(R|O) = \Pr(O|R) = 0$ since red and orange don't overlap at all.

Let's try a trickier example.

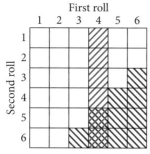

First roll
1 2 3 4 5 6

Second roll
1
2
3
4
5
6

Figure 27.3. The sample space for a pair of die rolls. The squares in the fourth column represent rolling a 4 as the first roll, and squares in the bottom right triangle represent outcomes summing to at least 9.

Example 27.3. *If a die is rolled twice, what is the probability that the sum of the rolls is at least 9? If the first roll is a 4, does the probability increase, decrease, or stay the same?*

Solution to example. Let $S = \{1, \ldots, 6\}^2$ be the sample space, let N be the event that the rolls total at least 9, and let F be the event that the first roll is a 4 (Figure 27.3). Since all outcomes in S are equally likely, we can use (27.2) and just calculate the size of the various event sets.

First, let's calculate $|N|$. There are $6 \cdot 6 = 36$ total outcomes, and the elements of N are:

$$N = \{\langle 3, 6 \rangle, \langle 4, 5 \rangle, \langle 4, 6 \rangle, \langle 5, 4 \rangle, \langle 5, 5 \rangle, \langle 5, 6 \rangle, \langle 6, 3 \rangle, \langle 6, 4 \rangle, \langle 6, 5 \rangle, \langle 6, 6 \rangle\},$$

so $|N| = 10$, and $\Pr(N) = \frac{10}{36} = \frac{5}{18}$.

The set $N \cap F$ contains just the two elements $\langle 4, 5 \rangle$ and $\langle 4, 6 \rangle$, since $4 + k < 9$ if $k < 5$. Finally $|F| = 6$, since the first roll's value is 4 while the second roll can be any of the 6 possible values. So

$$\Pr(N|F) = \frac{|N \cap F|}{|F|}$$
$$= \frac{2}{6} = \frac{1}{3}$$
$$> \frac{5}{18} = \Pr(N).$$

So rolling a 4 as the first roll slightly improves the chances of totaling at least 9.   ■

Note that, for two events A and B, $\Pr(A|B)$ and $\Pr(B|A)$ are not usually the same: they share a numerator, $\Pr(A \cap B)$, but have different denominators. In the extreme case, one of these conditional probabilities can be 1 while the other is arbitrarily small. For example, the probability that two people have the same birthday, given that they were both born on September 16, is 1. The probability that those two people were both born on September 16, given that they have the same birthday, is $\frac{1}{365}$ (on the assumption that all birthdays are equally likely and that February 29 is not a possibility).

This is a common source of confusion in real-world applications of probability. See if you can spot the mistake in this next example, an instance of what is called the *prosecutor's fallacy*:

Example 27.4. *A driver is accused of speeding. The prosecutor notes that the defendant had a radar detector (a tool that is used to detect the police radar that identifies speeders). Citing data that 80% of speeders have radar detectors, the prosecutor argues that the driver is therefore 80% likely to be guilty.*
What is the flaw in this logic?

Solution to example. The prosecutor's figure of 80% represents $\Pr(\text{Det}|\text{Spd})$, the probability of owning the radar detector given that the defendant is a speeder. But the relevant value is instead the probability that the defendant is a speeder given that he or she owns the radar detector: $\Pr(\text{Spd}|\text{Det})$. The given information is insufficient to calculate this value, because we know nothing about the relative sizes of Spd and Det; that is, the relative values of |Spd| and |Det|. It is possible, for example, that 80% of *all* drivers own radar detectors, but almost nobody speeds, and those who do speed own radar detectors in the same proportion as those who don't speed. In that case, using possession of a radar detector as evidence of speeding would result in the prosecution of vast numbers of nonspeeders.   ■

In Example 27.3, conditional probability was used to analyze a restricted subspace of the sample space. Conditional probability is also useful for

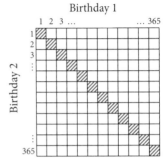

Figure 27.4. Each box represents a pair of birthdays: the horizontal position represents one birthday, and the vertical position represents the other. The boxes along the diagonal represent outcomes where the pair share a birthday; only one of those boxes represents the outcome that both have the birthday September 16.

problems that concern the entire sample space, but are best analyzed by splitting it into several cases. On page 144 we defined a *partition* of a set as a collection of nonempty subsets that are mutually disjoint and such that their union is the entire set. The *law of total probability* is a technique for computing a probability as a weighted average of the event's conditional probabilities over each block of a partition:

Theorem 27.5. Total Probability. *Let A be an event in a sample space S, and let events $S_1, \ldots, S_n$ form a partition of S. Then*

$$\Pr(A) = \sum_{i=1}^{n} \Pr(A|S_i) \cdot \Pr(S_i). \tag{27.6}$$

Proof. First, let's rewrite A:

$$A = A \cap S \qquad \text{(since } A \text{ is a subset of } S\text{)}$$

$$= A \cap \left(\bigcup_{i=1}^{n} S_i \right)$$

$$= \bigcup_{i=1}^{n} (A \cap S_i) \qquad \text{(distributing intersection over union).}$$

Therefore

$$\Pr(A) = \Pr \left(\bigcup_{i=1}^{n} (A \cap S_i) \right)$$

$$= \sum_{i=1}^{n} \Pr(A \cap S_i) \qquad \text{(since the } S_i \text{ are disjoint)}$$

$$= \sum_{i=1}^{n} \Pr(A|S_i) \cdot \Pr(S_i) \qquad \text{(by the definition of conditional probability).} \qquad \blacksquare$$

The overall probability of A is essentially the weighted average of the probabilities $\Pr(A|S_i)$, weighted according to the probabilities of the S_i.

The following example applies the law of total probability to a simple partition: an event and its complement.

Example 27.7. *There are two jars of marbles. The first jar contains 40 red marbles and 60 blue marbles, and the second contains 15 red and 35 blue. One marble is drawn from each jar. What is the probability that the two marbles are the same color? (See Figure 27.5.)*

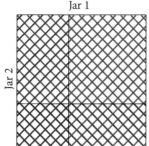

Jar 1

Jar 2

Figure 27.5. A partition of the sample space of Example 27.7. The horizontal axis represents Jar 1, the vertical axis Jar 2. Lines oriented in this direction ╱ represent the color of the marble drawn from Jar 1, so the left portion of the figure has these ╱ lines in red and the right portion has these ╱ lines in blue. Lines oriented in this direction ╲ represent the color of the marble drawn from Jar 2, so the bottom portion of the figure has those ╲ lines in red and the top portion has those ╲ lines in blue. The event that the two marbles are the same color is the union of the two rectangles (the bottom left and top right) in which both sets of diagonal lines are the same color. The combined area of those two rectangles, as a percentage of the area of the square, is the probability that the two marbles drawn are the same color.

Solution to example. It is easiest to split the problem into two cases: either the marble from the first jar is red (event R, with probability $\frac{40}{100} = \frac{2}{5}$) or it is not (event $\overline{R}$, with probability $\frac{60}{100} = \frac{3}{5}$). (See Figure 27.6.) Let M be the event that the two marbles match. Then $\Pr(M|R)$ is the probability that M occurs when the marble from the first jar is red—so the marble from the second jar is red, which has probability $\frac{15}{50} = \frac{3}{10}$. And $\Pr(M|\overline{R})$ is the probability that M occurs when the marble from the first jar is blue—so the marble from the second jar is blue, which has probability $\frac{35}{50} = \frac{7}{10}$. Using Theorem 27.5,

$$
\begin{aligned}
\Pr(M) &= \Pr(M|R) \cdot \Pr(R) + \Pr(M|\overline{R}) \cdot \Pr(\overline{R}) \\
&= \frac{3}{10} \cdot \frac{2}{5} + \frac{7}{10} \cdot \frac{3}{5} \\
&= 0.54.
\end{aligned}
$$
∎

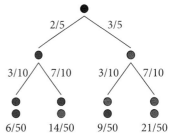

Figure 27.6. This probability tree is a visualization of the calculation for Example 27.7. The tree has branches representing each possibility at each step of the procedure. The probability of a "leaf" (in the bottom row) is the product of the probabilities of each branch along the way, and the probability of a given event is the sum of the probabilities for all leaves that belong to that event.

A famous application of the law of total probability is to the question of strategy in the game show *Let's Make A Deal*. The host, Monty Hall, offers contestants a choice between three doors, where one conceals a car and each of the other two conceals a goat. After the contestant chooses, Monty, who knows where the car is, opens one of the other two doors—one that he knows is hiding a goat. He then offers the contestant the option to switch from the chosen door to the other unopened door. The contestant gets to keep whatever is behind the chosen door.

Example 27.8. *Should the contestant stick to the original choice or switch? (Assume the contestant prefers to win the car.)*

Solution to example. The sample space can be partitioned into three events: the car is behind door 1 (event C_1), or it is behind door 2 (event C_2), or it is behind door 3 (event C_3), each with equal probability $\frac{1}{3}$. Without loss of generality (since the doors can be labeled as we wish), let's say the contestant initially chose door 1. (See Figure 27.7.)

If the contestant keeps the original choice of door 1, the chance of winning the car is just $\Pr(C_1) = \frac{1}{3}$.

If the contestant chooses to switch, we can compute the probability of winning the car (event W) by the law of total probability, using the partition of the sample space into the events C_1, C_2, and C_3:

$$
\Pr(W) = \Pr(W|C_1)\Pr(C_1) + \Pr(W|C_2)\Pr(C_2) + \Pr(W|C_3)\Pr(C_3).
$$

In the event C_1, switching guarantees a loss: the contestant started off with the car, so whether Monty opens door 2 or door 3, he switches to a door that conceals a goat. So $\Pr(W|C_1) = 0$.

But in the event C_2, switching guarantees a win: the contestant started with a goat behind door 1, so Monty had only one choice, which is to open door 3, leaving the contestant to switch to the winning door 2. So $\Pr(W|C_2) = 1$.

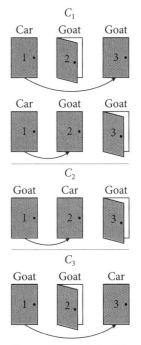

Figure 27.7. Analysis of the switching strategy: In case C_1, no matter which door Monty opens, the contestant always loses. In cases C_2 and C_3, Monty has only one choice of door to open, and the contestant always wins.

Similar logic holds for event C_3: after the contestant chooses door 1, Monty is forced to open door 2, and the contestant switches to the winning door 3. So $\Pr(W|C_3) = 1$.

Substituting these values into (27.6):

$$\Pr(W) = 0 \cdot \frac{1}{3} + 1 \cdot \frac{1}{3} + 1 \cdot \frac{1}{3} = \frac{2}{3}.$$

By switching, the contestant wins with probability $\frac{2}{3}$. A contestant who sticks to the original choice wins with probability only $\frac{1}{3}$. This counterintuitive and much debated result is known as the *Monty Hall paradox*.[1]

The intuitive reason that switching is the right strategy becomes clearer if we imagine a more extreme version of the game, where there are a hundred doors rather than three (and again only one door has a car behind it). The contestant makes an initial choice, and Monty then opens a door, reveals a goat, and invites the contestant to switch. Already, switching would provide a slight advantage, though the advantage grows with each additional door that Monty opens. Suppose the contestant does not switch yet, and Monty opens another door, and another, and another, and each time the contestant declines to switch. Finally, all but two doors have been opened, the prize has not been revealed, and Monty pauses one last time to allow the contestant the opportunity to switch. Now 98 doors have been opened, and the only unopened doors are the one the contestant chose originally and another door that Monty seems to have been avoiding! A contestant would be mad not to switch. (Problem 27.9 is about the odds of winning by switching in this version of the game.) ∎

Another surprising phenomenon in total probability is *Simpson's paradox*: when comparing the total probabilities of two events, one may be more likely overall even when the other is more likely within each event of the partition. Consider the following example:

Example 27.9. *Alexis and Bogdan are students. Each is required to complete a total of 100 problems, but the problems relate to two different topics, and they can choose how many problems to do for each topic. Figure 27.8 shows how many problems they try for each topic, and how many they get right.*

Who is the better student?

Solution to example. The answer is different depending on whether the measure is overall performance, or performance on each individual topic.

Both students completed 100 problems between the two topics. Alexis got correct answers to 63% of them, while Bogdan was right only 47% of the time, so it looks like Alexis is the winner. But we can also compare the students' success rates for each topic (Figure 27.9). Bogdan does better than Alexis in both, even though Alexis has the higher score overall. How is this possible?

[1] Indeed, Monty Hall himself seems not to have understood the paradox. In a letter dated September 10, 1990, responding to an inquiry about using this paradox as an exercise in another Harry Lewis textbook, Mr. Hall wrote, "Now, I am not well versed in algorithms; but as I see it, it wouldn't make any difference after the player has selected Door A, and having been shown Door C—why should he then attempt to switch to Door B? ... The chances of the major prize being behind Door A have not changed. He still has one of the two remaining doors. What makes Door B such an attraction?"

	Alexis	Bogdan
Topic 1	3/8	40/90
Topic 2	60/92	7/10
Total	63%	47%

Figure 27.8. Alexis and Bogdan each completed 100 problems split between two topics, but Alexis did more on the second topic and Bogdan did more on the first. Overall, Alexis got more correct answers.

	Alexis	Bogdan
Topic 1	38%	44%
Topic 2	65%	70%

Figure 27.9. If scores are computed per topic, it looks like Bogdan is the better student. Both students did better on the second topic, and the score on the second topic comprised most of Alexis's total score but little of Bogdan's total score.

The answer is that the students did different numbers of problems on each topic. In the cases where they did only a few problems on a topic, performing very well (as Bogdan does for Topic 2) or very badly (as Alexis does for Topic 1) has a dramatic effect on the score for that topic, but little effect on the overall score.

Perhaps Alexis's poor performance on Topic 1 and Bogdan's good performance on Topic 2 are not reliable indicators of their abilities, since those represent so few problems. In that case, the fairer measure is to judge the students on their performance over all 100 problems.

On the other hand, perhaps Topic 1 was difficult while Topic 2 was easy. That would have hurt Bogdan's overall score, since he did more problems on Topic 1, while helping Alexis, since she gained most of her points in Topic 2. In that case, it would make sense to compare each topic separately, so that Alexis would not be rewarded for attempting more easy problems and Bogdan would not be penalized for attempting more hard problems. ∎

To drive home the point that a table like Figure 27.9 should not be taken at face value, we'll use exactly the same data but change the description. Now the question is which of two diets promotes longevity. The populations of two lands (Oz and Xanadu) are studied for survival to age 75, and the population of each land is broken down into those on Diet 1 and those on Diet 2.

If the only result published is Figure 27.10, it looks like Diet 2 is better for the residents of both Oz and Xanadu, though the Xanadu folks have significantly greater longevity on either diet. But the underlying data tell a different story (Figure 27.11). Now Diet 1 seems superior, though it's hard to tell. Almost no one in Oz is on Diet 1 and almost no one in Xanadu is on Diet 2. Perhaps Diet 1 seems superior only because most of its followers are the already long-lived residents of Xanadu; or perhaps Xanadu residents live longer because they mostly follow the superior Diet 1. Before attributing anything to diet, it would be important to figure out if there is some reason other than diet for the shorter lifespans of people from Oz.

❊

The concept of independent events (page 302) is closely related to conditional probability. *A* and *B* were said to be independent just in case

$$\Pr(A \cap B) = \Pr(A) \cdot \Pr(B).$$

Independence can also be defined in terms of conditional probability:

Theorem 27.10. *For events A and B that both have nonzero probability, A and B are independent if and only if*

$$\Pr(A|B) = \Pr(A). \tag{27.11}$$

	Diet 1	Diet 2
Oz	38%	44%
Xanadu	65%	70%

Figure 27.10. Survival to age 75 by residents of two lands, broken down by their diets.

	Diet 1	Diet 2
Oz	3/8	40/90
Xanadu	60/92	7/10
Total	63%	47%

Figure 27.11. No conclusion about diet can be drawn from the underlying data, since residents of Oz and Xanadu may differ in some other important regard.

(If either A or B has probability 0, *then A and B are independent by the original definition.)*

This makes intuitive sense: A and B are independent if knowing that B occurred doesn't change the likelihood of A occurring (Figure 27.12).

Proof. By the definition of conditional probability,

$$\Pr(A|B) = \frac{\Pr(A \cap B)}{\Pr(B)},$$

provided that $\Pr(B) > 0$. But if A and B are independent, we can replace $\Pr(A \cap B)$ with $\Pr(A) \cdot \Pr(B)$:

$$\Pr(A|B) = \frac{\Pr(A) \cdot \Pr(B)}{\Pr(B)} = \Pr(A). \qquad \blacksquare$$

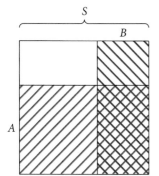

Figure 27.12. *A and B are independent events of a sample space S. Metaphorically, A depends on the vertical dimension and B on the horizontal dimension; these events are independent since the dimensions are uncorrelated. Mathematically, A takes up the same proportion of S as A ∩ B takes of B, and B takes up the same proportion of S as B ∩ A takes of A. So knowing whether an outcome falls within one event gives no additional information about the probability that it falls within the other.*

Although the formula (27.11) is asymmetric, independence is a symmetric relationship. Reversing the roles of A and B in (27.11), $\Pr(B|A) = \Pr(B)$ is a third way to state the same condition. When determining whether two events are independent, it is a matter of convenience whether to use either version of (27.11), or to use (26.12) from page 302 instead. The following example demonstrates all three approaches.

Example 27.12. *Consider a company with* 100 *employees:* 60 *women and* 40 *men. There are* 15 *managers,* 10 *of whom are women. An employee is chosen at random. Is the event that the employee is a woman independent of the event that the employee is a manager?*

Solution to example. Using Theorem 27.10, we compare:

$$\Pr(\text{Wmn}|\text{Mgr}) = \frac{\Pr(\text{Wmn} \cap \text{Mgr})}{\Pr(\text{Mgr})}$$

$$= \frac{10/100}{15/100}$$

$$= \frac{2}{3}$$

$$\neq \frac{60}{100} = \Pr(\text{Wmn}).$$

These are not equal, so the events are not independent. Alternatively, we can come to the same conclusion by comparing $\Pr(\text{Mgr}|\text{Wmn})$ to $\Pr(\text{Mgr})$.

$$\Pr(\text{Mgr}|\text{Wmn}) = \frac{\Pr(\text{Mgr} \cap \text{Wmn})}{\Pr(\text{Wmn})}$$

$$= \frac{10/100}{60/100}$$

$$= \frac{1}{6}$$

$$\neq \frac{15}{100} = \Pr(\mathrm{MGR}).$$

As a final check, (26.12) gives the same result, that they are not independent:

$$\Pr(\mathrm{WMN} \cap \mathrm{MGR}) = \frac{10}{100}$$

$$\neq \frac{60}{100} \cdot \frac{15}{100} = \Pr(\mathrm{WMN}) \cdot \Pr(\mathrm{MGR}). \qquad \blacksquare$$

Whether or not two events are independent can only be determined by carrying out the calculation—there is no way to tell without checking the numbers. For example, consider an altered version of this scenario:

Example 27.13. *Suppose the scenario is as in Example 27.12, except that 9 of the 15 managers are women, instead of 10. Are WMN and MGR independent?*

Solution to example. Now

$$\Pr(\mathrm{WMN}|\mathrm{MGR}) = \frac{\Pr(\mathrm{WMN} \cap \mathrm{MGR})}{\Pr(\mathrm{MGR})}$$

$$= \frac{9/100}{15/100}$$

$$= \frac{3}{5}$$

$$= \frac{60}{100} = \Pr(\mathrm{WMN}),$$

so in this case the events are independent. $\qquad \blacksquare$

Chapter Summary

- A *conditional probability* is the probability of an event conditioned on the fact that some other event occurs. The conditional probability of A given B, assuming $\Pr(B) > 0$, is

$$\Pr(A|B) = \frac{\Pr(A \cap B)}{\Pr(B)}.$$

- Usually $\Pr(A|B) \neq \Pr(B|A)$. The mistake of confusing $\Pr(A|B)$ with $\Pr(B|A)$ is called the *prosecutor's fallacy*.

- The *law of total probability* states that an event's probability can be computed by *partitioning* the sample space into disjoint blocks such that their union is the entire sample space, and computing the weighted average of the event's conditional probabilities on each block: for blocks S_i forming a partition of the sample space,

$$\Pr(A) = \sum_{i=1}^{n} \Pr(A|S_i) \cdot \Pr(S_i).$$

- For two events A and B, it is possible for A to have a higher total probability even if B has a higher conditional probability over every block in a given partition. This surprising phenomenon is called *Simpson's paradox*.

- Independence can be redefined in terms of conditional probability: if events A and B both have nonzero probability, they are independent just in case $\Pr(A|B) = \Pr(A)$, or equivalently, just in case $\Pr(B|A) = \Pr(B)$.

Problems

27.1. In Figure 27.2, what is $\Pr(O|G)$? $\Pr(G|O)$? $\Pr(B|R)$? $\Pr(R|B)$?

27.2. Suppose that a family has two children. Assume that each child is equally likely to be a boy or a girl, and that their genders are independent (no identical twins).
  (a) What is the probability that both children are girls?
  (b) If the eldest is a girl, what is the probability that both children are girls?
  (c) If at least one child is a girl, what is the probability that both are girls?

27.3. When is $\Pr(A|B) = \Pr(B|A)$?

27.4. Suppose you have 50 black marbles, 50 white marbles, and 2 jars. You can distribute the marbles between the 2 jars in any way, as long as neither jar is empty. Then you will pick a jar at random, and then pick a random marble from that jar. How should you distribute the marbles to maximize the probability of choosing a black marble?

27.5. A man is accused of robbing a bank. Eyewitnesses testify that the robber was 6 feet tall and had red hair and green eyes; the suspect matches this description. Suppose that only 100 of the 100000 people in the town are men who are 6 feet tall with red hair and green eyes, and assume that one of them robbed the bank.
  (a) What is the probability that the suspect is innocent, given that he matches the description?
  (b) What is the probability that the suspect would match the description, given that he is innocent?

(c) Suppose the police reviewed images of 1000 potential suspects when investigating the case, and every one of the 100000 residents was equally likely to appear in this review. What is the probability that at least one of the 1000 would match the description?

27.6. In the last chapter (Problem 26.4), we considered a professor who wants to know whether her students completed a reading assignment. Instead of asking directly, the professor asks the students to each flip a coin, and to raise their hands if either their coin lands heads up or they did not do the assignment. Suppose, as in Problem 26.4, that $\frac{3}{4}$ of the class raise their hands. Given that a student raised his or her hand, what is the probability that he or she did not do the reading?

27.7. Consider again the spam classifier from Problem 26.10. Now suppose that 80% of all emails are sent from email addresses that are unknown to the receiver (event U), that 10% of emails include a request for money (event M), and that 90% of emails that include a request for money are sent from an unknown email address. Are U and M independent?

27.8. A professor suspects a student of cheating on an exam, based on the fact that he achieved a perfect score. Which of the following would be relevant evidence? There are multiple correct answers.

- The probability of getting a perfect score, given that he cheated.
- The probability of getting a perfect score, given that he did not cheat.
- The probability of not getting a perfect score, given that he cheated.
- The probability of not getting a perfect score, given that he did not cheat.
- The probability that he cheated, given that he got a perfect score.
- The probability that he cheated, given that he did not get a perfect score.
- The probability that he did not cheat, given that he got a perfect score.
- The probability that he did not cheat, given that he did not get a perfect score.

27.9. In the version of the Monty Hall problem (Example 27.8) with 100 doors, what are the odds of winning:
(a) If the player does not switch?
(b) If the player switches after Monty opens 1 door?
(c) If the player switches after Monty opens 98 doors?

Chapter 28

Bayes' Theorem

The prosecutor in the speeding trial (page 313) misused the statistic $\Pr(\text{DET}|\text{SPD})$ as though it was $\Pr(\text{SPD}|\text{DET})$. The prosecutor claimed that 80% of those who speed owned radar detectors, and incorrectly inferred that the driver was 80% likely to be guilty. But this 80% was actually the probability that the suspect would own a radar detector given that he or she was guilty, not the probability of guilt given that he or she owned the radar detector.

These two conditional probabilities are not the same, but they are related—one can be calculated from the other with the aid of two additional pieces of information: $\Pr(\text{SPD})$, the percent of all drivers who speed, and $\Pr(\text{DET})$, the percent of all drivers who own radar detectors. The more drivers who speed, the more likely it is that the suspect is one of them. But the more drivers who own radar detectors, the less suspicious it is to have one. From $\Pr(\text{SPD})$, $\Pr(\text{DET})$, and $\Pr(\text{DET}|\text{SPD})$, it is possible to calculate $\Pr(\text{SPD}|\text{DET})$ by means of *Bayes' theorem*:[1]

Theorem 28.1. Bayes' Theorem. *Provided that* $\Pr(A)$ *and* $\Pr(B)$ *are both nonzero,*

$$\Pr(A|B) = \frac{\Pr(B|A)\,\Pr(A)}{\Pr(B)}.$$

Proof. By the definition of conditional probability (page 311),

$$\Pr(A \cap B) = \Pr(A|B)\,\Pr(B),$$

and symmetrically,

$$\Pr(B \cap A) = \Pr(B|A)\,\Pr(A).$$

Combining these equations,

$$\Pr(A|B)\,\Pr(B) = \Pr(A \cap B) = \Pr(B \cap A) = \Pr(B|A)\,\Pr(A),$$

from which the result follows by dividing by $\Pr(B)$. ∎

[1] Thomas Bayes (1701–61) was a British mathematician and minister. In notes that were edited and published only after his death, he developed a restricted version of the theorem that now bears his name.

So $\Pr(A|B)$ and $\Pr(B|A)$ differ by a factor of exactly $\frac{\Pr(A)}{\Pr(B)}$.

Let's plug in some numbers to correct the argument in the speeding trial, assuming various probabilities of speeding and radar-detector ownership in the population:

Example 28.2. *If 80% of drivers who speed own radar detectors, calculate the probability that a driver who owns a radar detector speeds, if*

- *5% of all drivers speed, and 40% of all drivers own radar detectors;*
- *20% of all drivers speed, and 20% of all drivers own radar detectors;*
- *20% of all drivers speed, and 16% of all drivers own radar detectors.*

Solution to example.

- In the first scenario:

$$\Pr(\text{Spd}|\text{Det}) = \frac{\Pr(\text{Det}|\text{Spd})\,\Pr(\text{Spd})}{\Pr(\text{Det})}$$

$$= \frac{0.8 \cdot 0.05}{0.4} = 0.1 = 10\%.$$

5% speed, 40% own

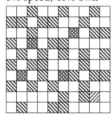

Not many drivers speed, which makes it less likely that the suspect is a speeder, and many drivers—including many who don't speed—own radar detectors, so the fact that the driver owns one is not very compelling evidence.

- In the second scenario:

$$\Pr(\text{Spd}|\text{Det}) = \frac{\Pr(\text{Det}|\text{Spd})\,\Pr(\text{Spd})}{\Pr(\text{Det})}$$

$$= \frac{0.8 \cdot 0.2}{0.2} = 0.8 = 80\%.$$

20% speed, 20% own

In this case, the prosecutor's figure actually happened to be right. There are equal numbers of drivers who speed and drivers who own radar detectors, so their intersection comprises an equal proportion of both populations.

- In the third scenario:

$$\Pr(\text{Spd}|\text{Det}) = \frac{\Pr(\text{Det}|\text{Spd})\,\Pr(\text{Spd})}{\Pr(\text{Det})}$$

$$= \frac{0.8 \cdot 0.2}{0.16} = 1.0 = 100\%.$$

20% speed, 16% own

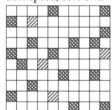

Figure 28.1. Drivers who speed are indicated in blue; drivers who own radar detectors are indicated in red. In each scenario, 80% of the speeders own radar detectors; but the percentage of radar-detector owners who speed varies.

The numbers here work out so that every driver who speeds owns a radar detector, and every driver who doesn't speed doesn't own one, so the fact that the suspect owns one is definite proof of guilt. (At least in the world of discrete mathematics. In reality, probabilities are never numerically precise.)

So the original figure of 80% is quite uninformative by itself—keeping $\Pr(\textsc{Det}|\textsc{Spd})$ constant, very different probabilities of guilt arise from different values of the other parameters. ■

Mathematically, Bayes' theorem is a formula for converting between $\Pr(A|B)$ and $\Pr(B|A)$. A more utilitarian way to think about Bayes' theorem is that it provides a mechanism for updating the probability of a *hypothesis* in light of new *evidence*.

Example 28.3. *A given candidate in an election has a* 40% *chance of winning, according to pre-election polls. On election night, early reports show that this candidate has won a swing district, considered a "must win" for her—without that district she would certainly lose the election. Originally it had been estimated that she had a* 50% *of winning the swing district. What is the updated probability that she will win the election?*

Solution to example. Let $\textsc{El}$ be the event that the candidate wins the election (the hypothesis), and $\textsc{Sw}$ the event that she wins the swing district (the evidence). The candidate can win the election only if she wins the swing district, so $\Pr(\textsc{Sw}|\textsc{El}) = 1$. The objective is to calculate the probability $\Pr(\textsc{El}|\textsc{Sw})$ that she will win the election, given the new information that she has won the swing district.

$$\Pr(\textsc{El}|\textsc{Sw}) = \frac{\Pr(\textsc{Sw}|\textsc{El}) \, \Pr(\textsc{El})}{\Pr(\textsc{Sw})}$$

$$= \frac{1 \cdot 0.4}{0.5} = 0.8.$$

The candidate is now expected to win the election, with a chance of 80%. ■

Because Bayes' theorem can be used to "update" $\Pr(A)$ to $\Pr(A|B)$, based on the new information that B occurred, it is sometimes referred to as *Bayesian updating*.

✳

Bayes' theorem provides the basis for certain approaches to *machine learning*, in which algorithms modify their behavior based on experience rather than relying purely on pre-programmed instructions. Machine learning is commonly used to solve classification problems, where inputs must be sorted into categories. Optical character recognition, image recognition, and spam filtering are typical examples.

Example 28.4. *How might email be classified automatically as spam or not spam, based on some examples that have been provided of emails that are and are not spam?*

Solution to example. We'll describe a very simple set of criteria that might be used to automate such decisions. The first step is to identify features that may be associated with spam, say

- whether the sender's email is unknown to the recipient (event U);
- whether the email includes a request for money (event M);
- whether the email uses special characters like @ and $ within words, in place of letters (event C).

The next step is to create a training data set, where some emails are manually labeled as spam or not spam, and categorized according to each of the features. For example, suppose a training set includes 100 emails, of which 60 are spam and 40 are not spam, with the following characteristics:

	Unknown	Money	Characters
Spam	50/60	20/60	35/60
Not spam	10/40	5/40	15/40

Now let's consider the problem of predicting the correct label for a new email, based on this data. Suppose the new email is from an unknown sender and uses special characters, but does not contain a request for money. We want to know whether the email is spam (event S_P), conditioned on the event $U \cap \overline{M} \cap C$. By Bayes' theorem,

$$\Pr(S_P | (U \cap \overline{M} \cap C)) = \frac{\Pr((U \cap \overline{M} \cap C) | S_P) \Pr(S_P)}{\Pr(U \cap \overline{M} \cap C)}$$

$$= \frac{\Pr(U \cap \overline{M} \cap C \cap S_P)}{\Pr(U \cap \overline{M} \cap C)}. \tag{28.5}$$

Let's rewrite $\Pr(U \cap \overline{M} \cap C \cap S_P)$, repeatedly using the definition of conditional probability.

$$\Pr(U \cap \overline{M} \cap C \cap S_P) = \Pr(U | (\overline{M} \cap C \cap S_P)) \Pr(\overline{M} \cap C \cap S_P)$$

$$= \Pr(U | (\overline{M} \cap C \cap S_P)) \Pr(\overline{M} | (C \cap S_P)) \Pr(C \cap S_P)$$

$$= \Pr(U | (\overline{M} \cap C \cap S_P)) \Pr(\overline{M} | (C \cap S_P)) \Pr(C | S_P) \Pr(S_P).$$

We now make an important assumption, called the *conditional independence* of the features. The conditional independence assumption is that, conditioned on S_P, the features U, $\overline{M}$, and C (or equivalently, U, M, and C)

are independent of each other. The general definition is given at the end of this example, but in the case at hand, conditional independence means specifically that

$$\Pr(U|(\overline{M} \cap C \cap \text{Sp})) = \Pr(U|\text{Sp}) \text{ and}$$

$$\Pr(\overline{M}|(C \cap \text{Sp})) = \Pr(\overline{M}|\text{Sp}).$$

Conditional independence of the features expresses an assumption that the various features are measures of different things. Though it is unlikely to be precisely true in practice, it has proved remarkably useful in a variety of real-world situations. On the assumption of conditional independence, (28.5) can be rewritten as

$$
\begin{aligned}
\Pr(\text{Sp}|(U \cap \overline{M} \cap C)) &= \frac{\Pr(U \cap \overline{M} \cap C \cap \text{Sp})}{\Pr(U \cap \overline{M} \cap C)} \\
&\approx \frac{\Pr(U|\text{Sp})\Pr(\overline{M}|\text{Sp})\Pr(C|\text{Sp})\Pr(\text{Sp})}{\Pr(U \cap \overline{M} \cap C)} \\
&= \frac{(50/60) \cdot (40/60) \cdot (35/60) \cdot (60/100)}{\Pr(U \cap \overline{M} \cap C)} \\
&= \frac{7/36}{\Pr(U \cap \overline{M} \cap C)}.
\end{aligned}
\tag{28.6}
$$

From the given information, we can't calculate the value of the denominator directly, since U, M, and C are assumed to be *conditionally* independent, but not unconditionally independent. Instead, we can calculate the probability of the complement of the above event—that is, $\Pr(\overline{\text{Sp}}|(U \cap \overline{M} \cap C))$—and then use the fact that these two probabilities sum to 1. Using the same assumption of conditional independence,

$$
\begin{aligned}
\Pr(\overline{\text{Sp}}|(U \cap \overline{M} \cap C)) &\approx \frac{\Pr(U|\overline{\text{Sp}}) \cdot \Pr(\overline{M}|\overline{\text{Sp}}) \cdot \Pr(C|\overline{\text{Sp}}) \cdot \Pr(\overline{\text{Sp}})}{\Pr(U \cap \overline{M} \cap C)} \\
&= \frac{(10/40) \cdot (35/40) \cdot (15/40) \cdot (40/100)}{\Pr(U \cap \overline{M} \cap C)} \\
&= \frac{21/640}{\Pr(U \cap \overline{M} \cap C)}.
\end{aligned}
\tag{28.7}
$$

The values (28.6) and (28.7) sum to 1:

$$
\begin{aligned}
\Pr(\text{Sp}|(U \cap \overline{M} \cap C)) + \Pr(\overline{\text{Sp}}|(U \cap \overline{M} \cap C)) &= \frac{7/36}{\Pr(U \cap \overline{M} \cap C)} \\
&\quad + \frac{21/640}{\Pr(U \cap \overline{M} \cap C)} \\
&= 1,
\end{aligned}
$$

so we can solve for the denominator:

$$\Pr(U \cap \overline{M} \cap C) = \frac{7}{36} + \frac{21}{640} = \frac{1309}{5760},$$

and then

$$\Pr(\text{Sp}|(U \cap \overline{M} \cap C)) = \frac{7/36}{1309/5760} \approx 0.86,$$

$$\Pr(\overline{\text{Sp}}|(U \cap \overline{M} \cap C)) = \frac{21/640}{1309/5760} \approx 0.14.$$

It is therefore likely that this message is spam. ■

Applications such as spam filtering call for conservative estimates, and demand a high threshold to avoid false positives—it is much worse to classify nonspam as spam than to classify spam as legitimate email. So it would make sense to classify a message as spam only if the estimated probability is, say, at least 80%. In other applications—for example, optical character recognition, where there are many possible labels, and errors in any direction are equally acceptable—we can simply assign the label that has the highest likelihood.

Returning to the definition of conditional independence, we can simply update either of our previous definitions of independence to use conditional probabilities. In either case, we are essentially asking about the events' independence within a restricted sample space. As before, these definitions apply only when the event that is conditioned on has nonzero probability.

By analogy to the definition that A and B are independent if

$$\Pr(A \cap B) = \Pr(A) \Pr(B)$$

from page 302, A and B are conditionally independent given E if

$$\Pr(A \cap B|E) = \Pr(A|E) \Pr(B|E),$$

provided that $\Pr(E) > 0$.

Or, by analogy to the definition that A and B are independent if

$$\Pr(A|B) = \Pr(A)$$

from page 318, A and B are conditionally independent given E if

$$\Pr(A|B \cap E) = \Pr(A|E),$$

provided that $\Pr(B \cap E) > 0$.

Recalling mutual independence (defined on page 305), we can generalize this latter formula to define *conditional mutual independence*. A set of events

X is conditionally mutually independent relative to an event E if and only if, for every event $A \in X$ and every $Y \subseteq X - \{A\}$,

$$\Pr\left(A \,\middle|\, \left(\bigcap_{B \in Y} B\right) \cap E\right) = \Pr(A|E),$$

provided that $\Pr((\bigcap_{B \in Y} B) \cap E) > 0$. That is, the probability of A given E doesn't depend on whether the sample space is restricted to a subset specified by any number of the other events in X, provided that the restricted sample space has nonzero probability. In Example 28.4, X was the set $\{U, \overline{M}, C\}$ and E was the event Sp, and the assumption was applied first with $A = U$, $Y = \{\overline{M}, C\}$ and then with $A = \overline{M}$, $Y = \{C\}$.

A *naïve Bayes classifier* is one, like that of Example 28.4, that assumes all features are conditionally mutually independent. Naïve Bayes classifiers are commonly used because they are relatively easy to implement, and give satisfactory results in a variety of circumstances.

✳

In the preceding examples, the probability of the evidence was a given. But in many applications of Bayes' theorem, this value needs to be computed—perhaps using the law of total probability (Theorem 27.5).

Example 28.8. *Two bowls contain marbles. The first contains only blue marbles; the second contains 4 blue and 12 red. A bowl is chosen at random, and a marble is chosen at random from that bowl. If the marble selected is blue, what is the probability that it was chosen from the first bowl?*

Solution to example. Let F be the event that the marble came from the first bowl, and B the event that the chosen marble is blue.

By Bayes' theorem,

$$\Pr(F|B) = \frac{\Pr(B|F)\Pr(F)}{\Pr(B)}.$$

The denominator can be expanded into F and $\overline{F}$ subcases, using the law of total probability:

$$\Pr(F|B) = \frac{\Pr(B|F)\Pr(F)}{\Pr(B|F)\Pr(F) + \Pr(B|\overline{F})\Pr(\overline{F})}$$

$$= \frac{1 \cdot \frac{1}{2}}{1 \cdot \frac{1}{2} + \frac{1}{4} \cdot \frac{1}{2}}$$

$$= \frac{4}{5}. \qquad \blacksquare$$

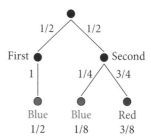

Figure 28.2. This tree diagram shows branching possibilities: to get the probability of a "leaf" (in the bottom row), multiply the probabilities along the path from the root. To find the probability of the first-bowl/ blue-marble branch, given that the marble is blue, divide by the sum of the probabilities over all blue-marble branches.

Armed with this technique, let's revisit the Monty Hall paradox from last chapter (Example 27.8, page 315). This problem is a natural candidate for Bayes' theorem: we start with an initial probability for an event; then we incorporate new information when we learn it, to find an updated probability.

Example 28.9. *To again set the stage: the contestant must choose between three doors. Behind one door is a car, and behind the other two, a goat. The contestant makes an initial choice; the host, Monty Hall, opens one of the other doors to reveal a goat; Monty asks if the contestant wants to switch to the remaining closed door. Should the contestant switch?*

Solution to example. As before, without loss of generality we can assume the contestant started by choosing door 1. If the contestant sticks with door 1, the chance of winning the car is the probability that the car was behind door 1: $\Pr(C_1) = \frac{1}{3}$.

Again without loss of generality, assume Monty opens door 2 (since the analysis is the same if we switch the roles of door 2 and door 3). We want to find the probability of winning if the contestant switches—that is, the probability that the car is behind door 3 (event C_3), knowing that Monty opened door 2 (event M_2). Using Bayes' theorem,

$$\Pr(C_3|M_2) = \frac{\Pr(M_2|C_3)\,\Pr(C_3)}{\Pr(M_2)}.$$

Rewriting the denominator using the law of total probability, conditioning on which door the car lies behind,

$$\Pr(C_3|M_2) = \frac{\Pr(M_2|C_3)\,\Pr(C_3)}{\Pr(M_2|C_1)\,\Pr(C_1) + \Pr(M_2|C_2)\,\Pr(C_2) + \Pr(M_2|C_3)\,\Pr(C_3)}.$$

The contestant picked door 1. If the car was behind door 1, Monty could have chosen either door 2 or 3: $\Pr(M_2|C_1) = \frac{1}{2}$. If it was behind door 2, Monty could not have opened door 2: $\Pr(M_2|C_2) = 0$. And if it was behind door 3, Monty had to open door 2: $\Pr(M_2|C_3) = 1$. Substituting these values,

$$\Pr(C_3|M_2) = \frac{1 \cdot \frac{1}{3}}{\frac{1}{2} \cdot \frac{1}{3} + 0 \cdot \frac{1}{3} + 1 \cdot \frac{1}{3}} = \frac{2}{3}. \qquad \blacksquare$$

Chapter Summary

- The conditional probabilities $\Pr(A|B)$ and $\Pr(B|A)$ are related: if $\Pr(A)$ and $\Pr(B)$ are both known, either conditional probability can be calculated from the other using *Bayes' theorem*:

$$Pr(A|B) = \frac{Pr(B|A)\,Pr(A)}{Pr(B)}.$$

- *Bayesian updating* refers to the use of Bayes' theorem to update the unconditional probability of a *hypothesis* $Pr(A)$ to a conditional probability $Pr(A|B)$, in light of the *evidence* that B occurred.

- Two events A and B are *conditionally independent* relative to a third event E if they are independent given E. Equivalently:

$$Pr(A \cap B|E) = Pr(A|E)\,Pr(B|E) \text{ or}$$
$$Pr(A|B \cap E) = Pr(A|E),$$

  provided that the relevant denominators are nonzero.

- A set of events X is *conditionally mutually independent* relative to an event E if it is mutually independent given E: for every $A \in X$ and $Y \subseteq X - \{A\}$,

$$Pr\left(A \,\middle|\, \left(\bigcap_{B \in Y} B\right) \cap E\right) = Pr(A|E),$$

  provided that the denominators are nonzero.

- A *naïve Bayes classifier* is a simple type of machine-learning algorithm that uses Bayes' theorem with the assumption that all features are conditionally mutually independent.

- If $\{A_1, \ldots, A_n\}$ is a partition of the event A, then Bayes' theorem can be rewritten using the law of total probability:

$$Pr(A|B) = \frac{Pr(B|A)\,Pr(A)}{Pr(B|A_1)\,Pr(A_1) + \ldots + Pr(B|A_n)\,Pr(A_n)}.$$

  This formulation can be helpful when $Pr(B)$ is not explicitly given.

Problems

28.1. Suppose a die is chosen at random from two dice; one is fair, with all values equally likely, while the other is weighted such that a 6 is twice as likely as any other value. The chosen die is rolled and comes up as 6; what is the probability that it was the unfairly weighted die?

28.2. This problem explores how two people with different initial beliefs about the probability of an event can converge to the same updated probability, if given sufficient evidence, through Bayesian updating.

Adam has a collection of 10 coins and Beatriz has a collection of 3 coins. Each collection contains 1 coin that is biased to land on heads with probability $\frac{9}{10}$, while the remaining coins are fair.

(a) Adam and Beatriz each pick a coin at random from their own collections. What is the probability that each person chose the biased coin?

(b) If Adam and Beatriz each flip their chosen coin once, and it lands on heads, what updated probability do they each assign to the event that their coin is biased?

(c) If Adam and Beatriz each flip their chosen coin 10 times, and it lands on heads on all 10 times, what updated probability do they each assign to the event that their coin is biased?

28.3. Suppose a jar contains 100 coins. Some of the coins are real, and have heads on one side and tails on the other; others are fake, and have heads on both sides.

(a) Assume 99 of the coins are real and 1 is fake, with two heads. A coin is chosen at random and flipped twice; both times it lands heads up. What is the chance it is the fake?

(b) Some unknown number of the coins are fakes with two heads. A coin is chosen at random and flipped twice; both times it lands heads up. How many of the 100 coins must be fake to give at least a 50% chance that the chosen coin is a fake?

28.4. Two jars are filled with red and white marbles; one contains $\frac{2}{3}$ red and $\frac{1}{3}$ white, and the other $\frac{2}{3}$ white and $\frac{1}{3}$ red. Ann chooses one of the jars and draws 5 marbles, with replacement, all of which are red. From the same jar, Betty draws 20 marbles, with replacement, and gets 15 red and 5 white. Whose experiment provides the stronger evidence that the chosen jar was $\frac{2}{3}$ red rather than $\frac{2}{3}$ white?

28.5. A drug test has a false positive rate of 2% (2% of people who do not use the drug will falsely test positive), and a false negative rate of 5% (5% of people who do use the drug will falsely test negative). If 1% of the population uses the drug, what is the probability that a given person who tests positive is actually a user? See Figure 28.3.

28.6. Suppose an algorithm forecasts the chance of rain using a naïve Bayes classifier. The training data is as follows, representing the weather in a given location for the same month in the previous year, when it rained 7 days out of 30:

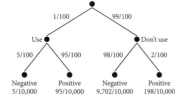

Figure 28.3. A tree diagram showing the numbers of users and non-users that test positive for the drug.

	Cloudy	**Humid**	**Low Air Pressure**
Rain	7/7	5/7	6/7
No rain	16/23	12/23	8/23

What chance of rain does the algorithm predict, if:

(a) It is cloudy and humid, but the air pressure is not low?

(b) It is cloudy and the air pressure is low, but it is not humid?

(c) It is humid and the air pressure is low, but it is not cloudy?

28.7. Alexei Romanov, the only son of Tsar Nicholas II of Russia, suffered from the genetic disease hemophilia. Hemophilia is inherited via the X-chromosome: women have two versions of the allele, one inherited from each parent; while men have one, inherited from the mother. A person with at least one version of the dominant allele, H, does not have the disease, while a person who has only the recessive allele, h, has the disease. That is, women may have alleles HH (no hemophilia), Hh (carrier of hemophilia), or hh (hemophilia); men may have H (no hemophilia) or h (hemophilia).

Neither Tsar Nicholas nor his wife Tsarina Alexandra had hemophilia.

(a) Given that Alexei had hemophilia, what is the probability that his sister Anastasia was a carrier?

(b) During the Bolshevik revolution, the entire Romanov family was assassinated by revolutionaries; however, rumors abounded that some of the women had escaped. Several years later, a woman named Anna Anderson surfaced, claiming to be Anastasia. According to her story, Anna had escaped from Russia, married, and had a son, whom she left in an orphanage.

Suppose it was initially believed to be only 5% likely that Anna was truly Anastasia, and suppose that 1% of men at the time had hemophilia. If Anna's son were found to have hemophilia, how would that change the probability that Anna was Anastasia?

(c) If instead, Anna's son were found not to have hemophilia, what would be the updated probability that Anna was Anastasia?

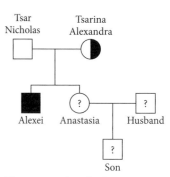

Figure 28.4. An inheritance diagram illustrating the Romanov family members in question. Men are represented as squares, women as circles. A white interior indicates no hemophilia, black indicates hemophilia, and half white/half black indicates carrier of hemophilia. Question marks indicate hemophilia status unknown.

Chapter 29

Random Variables and Expectation

Consider an experiment that depends on a random event and results in a numerical value. An example might be the value of a single roll of a die, or the number of rolls of a die until a result of 6 appears. Over multiple runs of the experiment this number takes on different values, each occurring with a certain probability. A *random variable* is a way of describing such a number: its value is not certain, but something is known about the range of possible values and the likelihood of each.

Formally, a random variable is a function that maps outcomes to numerical values. That is, its domain is the sample space of an experiment, and its codomain is a numerical set, such as $\{0,1\}$, $\mathbb{N}$, or $\mathbb{R}$. A random variable is usually denoted with a capital letter, such as X. The sample space is typically denoted by S and the set of values by T, so $X : S \to T$.

For example, in the experiment of flipping a coin, if s is an outcome in the sample space $S = \{\text{Heads}, \text{Tails}\}$, then we might define

$$X(s) = \begin{cases} 1, & \text{if } s = \text{Heads}, \\ 0, & \text{if } s = \text{Tails}. \end{cases}$$

A random variable, like this one, that takes on only the values 0 and 1 is called a *Bernoulli variable*,[1] and can be described by just one parameter: the probability p with which it has the value 1. Such a random variable is also called an *indicator variable*, since it indicates whether an event happened (value 1, sometimes called a *success*) or did not (value 0, sometimes called a *failure*).

[1] Named after the Swiss mathematician Jacob Bernoulli, 1654–1705.

Random variables may indicate more than just success or failure. Suppose we perform the experiment of flipping a coin 10 times; then we might define $X(s)$ to be the number of heads in outcome s. That is, $X : \{\text{Heads}, \text{Tails}\}^{10} \to \{0, \ldots, 10\}$, where $X(s)$ is the number of heads in the sequence s.

The codomain of a random variable may be infinite.[2] Suppose we run the experiment of flipping a coin until it lands on heads, and let X be the number of flips required. The value of X could be any positive integer—though larger values are less probable than smaller values.

[2] This book discusses only *discrete* random variables, which may have only a countable number of possible values—finite or infinite. Continuous random variables may take on any of an uncountable infinity of values—real numbers, for example.

What exactly is the probability distribution of this random variable? The probability of a fair coin landing heads up is $\frac{1}{2}$, so

$$\Pr(\{s \in S : X(s) = 1\}) = \frac{1}{2}.$$

The probability that any given flip will be the first to land on heads is the probability that all of the previous flips landed on tails, and that the given flip landed on heads; that is,

$$\Pr(\{s \in S : X(s) = n\}) \left(\frac{1}{2}\right)^{n-1} \cdot \frac{1}{2}$$

$$= \frac{1}{2^n}. \tag{29.1}$$

In this case, the probability of heads was equal to the probability of tails; let's generalize to the case of unequal probabilities. Suppose the coin is weighted so that it lands on heads only $\frac{1}{3}$ of the time. Again letting X be the number of flips until getting heads, the probability of $n-1$ tails followed by a heads would be

$$\Pr(\{s \in S : X(s) = n\}) = \left(\frac{2}{3}\right)^{n-1} \cdot \frac{1}{3}. \tag{29.2}$$

Each of these last two coin experiments can be thought of as a sequence of trials, where each trial's outcome is represented by the same Bernoulli variable—that is, each trial has the same probability of success. Such trials are called *Bernoulli trials*. In general, for a sequence of Bernoulli trials, each with probability p of success, the probability that the first success will occur after n trials is $(1 - p)^{n-1} \cdot p$. A random variable with a distribution given by this formula is called a *geometric random variable*, so named because the probabilities of successive values form a *geometric sequence*, where the ratio between any two consecutive terms is constant (analogous to a geometric series, defined in Chapter 24).

We can define a function like (29.1) or (29.2) not just for geometric random variables but for any random variable. The *probability mass function* for random variable X maps each possible value x to the probability that X takes on value x. The probability mass function is denoted $\mathrm{PMF}_X(x)$ to make clear that it is a function of the value x of a random variable X, but is more conveniently abbreviated as $\Pr(X = x)$, "the probability that the value of X is x":

$$\mathrm{PMF}_X(x) = \Pr(X = x)$$

$$= \Pr(\{s \in S : X(s) = x\}). \tag{29.3}$$

Since the values of a random variable are real numbers, those values can be ordered numerically. The *cumulative distribution function* describes the probability that the value of a random variable is below a specified number:

$$\mathrm{CDF}_X(x) = \Pr(\{s \in S : X(s) \le x\}),$$

which is also often abbreviated:

$$\mathrm{CDF}_X(x) = \Pr(X \le x).$$

The cumulative distribution function is so named because is it the accumulation of the distribution described by the probability mass function, for values up to x:

$$\mathrm{CDF}_X(x) = \sum_{y \le x} \mathrm{PMF}_X(y).$$

The cumulative distribution function answers questions like "How likely is it that it will take no more than 3 flips for a coin to land on heads?" For a fair coin, that would be

$$\mathrm{CDF}_X(3) = \sum_{y=1}^{3} \mathrm{PMF}_X(y)$$

$$= \sum_{y=1}^{3} \frac{1}{2^y}$$

$$= \frac{1}{2} + \frac{1}{4} + \frac{1}{8} = \frac{7}{8}.$$

✳

The *expected value* of a random variable X, also called its *expectation* or *mean* and denoted by $E(X)$, is the weighted average of its possible values, weighted according to their probabilities. If X is a random variable with codomain T, then its expected value is

$$E(X) = \sum_{x \in T} \Pr(X = x) \cdot x. \tag{29.4}$$

Let's try out this definition on a few simple experiments:

Example 29.5. *What is the expected value of a roll of a standard six-sided die?*

Solution to example. The possible values are the integers 1 through 6 with equal probability, so

$$E(X) = \sum_{i=1}^{6} \left(\frac{1}{6} \cdot i \right)$$

$$= \frac{1}{6} \cdot \frac{6 \cdot 7}{2} = 3.5 \quad \text{(using (3.12))}.$$

Note that the expected value of a random variable need not be one of the possible values of that variable! ∎

Example 29.6. *If a fair six-sided die is rolled three times, how many distinct values are expected to appear?*

Solution to example. There are 6^3 outcomes of the experiment. An outcome might have 3 distinct values (for example, rolls of 5, 1, and 6), or 2 distinct values (for example, 3, 4, and 3), or 1 distinct value (for example, 2, 2, and 2). Let X be the number of distinct values; then $\Pr(X=1)$ and $\Pr(X=3)$ are relatively straightforward to calculate, and $\Pr(X=2)$ represents the remaining outcomes:

$$\Pr(X=1) = \frac{6}{6^3} \qquad\qquad = \frac{1}{36}$$

$$\Pr(X=3) = \frac{6!/3!}{6^3} \qquad\qquad = \frac{20}{36}$$

$$\Pr(X=2) = 1 - \Pr(X=1) - \Pr(X=3) \qquad = \frac{15}{36}.$$

Applying the definition of expected value,

$$E(X) = \Pr(X=1) \cdot 1 + \Pr(X=2) \cdot 2 + \Pr(X=3) \cdot 3$$

$$= \frac{1 \cdot 1}{36} + \frac{15 \cdot 2}{36} + \frac{20 \cdot 3}{36} \approx 2.53. \qquad\qquad ∎$$

Example 29.7. *Let X be the number of heads that appear in a sequence of 10 flips of a fair coin. What is $E(X)$?*

Solution to example. X can have any integer value from 0 to 10, and for any integer i there are $\binom{10}{i}$ sequences of 10 flips that contain exactly i heads. Each such sequence occurs with probability $(\frac{1}{2})^{10}$. So

$$E(X) = \sum_{i=0}^{10} \Pr(X=i) \cdot i$$

$$= \sum_{i=1}^{10} \Pr(X=i) \cdot i \qquad\qquad \text{(since the } i=0 \text{ term is 0)}$$

$$= \sum_{i=1}^{10} \binom{10}{i} \left(\frac{1}{2}\right)^{10} \cdot i$$

$$= \left(\frac{1}{2}\right)^{10} \cdot 10 \cdot 2^9 \qquad \text{(see Problem 23.8)}$$

$$= 5. \qquad \blacksquare$$

This result makes intuitive sense: the number of sequences that have 4 heads is equal to the number that have 6, and the average number of heads between all of those sequences together is 5; and similarly for 3 and 7, 2 and 8, 1 and 9, and 0 and 10.

There are simple, useful formulas for the expected values of Bernoulli and geometric random variables.

Theorem 29.8. *For $0 \le p \le 1$,*

- *the expected value of the Bernoulli variable with probability p of success is p, and*
- *the expected value of the geometric random variable for which each trial has probability p of success is $\frac{1}{p}$.*

Proof. Let B_p be the Bernoulli variable. Then

$$E(B_p) = \sum_{x \in \{0,1\}} \Pr(B_p = x) \cdot x$$

$$= (1 - p) \cdot 0 + p \cdot 1 = p.$$

Let G_p be the geometric random variable. Then

$$E(G_p) = \sum_{i=1}^{\infty} \Pr(G_p = i) \cdot i$$

$$= \sum_{i=1}^{\infty} (1 - p)^{i-1} \cdot p \cdot i$$

$$= \frac{p}{1 - p} \cdot \sum_{i=1}^{\infty} i \cdot (1 - p)^i$$

$$= \frac{p}{1 - p} \cdot \frac{1 - p}{p^2} \qquad \text{(by (24.13))}$$

$$= \frac{1}{p}. \qquad \blacksquare$$

Expectation is important to computer science because it gives a way of calculating the *average runtime* of an algorithm, as opposed to the worst-case runtime discussed in Chapter 21. Worst-case analysis gives an upper bound, while average-case analysis may more realistically estimate the runtime of an algorithm on typical inputs.

As an example, consider the *binary search* algorithm for finding an element in a sorted list, described and analyzed on page 224. The runtime of this algorithm was found to be $\Theta(\log n)$, but that is an asymptotic bound for the worst case. The search might succeed after probing only one or two elements. How many steps does binary search take on average? To formalize the question:

Example 29.9. *Suppose s is an element in a sorted list L, and let $X(s)$ be the number of elements that the binary search algorithm must inspect in order to find s in L. What is $E(X)$?*

Solution to example. First, to simplify the arithmetic, we'll assume that L is a list of length $2^k - 1$, for some integer k. This ensures that at each stage of the algorithm, the size of the current sublist is odd—so we can inspect the middle element and then remove it, leaving left and right sublists of the same length, which is also odd.

The minimum number of steps is 1, which happens if s is the middle element of L; and the maximum number is k, if s is not found until we reach a sublist of size 1. The expected value will be somewhere in between.

We need to find $\Pr(X = i)$, for each value $1 \leq i \leq k$. Let L_i be the set of elements in L that would be found after inspecting exactly i elements. Since s is equally likely to be any of the elements,

$$\Pr(X = i) = \frac{|L_i|}{|L|}.$$

L_1 contains just the middle element of L, so $|L_1| = 1$. L_2 contains the middle elements of both the left sublist and the right sublist, so $|L_2| = 2$. At each successive stage, L_i contains the middle elements of both the left and right sublists that would remain after inspecting an element from L_{i-1}. That is, $|L_i|$ doubles at each stage, starting with $|L_1| = 1$, so $|L_i| = 2^{i-1}$. (Figure 29.1 shows an example of the $k = 4$ case; that is, the list L has length 15. In row $i = 1, \ldots, 4$, the elements boxed in red make up L_i.) So

$$E(X) = \sum_{i=1}^{k} \Pr(X = i) \cdot i$$

$$= \sum_{i=1}^{k} \frac{|L_i|}{|L|} \cdot i$$

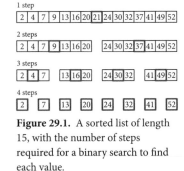

Figure 29.1. A sorted list of length 15, with the number of steps required for a binary search to find each value.

$$= \sum_{i=1}^{k} \frac{2^{i-1}}{2^k - 1} \cdot i$$

$$= \left(\frac{1}{2^k - 1} \right) \sum_{i=1}^{k} 2^{i-1} \cdot i$$

$$= \left(\frac{1}{2^k - 1} \right) \cdot \left(2^k \cdot (k-1) + 1 \right) \qquad \text{(see Problem 3.8).} \qquad \blacksquare$$

So it turns out that the average case runtime for binary search isn't much better than the worst case. For a list of length $2^k - 1$, the worst case is k, as noted above; and taking the limit of the expected value as $k \to \infty$,

$$\lim_{k \to \infty} E(X) = \lim_{k \to \infty} \frac{2^k \cdot k - 2^k + 1}{2^k - 1}$$

$$= k - 1;$$

that is, just 1 step fewer than in the worst case. Although the number of steps can be any integer between 1 and k, there are many more elements that take more steps than the few elements that take few steps, so on balance binary search is expected to take close to the maximum number of steps.

Note that this analysis was the expected runtime only for elements actually found in L. An element that is not in L would always require the binary search algorithm to make k probes into the list before returning empty-handed. To analyze binary search for the case in which the item s might or might not be in the list, we would need to know what proportion of the searches were for items in the list. If most searches were for items not in the list, the average runtime of the algorithm would be even closer to the worst-case runtime.

The average case runtime of an algorithm depends both on the algorithm and on assumptions about the distribution of inputs. In our analysis of binary search, we assumed that the searched item was equally likely to be in any position within the list. If the middle element of the list was for some reason a more frequent search target than others, the average-case search time would be decreased. For some algorithms, the average case runtime is much less than the worst case. *Quicksort* is a sorting algorithm that has this property, under reasonable assumptions about the lists being sorted. In the worst case the runtime is $\Omega(n^2)$, but the average-case runtime is $\Theta(n \log n)$.

✳

The *variance* of a random variable is a measure of how spread out the possible values are. It is defined as the expected value of the square of the variable's difference from its mean:

$$\text{Var}(X) = E((X - E(X))^2)$$

$$= \sum_x \Pr(X = x) \cdot (x - E(X))^2.$$

So the variance is 0 for a random variable that assumes only a single value, and the variance of a random variable is never negative. The more frequently extreme values (that is, values far from the mean) occur, the higher the variance, as the following example illustrates.

Example 29.10. *Calculate the variance of the following random variables, each of which has expected value $\frac{1}{2}$:*

- *X: has value $\frac{1}{2}$ always;*
- *Y: has value 0 or 1, each with equal probability;*
- *Z: has value 0, $\frac{1}{4}$, $\frac{1}{2}$, $\frac{3}{4}$, or 1, each with equal probability.*

Solution to example.

$$\text{Var}(X) = \left(\frac{1}{2} - \frac{1}{2}\right)^2 = 0$$

$$\text{Var}(Y) = \frac{\left(0 - \frac{1}{2}\right)^2 + \left(1 - \frac{1}{2}\right)^2}{2} = \frac{1}{4}$$

$$\text{Var}(Z) = \frac{\left(0 - \frac{1}{2}\right)^2 + \left(\frac{1}{4} - \frac{1}{2}\right)^2 + \left(\frac{1}{2} - \frac{1}{2}\right)^2 + \left(\frac{3}{4} - \frac{1}{2}\right)^2 + \left(1 - \frac{1}{2}\right)^2}{5} = \frac{1}{8}.$$

∎

The variance of flips of a coin (counting tails as 0 and heads as 1) is highest when the coin is fair. To be precise:

Example 29.11. *What is the variance of a Bernoulli variable with probability of success p?*

Solution to example. Let B_p be the variable, and recall from Theorem 29.8 that $E(B_p) = p$. Then

$$\text{Var}(B_p) = \sum_x \Pr(B_p = x) \cdot (x - E(B_p))^2$$

$$= (1 - p) \cdot (0 - p)^2 + p \cdot (1 - p)^2$$

$$= p \cdot (1 - p).$$

∎

So for a fair coin, $\text{Var}(B_{0.5}) = 0.25$, while a coin that has a 90% chance of landing on tails has $\text{Var}(B_{0.1}) = 0.09$, reflecting the fact that the mean is close to the frequently occurring value and far from the rarely occurring value.

What is the variance in the number of probes needed to find an item in an ordered list using binary search? As we noted in Example 29.9, if X represents the number of steps required in a binary search, then X ranges over values from 1 to k (where k is approximately the base-2 log of the length of the list), with the values weighted more toward k than toward 1 (see Figure 29.2). Since the values are clustered close to the mean, $\text{Var}(X)$ should be relatively low; for instance, it is less than that of a random variable Y that takes on the same values 1 to k, but with equal probability. Problem 29.8 verifies this.

The expectation and variance of a random variable can interact in surprising ways. The evolutionary biologist Stephen Jay Gould proposed a striking theory to explain the disappearance of the .400 hitter in baseball.[3] To set the stage for readers unfamiliar with the game, the *batting average* of a player serves as a measure of offensive skill, the higher the better: it is the player's number of hits divided by the number of times at bat, a fraction between .000 and 1.000. Though .400 hitters were not uncommon in baseball's early days, the last .400 hitter was Ted Williams, who hit .406 in 1941. Since then, not a single player has broken the .400 threshold.

Example 29.12. *Why are there no .400 hitters any more?*

Solution to example. Of course this is not purely a mathematical question, but according to Gould, it has an important mathematical angle. Gould contended that, paradoxically, *the disappearance of the .400 hitter is a consequence of the general improvement in the quality of the game.* It used to be much easier for mediocre players to play at the Major League level, when the competitive pressures were lower. As a result, the variance in the batting averages of players was high: the best players were as good then as they are today, but the poorer players of past years would not be on any team today. But this concentration of the talent pool has not resulted in a general increase in batting averages, because those responsible for the game have tweaked the rules to keep the mean (or expected) batting average roughly constant, at around .260. (For example, the rules committee might raise the pitching mound or expand the strike zone to disadvantage batters, and therefore lower batting averages.) Because the expected batting average has been held roughly constant while its variance has decreased, values far from the mean—including values over .400—have become rare or entirely nonexistent. ∎

✳

Applications of expectation and variance often involve functions of a random variable, or combinations of more than one random variable using

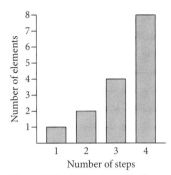

Figure 29.2. A histogram for the PMF of the number of steps required in a binary search over a list of length 15.

[3] Originally published in *Discover* magazine under the title "Entropic Homogeneity Isn't Why No One Hits .400 Any More," Gould's theory appears as a chapter of his book *Full House: The Spread Of Excellence From Plato To Darwin* (Harmony Books, 1996).

addition and multiplication. Some simple mathematical rules govern how expectation and variance behave when combined algebraically.

Theorem 29.13. Basic Properties of Expectation. *Let c be a constant and X a random variable. Then*

(a) $E(c) = c$;

(b) $E(E(X)) = E(X)$;

(c) $E(cX) = cE(X)$.

Proof.

(a) The only value c can take is c, with probability 1, so by the definition of expectation
$$E(c) = \sum_{c} \Pr(c = c) \cdot c = 1 \cdot c = c.$$

(b) $E(X)$ is a number, not a random variable, so $E(E(X)) = E(X)$ by part (a).

(c) Let T be the codomain of X:
$$E(cX) = \sum_{x \in T} \Pr(X = x) \cdot cx$$
$$= c \sum_{x \in T} \Pr(X = x) \cdot x = cE(X). \qquad \blacksquare$$

For example, suppose that X is the random variable representing the value of a roll of a six-sided die. Example 29.5 showed that $E(X) = 3.5$. Then the expected value of double the roll is just

$$E(2X) = 2E(X) = 2 \cdot 3.5 = 7.$$

The value "double the roll of a die" can also be regarded as the sum of two random variables X and Y, where both variables are just the value of the same single roll. More generally, for any two random variables X and Y—whether they are equal, completely unrelated, or have some relationship in between—we can simply add their expectations:

Theorem 29.14. *The expectation of the sum of two random variables is equal to the sum of their expectations. That is, $E(X + Y) = E(X) + E(Y)$.*

This result—that the expectation of a sum of random variables is the sum of the expectations of those random variables—is one of several related results for finding the expectation of a complex expression in terms of the expectations of the random variables mentioned in that expression. It is helpful in proving these to back up to the definition of expectation and to reformulate it in terms of the probabilities of various outcomes, rather than in terms of the probabilities of the random variable having various values. That is, (29.4) defines the expectation of a random variable X as a sum over the values $x \in T$, where T is the codomain of the random variable:

$$E(X) = \sum_{x \in T} \Pr(X = x) \cdot x.$$

The following lemma sums instead over all possible outcomes $s \in S$.

Lemma 29.15. *If X is a random variable for which the domain is the sample space S, then its expected value $E(X)$ is equal to the weighted average, over the outcomes in S, of the value of X on each outcome, weighted according to the outcomes' probabilities:*

$$E(X) = \sum_{s \in S} \Pr(s) \cdot X(s).$$

The proof is left to Problem 29.6. In essence, we can change the order of a double summation over the possible values of X and the outcomes that produce those values, so that it is instead a single summation over the possible outcomes.

This alternative formula leads to a simple proof of Theorem 29.14:

Proof of Theorem 29.14. Let the sample space S be the domain of X and Y.[4] Using Lemma 29.15,

$$E(X + Y) = \sum_{s \in S} \Big(\Pr(s) \cdot (X(s) + Y(s)) \Big)$$
$$= \sum_{s \in S} \Big(\Pr(s) \cdot X(s) \Big) + \sum_{s \in S} \Big(\Pr(s) \cdot Y(s) \Big)$$
$$= E(X) + E(Y) \qquad \text{(by the definition of expectation).}$$

■

[4] We can say X and Y share a sample space S even if X and Y refer to different experiments—for instance, X is a function on the outcome of a coin toss and Y is a function on the outcome of a die roll. In that case S is just the cross product of their sample spaces.

Theorem 29.14 can be extended, by induction, to any finite sum of variables. Combining this extension with Theorem 29.13(c), we can prove a more general rule, called *linearity of expectation*:

Theorem 29.16. *For random variables X_i and constants c_i,*

$$E\left(\sum_i c_i X_i \right) = \sum_i c_i E(X_i).$$

The proof is left as Problem 29.1.

Linearity of expectation makes it easy to calculate the variance of X given the expectations of X and X^2:

Theorem 29.17. $\mathrm{Var}(X) = E(X^2) - E(X)^2$.

Proof. Starting from the definition,

$$\mathrm{Var}(X) = E((X - E(X))^2)$$

$$= E(X^2 - 2XE(X) + E(X)^2)$$

$$= E(X^2) - E(2XE(X)) + E(E(X)^2), \qquad (29.18)$$

where the last step follows from Theorem 29.16. The middle term of (29.18) can be simplified since 2 and $E(X)$ are both constants, so $E(2XE(X)) = 2E(X)^2$. By the same token, $E(E(X)^2) = E(X)^2$. But then (29.18) simplifies to $E(X^2) - E(X)^2$, as desired. ∎

We are now prepared to determine the variance of a geometric random variable.

Theorem 29.19. *The variance of the geometric random variable with probability p of success on each trial is $\frac{1-p}{p^2}$.*

For example, consider the count of coin flips until the first heads. This is a geometric random variable with $p = 0.5$ for a fair coin, or with $p = 0.1$ for a coin that comes up heads only with probability 0.1. In the first case the variance is 2, and in the second case the variance is 90.

Proof. We start with the formula for variance from Theorem 29.17, and rewrite it with a trick using linearity of expectation:

$$\mathrm{Var}(X) = E(X^2) - E(X)^2$$

$$= E(X(X-1)) + E(X) - E(X)^2. \qquad (29.20)$$

We now transform the term $E(X(X-1))$ into a recognizable series:

$$E(X(X-1)) = \sum_{i=1}^{\infty} \Pr(X = i) \cdot i(i-1)$$

$$= \sum_{i=1}^{\infty} (1-p)^{i-1} p \cdot i(i-1)$$

$$= \sum_{i=2}^{\infty}(1-p)^{i-1}p \cdot i(i-1) \qquad \text{(the } i = 1 \text{ term is 0)}$$

$$= (1-p)p\sum_{i=2}^{\infty} i(i-1)(1-p)^{i-2}. \qquad (29.21)$$

Now we require a bit of calculus—this sum is the second derivative of a geometric series:

$$\sum_{i=2}^{\infty} i(i-1)(1-p)^{i-2} = \frac{d^2}{dp^2}\left(\sum_{i=0}^{\infty}(1-p)^i\right)$$

$$= \frac{d^2}{dp^2}\left(\frac{1}{1-(1-p)}\right) \text{ (by (24.8))}$$

$$= \frac{2}{p^3}. \qquad (29.22)$$

Substituting (29.22) back into (29.21),

$$E(X(X-1)) = (1-p)p \cdot \frac{2}{p^3} = \frac{2(1-p)}{p^2}. \qquad (29.23)$$

Finally, substituting (29.23) back into (29.20), and using the fact that $E(X) = \frac{1}{p}$ (Theorem 29.8),

$$\text{Var}(X) = E(X(X-1)) + E(X) - E(X)^2$$

$$= \frac{2(1-p)p}{p^3} + \frac{1}{p} - \frac{1}{p^2} = \frac{1-p}{p^2}. \qquad \blacksquare$$

✳

Theorem 29.14, which says that the expectation of a sum is the sum of the expectations, suggests further questions: does a similar relationship hold for other operations? For example, is the expectation of a product the product of the expectations? Is the expectation of a reciprocal the reciprocal of the expectation? Combining those questions, what about the expectation of a quotient?

The answer for reciprocals is no, and therefore the answer about quotients is also no. The answer about products is that the expectation of a product of two random variables is the product of their expectations only if the two variables are independent—which we'll define shortly. The following example illustrates that these relationships do not generally hold for reciprocals and products.

Example 29.24. *Suppose a game at a casino offers a 50% chance of winning. A gambler decides to play this game repeatedly, stopping after the first win. Let G be the number of games played under this strategy. What is the expected value of G? What fraction of these games are expected to be wins, and what fraction are expected to be losses?*

Solution to example. G is a geometric random variable, where each trial results in a win with probability $\frac{1}{2}$, so by Theorem 29.8,

$$E(G) = \frac{1}{\left(\frac{1}{2}\right)} = 2.$$

If L is the number of losses preceding the first win, then

$$E(L) = E(G-1) = E(G) - 1 = 1.$$

Since only the last of the G games is a win, the expected fraction of wins among the G games is

$$E\left(\frac{1}{G}\right) = \sum_{i=1}^{\infty} \Pr(G=i) \cdot \frac{1}{i}$$

$$= \sum_{i=1}^{\infty} \frac{1}{2^i} \cdot \frac{1}{i}. \qquad (29.25)$$

It is easier to calculate the value of (29.25) if we generalize it by substituting a variable x for the value $\frac{1}{2}$:

$$\sum_{i=1}^{\infty} \frac{1}{i} \cdot x^i \qquad (29.26)$$

Now we can solve this with a bit of calculus: this sum is the integral of

$$\sum_{i=1}^{\infty} x^{i-1} = \sum_{i=0}^{\infty} x^i = \frac{1}{1-x} \quad \text{(by (24.8)).} \qquad (29.27)$$

But the integral of $\frac{1}{1-x}$ is $-\ln(1-x)$, so plugging in $x = \frac{1}{2}$, the value of (29.25) is

$$\sum_{i=1}^{\infty} \frac{1}{2^i} \cdot \frac{1}{i} = -\ln\left(1 - \tfrac{1}{2}\right)$$

$$= \ln 2 \approx 0.69.$$

We've now found $E(\frac{1}{G})$. Note that the expectation of the reciprocal of G is *not* the reciprocal of the expected value of G:

$$\ln 2 = E\left(\frac{1}{G}\right) \neq \frac{1}{E(G)} = \frac{1}{2}.$$

Now, we want to find $E(\frac{L}{G})$. Because L depends on G, it is *not* valid to simply multiply $E(L)$ by $E(\frac{1}{G})$! Since $E(L) = 1$, that would give the result $E(\frac{L}{G}) = E(\frac{1}{G}) = \ln 2$, which can't be right, since the fraction of losses plus the fraction of wins would then add up to more than 1.

Instead we use linearity of expectation:

$$E\left(\frac{L}{G}\right) = E\left(\frac{G-1}{G}\right) = 1 - E\left(\frac{1}{G}\right)$$

which is $1 - \ln 2 \approx 0.31$. ∎

✻

When can expectations be multiplied to find the expectation of the variables' product? Like events, random variables can be *independent*; in fact the definition of independence for random variables is based on the independence of the events that they take on each of their values. That is, X and Y are independent if the events $X = x$ and $Y = y$ are independent for all x and y.

Referring back to the definition of independent events given by (26.12), we can give the formal definition of independence of random variables. Suppose X and Y have codomains T_X and T_Y respectively. Then X and Y are independent if and only if for all $x \in T_X$ and $y \in T_Y$,

$$\Pr((X = x) \cap (Y = y)) = \Pr(X = x) \cdot \Pr(Y = y).$$

When two variables are independent, the expectation of their product is the product of their expectations:

Theorem 29.28. *If X and Y are independent,*

$$E(X \cdot Y) = E(X) \cdot E(Y).$$

Proof. Let T_X and T_Y be the codomains of X and Y respectively. Starting with the definition of expectation, and then applying the definition of independence, $E(X \cdot Y)$ is the following sum:

$$E(X \cdot Y) = \sum_{x \in T_X \text{ and } y \in T_Y} \left(\Pr((X = x) \cap (Y = y)) \cdot xy \right)$$

$$= \sum_{x \in T_X} \sum_{y \in T_Y} \left(\Pr(X = x) \cdot \Pr(Y = y) \cdot xy \right).$$

From the perspective of the inner summation, in which y is the variable, any quantity defined in terms of x is a constant. So we can factor out $\Pr(X = x)$

and x from the inner summation:

$$E(X \cdot Y) = \sum_{x \in T_X} \left(\Pr(X = x) \cdot x \cdot \left(\sum_{y \in T_Y} \Pr(Y = y) \cdot y \right) \right).$$

Similarly, the summation that just involves y is a constant with respect to x, so the last expression can be rewritten as

$$E(X \cdot Y) = \left(\sum_{x \in T_X} \Pr(X = x) \cdot x \right) \cdot \left(\sum_{y \in T_Y} \Pr(Y = y) \cdot y \right),$$

which is just $E(X) \cdot E(Y)$, as was to be shown.   ∎

Independent random variables satisfy several properties that are not true of random variables in general. For instance, the variance of the sum of two independent variables is the sum of their variances. This is not true of arbitrary random variables, as the following example shows.

Example 29.29. *Let X be a random variable. What is* $\mathrm{Var}(X + X)$*?*

Solution to example. First, note that X is not generally independent of itself, since

$$\Pr(X = x \cap X = x) = \Pr(X = x),$$

which is not equal to

$$\Pr(X = x) \cdot \Pr(X = x) = \Pr(X = x)^2,$$

unless $\Pr(X = x)$ is 0 or 1.
   As for $\mathrm{Var}(X + X)$,

$$\mathrm{Var}(X + X) = \mathrm{Var}(2X)$$
$$= E((2X)^2) - E(2X)^2$$
$$= E(4X^2) - (2(E(X))^2$$
$$= 4E(X^2) - 4E(X)^2$$
$$= 4\mathrm{Var}(X),$$

which is not equal to $\mathrm{Var}(X) + \mathrm{Var}(X)$ unless $\mathrm{Var}(X) = 0$.   ∎

On the other hand, the variance of the sum is the sum of their variances if the two variables are independent:

Theorem 29.30. *If X and Y are independent, then* $\text{Var}(X + Y) = \text{Var}(X) + \text{Var}(Y)$.

Proof. We start with the formula for variance given by Theorem 29.17:

$$\text{Var}(X + Y) = E((X + Y)^2) - E(X + Y)^2$$
$$= E(X^2 + 2XY + Y^2) - (E(X) + E(Y))^2.$$

Then, applying linearity to the first term and multiplying out the square in the second term, we get

$$\text{Var}(X + Y) = E(X^2) + E(2XY) + E(Y^2) - (E(X)^2 + 2E(X)E(Y) + E(Y)^2).$$

But $E(2XY) = 2E(X)E(Y)$ by linearity and the independence of X and Y, so

$$\text{Var}(X + Y) = E(X^2) + E(Y^2) - (E(X)^2 + E(Y)^2)$$
$$= E(X^2) - E(X)^2 + E(Y^2) - E(Y)^2$$
$$= \text{Var}(X) + \text{Var}(Y). \qquad \blacksquare$$

✳

We'll conclude this chapter by applying these techniques to some questions of computer science. Linearity of expectation in particular is a powerful tool, and can simplify calculations that might otherwise seem intractable. Consider the following application, in which we must calculate the number of "successes" in a sequence of events when each such success affects the probability that there will be another success.

Example 29.31. The Hiring Problem. *A company is hiring for an open position. Over the course of n days, the company interviews one candidate per day. On each day, if the current candidate is better than the current employee, the candidate is hired immediately (and the employee is fired); if the current candidate is worse than the current employee, the employee keeps the job. At the end of this process, how many hires will the company have made, on average?*

Solution to example. First, let's restate the problem in algorithmic terms. Assume the candidates come in random order, and assign each a rank from 1 to n. So the process is to iterate through a randomly ordered list of n distinct numbers, at all times keeping track of the maximum value seen so far, and we want to find the expected number of times this maximum value is updated.

Let I_j be the indicator variable that is 1 just in case the jth element in the list is the maximum of the first j elements. Then $\Pr(I_j = 1) = \frac{1}{j}$, and so $E(I_j) = \frac{1}{j}$.

We want to find the expectation of the sum

$$E\left(\sum_{j=1}^{n} I_j\right).$$

Without linearity, this problem would seem tricky, because the I_j are not independent: if element j is greater than all the previous elements, it is less likely that element $j + 1$ is also greater than all the ones before it, since element j is known to be large. But because of linearity of expectation, this does not matter; to find the sum $E(\sum_{j=1}^{n} I_j)$ we can simply take the sum

$$\sum_{j=1}^{n} E(I_j) = \sum_{j=1}^{n} \frac{1}{j}.$$

This is just the harmonic series H_n, which we encountered as (24.24). We found that this can be approximated by the natural logarithm; by (24.26),

$$\ln n \leq H_n \leq \ln(n + 1).$$

So the expected number of hires is between $\ln n$ and $\ln(n + 1)$.   ∎

Note that the minimum number of hires is 1 (if the first candidate is the best), and the maximum number is n (if the candidates are interviewed in ascending order), so the average number of hires is significantly less than the worst-case number.

Let's close by analyzing an example familiar to computer programmers. A *hash table* is a data structure for storing pairs consisting of a *key* and an associated *value*. Key-value pairs are stored so that the value can be retrieved quickly given the key. A key-value pair (also known as a *record*) is stored in a "bucket" via a *hash function* (first mentioned in Problem 1.13), which maps keys to buckets. One common design for a hash table uses buckets that are capable of storing multiple records; linked lists, perhaps. If no two distinct records have keys that map to the same bucket, the value can be retrieved by computing the hash function on the key and taking the record out of the bucket. If the keys of multiple records map to the same bucket, a *collision* is said to occur and the bucket itself needs to be searched for the record with the right key. To keep retrieval times low, the hash function should minimize collisions and spread the keys evenly across the buckets. (See Figure 29.3.)

Example 26.9, which we presented as a question about duplicate birthdays in a group of people, can be viewed as analyzing the likelihood of collisions in a hash table with 365 buckets storing $n \leq 365$ records—the hash function maps people to their birthdays.

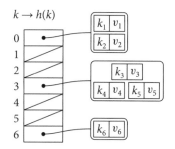

Figure 29.3. Schematic representation of a hash table. Hash function h is applied to the key to locate the bucket where the key-value pair is stored. This particular hash table has seven buckets; four are empty, one has two key-value records, one has three, and one has one. The records associated with keys k_1 and k_2 are both in bucket 0, because the hash values of k_1 and k_2 collide: $h(k_1) = h(k_2) = 0$.

When designing a hash table, the probability of hash collisions (which cost time, since buckets then have to be searched) has to be weighed against the likely number of empty buckets (which are costly in another way, because of the space they waste). So it is natural to ask: given that we are storing n records in a hash table of size k, how many of the k buckets are *expected* to be empty? Depending on the relative values of n and k, the number could be anywhere between 0 (if there are a great many more records than buckets) and $k - 1$ (if there is only one record).

Example 29.32. *Suppose that a hash table uses k buckets to store n records, and*

- *the hash function distributes records among the buckets randomly, each with equal probability;*
- *each record's location is independent of the location of all the other records.*

(These assumptions about the hash function constitute the "simple uniform hashing assumption.") What is the expected number of empty buckets?

Solution to example. Again, at first glance this problem seems difficult: if one bucket is empty, that makes it less likely that the next bucket is empty, since the records must end up somewhere. In other words, the events indicating whether each bucket is empty are not mutually independent.

But linearity of expectation makes the analysis easier. Let's label the buckets 0 through $k - 1$, and define I_j to be the indicator random variable for the event that bucket j is empty. The probability that j is empty is just the probability that all n records land in one of the other $k - 1$ buckets, that is:

$$\left(\frac{k-1}{k}\right)^n.$$

So I_j is 1 with probability $(\frac{k-1}{k})^n$, and 0 otherwise. The calculation is the same for each bucket j, $0 \leq j \leq k - 1$.

Now, the expected number of empty buckets is the expected value of the sum of the indicator variables:

$$E\left(\sum_{j=0}^{k-1} I_j\right) = \sum_{j=0}^{k-1} E(I_j)$$

$$= k \cdot \left(\frac{k-1}{k}\right)^n.$$

This formula gives the right results for the two extreme cases we mentioned ($k - 1$ if $n = 1$, and a number approaching 0 as n grows without bound). To take an intermediate case, if 1000 records were placed in a hash table of 1000 buckets, we would expect approximately 368 buckets to be empty, with the records spread over the remaining 632 buckets, since

$$k \cdot \left(\frac{k-1}{k} \right)^n = 1000 \cdot \left(\frac{999}{1000} \right)^{1000} \approx 367.70.$$

As we increase the size, say to 10000, more and more records will end up alone in their buckets, though there are still some collisions:

$$k \cdot \left(\frac{k-1}{k} \right)^n = 10000 \cdot \left(\frac{9999}{10000} \right)^{1000} \approx 9048.33,$$

so the 1000 records are expected to spread over 952 of the buckets, while 9048 buckets are empty. ∎

If collisions are resolved by storing all elements in a bucket in a list structure, then records are retrieved by computing the hash function and then linearly traversing the list at that bucket in search of the key. The expected number of records per bucket is just $\frac{n}{k}$, so assuming that computing the hash function takes time $O(1)$, the average-case lookup time is $\Theta(1 + \frac{n}{k})$. Of course, the worst case occurs when all records land in the same bucket, so looking up an element requires a linear traversal of a length-n list, which requires time $\Theta(n)$.

If the hash table is large compared to the number of records, the average lookup time $\Theta(1 + \frac{n}{k})$ reduces to $\Theta(1)$; that is, lookup takes constant time on average. In this analysis we have treated k as a variable rather than a constant: usually, hash table implementations will dynamically resize as values are added, so that k is always large compared to n. If the hash table did not resize, so that k really must be treated as a constant, then the runtime should be regarded as $\Theta(n)$, and lookup times would degrade as the number of records stored in the table approached the size of the table.

Chapter Summary

- A *random variable* describes a value that is not certain, but instead has a range of possible values, each with a certain probability.

- A *Bernoulli variable* or *indicator variable* is a random variable that indicates whether an event happened, taking on just the two values 1 (in case of *success*) and 0 (in case of *failure*).

- A *geometric random variable* represents the number of trials consisting of a sequence of *Bernoulli trials* that ends after the first success.

- The *probability mass function* of a random variable X describes the probability that the value of X is equal to a given value. It is written $PMF_X(x)$ or $\Pr(X = x)$.

- The *cumulative distribution function* of a random variable X describes the probability that the value of X is less than or equal to a given value. It is written $CDF_X(x)$ or $\Pr(X \le x)$.

- The *expected value* or *expectation* of a random variable X, denoted $E(X)$, is the weighted average of its possible values:

$$E(X) = \sum_{x \in T} \Pr(X = x) \cdot x.$$

- The *average runtime* of an algorithm is a function describing the expected value of the runtime, which depends on the algorithm itself as well as the distribution of inputs to the algorithm.

- The *variance* of a random variable is a measure of how spread out its possible values are, with the following two equivalent definitions:

$$\mathrm{Var}(X) = E((X - E(X))^2) = E(X^2) - E(X^2).$$

- Expectation is *linear*: $E\left(\sum_i c_i X_i\right) = \sum_i c_i E(X_i)$.

- A Bernoulli variable with probability p of success has expected value p and variance $p \cdot (1 - p)$.

- A geometric random variable with probability p of success for each trial has $\Pr(X = n) = (1 - p)^{n-1} \cdot p$, expected value $\frac{1}{p}$, and variance $\frac{1-p}{p^2}$.

- Two variables X and Y are *independent* if the events $X = x$ and $Y = y$ are independent for all values of x and y; that is,

$$\Pr((X = x) \cap (Y = y)) = \Pr(X = x) \cdot \Pr(Y = y).$$

- If two variables X and Y are independent, then

$$E(X \cdot Y) = E(X) \cdot E(Y)$$
$$\mathrm{Var}(X + Y) = \mathrm{Var}(X) + \mathrm{Var}(Y).$$

Problems

29.1. Prove that expectation is linear (Theorem 29.16); that is, for random variables X_i and constants c_i,

$$E\left(\sum_i c_i X_i\right) = \sum_i c_i E(X_i).$$

Use Theorem 29.13 and Theorem 29.14.

29.2. Let X represent the number of rolls for which the value is at least 5, in a sequence of 10 rolls of a fair six-sided die. What is $E(X)$?

29.3. Consider the following dice games:
(a) You roll a die once, and decide to either take its value as your score or roll again. If you roll again, you score the value of your second roll. What strategy maximizes the expected score, and what is that expected score?
(b) You roll a die repeatedly, stopping when you roll a 1. Your score is the sum of the values of all your rolls. What is the expected score?

29.4. What is the variance of:
(a) one roll of a fair six-sided die?
(b) double the value of one roll of a fair six-sided die?

29.5. Consider the following ways of producing a random number between 2 and 12. What are the expectation and variance of each? How do these compare to the expectation and variance of rolling one die and doubling the result? (See Problem 29.4.)
(a) Roll a fair six-sided die twice, and add the results.
(b) Roll a fair 11-sided die, where each face is a value between 2 and 12.

29.6. Prove Lemma 29.15. *Hint:* Start with (29.4). Then use (29.3) to substitute the definition of $\Pr(X = x)$ and achieve a double summation, over all possible values $x \in T$ and all possible outcomes $s \in S$ for which $X(s) = x$. Then change the order of summation to get the desired equality.

29.7. A number n can be represented with $\lg n$ binary digits, as we saw on page 213. What if a number is extremely large and we want a more compact representation, and an estimate is acceptable rather than an exact value—for example, when counting occurrences of a very frequent event? The following *approximate counting algorithm* provides an estimate with the correct expected value, and requires only $\lg \lg n$ bits:[5] Assign the counter's initial value as $c = 0$. At each occurrence of the event, increment c with probability $\frac{1}{2^c}$. At the end, output the value $2^c - 1$.

For example, if the current value of c is 0, increment with probability 1; or if c is 3, increment with probability $\frac{1}{8}$.

Let X_n be the value of the counter after n events.
(a) Write $\Pr(X_{n+1} = i)$ in terms of $\Pr(X_n = i)$ and $\Pr(X_n = i - 1)$.
(b) Prove by induction that $E(2^{X_n} - 1) = n$, using the relationship from part (a).

[5] This algorithm was devised by the cryptographer Robert Morris in 1977.

29.8. Let X represent the number of steps required in a binary search for an element in a list of length $2^k - 1$, where the element is known to be a member of the list, and let Y represent a number that is equally likely to be any of the integers 1 through k, as on page 343. What are $\text{Var}(X)$ and $\text{Var}(Y)$? *Hint:* To calculate $E(X^2)$, use the equality

$$\sum_{i=1}^{k} 2^{i-1} \cdot i^2 = 2^k(k^2 - 2k + 3) - 3$$

from Problem 3.8.

29.9. Suppose a hash table contains k buckets and holds n records, where $k, n \geq 2$, and suppose that the hash function obeys the simple uniform hashing assumption (page 353).

(a) What is the expected number of buckets that contain exactly 1 record? *Hint:* Define $E_{j,m}$ as the event that bucket j contains exactly the mth record, and no other records. Then define E_j as the event that bucket j contains exactly one record. Find $\Pr(E_j)$, using the fact that E_j is the union of the disjoint events $E_{j,m}$. Finally, define indicator variables I_j for each event E_j, and use linearity of expectation to find the answer.

(b) What is the expected number of buckets that contain 2 or more records? *Hint:* Use the values from the previous calculation and from Example 29.32, instead of calculating this directly.

29.10. In Example 29.32, what is the expected size of a bucket? (The "size" of a bucket is the number of records it contains.) What is the expected size of a *nonempty* bucket? Compare these quantities in the cases in which $n = k$, $n = 2k$, and $n = \frac{k}{2}$.

Chapter 30

Modular Arithmetic

Modular arithmetic is like prose: you may not realize that you've been speaking it your whole life. For example, if you have to take a pill every 6 hours and you took the last pill at 10pm, you have to take the next one at 4am. Mathematically, we would write

$$10 + 6 \equiv 4 \quad (\text{mod } 12), \tag{30.1}$$

since clocks (at least old analog clocks!) repeat themselves every 12 hours (Figure 30.1). In military notation, which uses 24-hour cycles instead, we would say that if your last pill was at 2200 hours, the next would be at 0400 hours; that is,

$$22 + 6 \equiv 4 \quad (\text{mod } 24). \tag{30.2}$$

Figure 30.1. Adding $10 + 6 \equiv 4$ (mod 12) on the face of a clock. By convention, the zero point on a traditional analog clock is labeled 12 rather than 0, but $12 \equiv 0$ (mod 12).

Of course we are ignoring details about stipulating "the next day" and so on.

The notation used in (30.1) and (30.2) is not ideal, but is now an unchangeable part of our cultural inheritance. The "(mod 12)" should be interpreted as part of the equivalence sign. So (30.1) says that $10 + 6$ and 4 are equivalent with respect to a particular equivalence relation, which might better have been called $\equiv_{12}$. What is that equivalence relation? It is the relation that holds between two integers if dividing them by 12 leaves the same remainder. Recall (page 155) our earlier use of the notation "p mod q" to denote the remainder, between 0 and $q - 1$ inclusive, when p is divided by q for positive integers p and q. We can then define $x \equiv y$ (mod m) in three equivalent ways.

For any integer $m > 0$, and any $x, y \in \mathbb{Z}$,

$$x \equiv y \quad (\text{mod } m)$$

1. if and only if

$$x \bmod m = y \bmod m; \tag{30.3}$$

2. if and only if there is an r, $0 \leq r < m$, and there exist integers i and j such that
$$x = i \cdot m + r \text{ and}$$

$$y = j \cdot m + r; \tag{30.4}$$

3. if and only if

$$x - y \text{ is divisible by } m. \tag{30.5}$$

We leave it as an exercise (Problem 30.1) to prove carefully that the three definitions are equivalent. Is "$x \equiv y \pmod{m}$" an equivalence relation on $\mathbb{Z}$ for any fixed m? This is easy to see from (30.3): equality of remainders when divided by m is reflexive, symmetric, and transitive.

For a fixed m, the equivalence classes of "$x \equiv y \pmod{m}$" are called the *congruence classes modulo m*. They are, for each nonnegative integer $r < m$, the set of integers that leave a remainder of r when divided by m. So there are exactly m such equivalence or congruence classes. We write that set of equivalence classes as

$$\mathbb{Z}_m = \{\{x \in \mathbb{Z} : x \bmod m = r\} : 0 \leq r < m\}. \tag{30.6}$$

Moreover, we'll write $[x]$ (the value of m is unstated, but is understood from the context) for the equivalence class containing integer x. So if $m = 12$, we might write $[10 + 6] = [4]$, for example, though it would also be true to write $[10 + 6] = [-20]$. We actually saw $\mathbb{Z}_2$ in disguise back in Chapter 9: the exclusive-or operator $\oplus$ is addition in $\mathbb{Z}_2$.

We can perform certain arithmetic operations, such as addition and subtraction, on the congruence classes themselves, yielding a congruence class as the result. For example, we can define $[x] + [y]$ as $[x + y]$—for example, we would define $[10] + [6]$ as the congruence class $[16]$; that is, $[4]$. But this definition is unambiguous only if it does not matter which elements of $[10]$ and $[6]$ are added together—each set has infinitely many members, and we need to be sure that the same congruence class results regardless of which representatives are chosen. That is, we need to be confident that

Theorem 30.7. *If $x' \in [x]$ and $y' \in [y]$, then $x' + y' \in [x + y]$.*

Proof. Suppose that the remainder when either x or x' is divided by m is r, and the remainder when either y or y' is divided by m is s. Then $[x + y] = [r + s] = [x' + y']$. ∎

The same holds true of subtraction: $[x] - [y]$ can consistently be defined as $[x - y]$. It works also for multiplication: $[x] \cdot [y]$ can consistently be defined as $[x \cdot y]$. It doesn't matter which representatives of $[x]$ and $[y]$ are chosen—if you pick x' and y' from $[x]$ and $[y]$ respectively, multiply them together, and take the remainder $\bmod\, m$, you get the same result as if you just multiply x and y together and take the remainder $\bmod\, m$.

This little calculus of congruence classes is starting to look a lot like ordinary arithmetic, but in a system that has only m "numbers." There are even additive and multiplicative identity elements: it is easy to see that for any $x \in \mathbb{Z}$,

$$[0] + [x] = [x] + [0] = [x] \text{ and} \qquad (30.8)$$

$$[1] \cdot [x] = [x] \cdot [1] = [x]. \qquad (30.9)$$

✻

We'll take a short diversion to discuss exponentiation before discussing division in $\mathbb{Z}_m$, which is more complicated.

First, it follows immediately from the fact that multiplication is well defined that the operation of raising a member of $\mathbb{Z}_m$ to an integer power is well defined:

$$[x]^n = [x^n]. \qquad (30.10)$$

This is simply a restatement of the fact that

$$\overbrace{[x \cdot x \cdot \ldots \cdot x]}^{n \text{ factors}} = \overbrace{[x] \cdot [x] \cdot \ldots \cdot [x]}^{n \text{ factors}}. \qquad (30.11)$$

Example 30.12. *Let $m = 7$. What is $[10^3]$?*

Solution to example. We can determine this two ways. If we calculate 10^3 and then divide by 7, we get that $1000 = 142 \cdot 7 + 6$, so $[1000] = [6]$. On the other hand, if we calculate $[10] = [3]$, then $[10]^3 = [3^3] = [27] = [6]$, since $27 = 3 \cdot 7 + 6$. ∎

One immediate caution. It is *not* true that $[x^y] = [x]^{[y]}$, or even that the expression $[x]^{[y]}$ has any obviously well-defined meaning. To make the arithmetic easy, let $m = 10$. Then $[2^{11}] = [2048] = [8]$, but if we try to calculate $[2^1]$, on the theory that $[11] = [1]$, we get $[2]$, not $[8]$.

So here is the Cardinal Rule of Modular Arithmetic: *When adding, subtracting, or multiplying elements of $\mathbb{Z}_m$, you may at any point replace a number x by x mod m. When computing x^n, you may replace x by x mod m, but you can't similarly change the exponent n.*

The cardinal rule speeds up certain computations quite a bit. If $m = 10$ and we need to compute $[12345 \cdot 54321]$, we could painstakingly do the operations in the order specified, first doing the multiplication to get the result

$$[12345 \cdot 54321] = [670592745]$$

and then reducing the result modulo 10 to get the answer $[5]$. But if we had done the reduction modulo 10 first, everything would have been so much easier!

$$[12345 \cdot 54321] = [12345] \cdot [54321]$$

$$= [5] \cdot [1]$$

$$= [5].$$

The second observation about modular exponentiation is that it can be done faster by repeated squaring. Whether or not the intermediate products are reduced modulo m as we go, if the exponent n is a large number it would take a long time to compute $[x^n]$ by actually carrying out the $n-1$ multiplications indicated in (30.11). The time can be reduced dramatically by computing x to an exponent that is about half as large and then squaring the result, which uses just one additional multiplication—or two if the exponent is odd:

$$x^n = \left(x^{\frac{n}{2}}\right)^2 \text{ if } n \text{ is even,}$$

$$x^n = \left(x^{\frac{n-1}{2}}\right)^2 \cdot x \text{ if } n \text{ is odd.}$$

Repeated squaring reduces the number of multiplications to compute x^n from $n-1$ to somewhere between $\lfloor \lg n \rfloor$ and $2\lfloor \lg n \rfloor$ (see Problem 30.6). When combined with the cardinal rule for modular arithmetic, repeated squaring can make computations go very quickly. For example, let's compute $3^{25} \bmod 7$:

$$3^{25} = \left(3^{12}\right)^2 \cdot 3$$

$$= \left(\left(3^6\right)^2\right)^2 \cdot 3$$

$$= \left(\left(\left(3^3\right)^2\right)^2\right)^2 \cdot 3$$

$$= \left(\left(\left(3^2 \cdot 3\right)^2\right)^2\right)^2 \cdot 3$$

$$\equiv \left(\left(\left(2 \cdot 3\right)^2\right)^2\right)^2 \cdot 3 \pmod 7$$

$$\equiv \left(\left(6^2\right)^2\right)^2 \cdot 3 \pmod 7$$

$$\equiv \left(1^2\right)^2 \cdot 3 \pmod 7$$

$$\equiv 3 \pmod 7. \tag{30.13}$$

❋

We'll say that y is a *multiplicative inverse* of x in $\mathbb{Z}_m$ just in case their product is equal to the multiplicative identity [1]; that is,

$$x \cdot y \equiv 1 \pmod m.$$

Depending on the value of m, some elements of $\mathbb{Z}_m$ have multiplicative inverses and some do not. Let's look at the multiplication tables for $\mathbb{Z}_4$ and $\mathbb{Z}_5$ (Figure 30.2) to see what is going on.

$\mathbb{Z}_4$	0	1	2	3
0	0	0	0	0
1	0	1	2	3
2	0	2	0	2
3	0	3	2	1

$\mathbb{Z}_5$	0	1	2	3	4
0	0	0	0	0	0
1	0	1	2	3	4
2	0	2	4	1	3
3	0	3	1	4	2
4	0	4	3	2	1

Figure 30.2. Multiplication tables for $\mathbb{Z}_4$ and $\mathbb{Z}_5$.

We can see that every nonzero member of $\mathbb{Z}_5$ has a multiplicative inverse: [2] and [3] are inverses of each other, and each of [1] and [4] is its own inverse. In $\mathbb{Z}_4$, however, [2] has no multiplicative inverse.

The difference is that 5 is a prime number and 4 is not.

Theorem 30.14. *If p is prime, then every nonzero member of $\mathbb{Z}_p$ has a multiplicative inverse.*

Proof. First note that if p is prime and $0 \leq a < p$, then

$$[0 \cdot a], [1 \cdot a], \ldots, [(p-1) \cdot a] \tag{30.15}$$

are all distinct. For suppose $0 \leq i < j \leq p - 1$ and $[i \cdot a] = [j \cdot a]$; that is, $i \cdot a \equiv j \cdot a \pmod{p}$. If we can prove that "canceling" the factor of a on both sides is legal, so that $i \equiv j \pmod{p}$, then we will have proved that $i = j$, since both are less than p. This would be a contradiction, since we assumed that $i < j$.

The rule that allows us to cancel in this case is the following lemma.

Lemma 30.16. *If $i \cdot a \equiv j \cdot a$ (mod p), where p is prime and a is not divisible by p, then $i \equiv j$ (mod p).*

Of course, in the case at hand a is not divisible by p, because a is a nonnegative integer less than p.

Let's prove the lemma. If $i \cdot a \equiv j \cdot a \pmod{p}$, then $a \cdot (j - i) \equiv 0 \pmod{p}$. That is, $a \cdot (j - i)$ is a multiple of p, and p is a factor of $a \cdot (j - i)$. We know that p is not a factor of a, so it must be a factor of $j - i$, which is to say, $i \equiv j \pmod{p}$. (By Theorem 1.7, page 6, if a prime number is a factor of a product of two numbers, it must be a factor of one or the other. This is where it's important that p is prime—the same would not be true if p was composite.)

That proves the lemma. So the congruence classes listed in (30.15) are just a permutation of [0], [1], ..., [p − 1] (you can see this for $\mathbb{Z}_5$ in Figure 30.2; each row except the first is a permutation of $\{0, \ldots, 4\}$). But one of them must therefore be [1]. That is, there is an $i < p$ such that $i \cdot a \equiv 1 \pmod{p}$. ∎

We have established that when p is prime, there is a multiplicative inverse for every nonzero element x, which we can call x^{-1}. So it makes sense to define division: $[x]/[y] = [z]$, where $[z] = [x] \cdot [y]^{-1}$. For example, in $\mathbb{Z}_5$,

$$[2]/[3] = [2] \cdot [3]^{-1}$$
$$= [2] \cdot [2]$$
$$= [4].$$

In group theory, it is common to use the name $\mathbb{Z}_n$ more narrowly, to refer to the group of congruence classes modulo n equipped with only the addition and subtraction operators. The related group $\mathbb{Z}_n^*$ contains only those congruence classes with elements relatively prime to n, equipped with the multiplication and division operators. The field of congruence classes mod n—when n is prime, so such a field exists—is then called F_n, and is equipped with all four operators. In this text, we simplify the notation by using $\mathbb{Z}_n$ to mean the set of all congruence classes modulo n, with all four operators, even though division is not defined on every element.

So when p is prime, the congruence classes obey all the standard rules of arithmetic with addition, subtraction, multiplication and division. A mathematical structure in which these four operations work as they do in $\mathbb{R}$ (the set of real numbers) is called a *field*. So we have shown that $\mathbb{Z}_p$ is a field.

✳

But can we actually *find* $[x]^{-1}$, the multiplicative inverse of x in $\mathbb{Z}_p$, in any other way than writing out the full multiplication table and looking for it? Yes! An old friend, Euclid's algorithm (page 154), comes to our assistance.

Let's restate Euclid's algorithm, being explicit about the division that happens on each iteration. We will then demonstrate the procedure on an example.

Euclid's algorithm for the greatest common divisor of m and n:

1. $\langle r, r' \rangle \leftarrow \langle m, n \rangle$

2. While $r' \neq 0$

    (a) $q \leftarrow \lfloor \frac{r}{r'} \rfloor$

    (b) $\langle r, r' \rangle \leftarrow \langle r', r \bmod r' \rangle$

3. Return r, which is the greatest common divisor of m and n.

Let's say that r_i is the value of r just prior to the beginning of the ith iteration of the "while" loop (counting from 0). So $r_0 = m$, and because for any $i > 0$ the value of r on the ith iteration is the same as the value of r' on the previous iteration, $r_1 = n$. For $i \geq 0$, let

$$q_{i+1} = \lfloor r_i / r_{i+1} \rfloor; \tag{30.17}$$

then q_{i+1} is the value of q computed during the ith iteration. Suppose the last iteration of the while loop is the kth. Then r_{k+1} is the greatest common divisor of m and n, and is also a divisor of all the r_i for $0 \leq i \leq k$.

The reason for introducing the variable q and the quantities q_i ($1 \leq i \leq k+1$) is so we can make explicit the way each r_i is related to the previous ones:

$$r_{i+1} = r_{i-1} - q_i r_i \tag{30.18}$$

$$r_0 = m$$

$$r_1 = n.$$

Now let's focus on the case at hand. We want to find the inverse of a, where $1 \leq a < p$, in $\mathbb{Z}_p$, where p is prime. We already know that the greatest common divisor of a and p is 1, but if we run the Euclidean algorithm on inputs $m = a$ and $n = p$, generating the q_i and r_i as byproducts, we will get some useful results! First off, we know that $r_{k+1} = 1$, since 1 is the only common divisor of a and p. So we can start with the equation

$$r_{k+1} = r_{k-1} - q_k r_k,$$

and work our way backward, repeatedly substituting the right side of (30.18) for instances of the left side. Only multiples of r_0 and r_1 will remain on the right side when all is said and done, and the left side will still be r_{k+1}. But we know that $r_{k+1} = 1$, so we will have derived an equation

$$r_{k+1} = 1 = c \cdot r_0 + d \cdot r_1 \text{ for some coefficients } c \text{ and } d$$
$$= c \cdot a + d \cdot p.$$

We don't care about the value of d, and we care only about the congruence class (mod p) of c. If we let $b = c \bmod p$, then the bottom line is that $a \cdot b \equiv 1$ (mod p); that is, $[b] = [a]^{-1}$.

Using Euclid's algorithm in this way to find multiplicative inverses is exponentially faster than searching through all possibilities, because the algorithm takes time proportional to the number of bits in its arguments, rather than proportional to their magnitude. (See page 287.) Moreover, the same method can be used to find integer coefficients c and d such that $c \cdot m + d \cdot n = s$ for other integer values of m, n, and s, or to determine that this is not possible—it turns out that such coefficients exist if and only if s is a multiple of the greatest common divisor of m and n. (This is proven in Problem 30.8.)

Let's use this method to find the multiplicative inverse of [2] in $\mathbb{Z}_5$, which we already know to be [3]. The algorithm computes the values shown in the table (Figure 30.3), starting with $r_0 = a = 2$ and $r_1 = p = 5$.

For example, $q_2 = 2$ because it is equal to $\lfloor r_1/r_2 \rfloor = \lfloor 5/2 \rfloor = 2$. Now using (30.18) starting from $i + 1 = k + 1 = 3$,

i	q_i	r_i	r_{i+1}
0		2	5
1	0	5	2
2	2	2	1
3	2	1	0

Figure 30.3. Trace of the algorithm starting with $a = 2$ and $p = 5$.

$$r_3 = r_1 - q_2 r_2$$
$$= r_1 - q_2(r_0 - q_1 r_1)$$
$$= -q_2 r_0 + (1 + q_1 q_2) r_1$$
$$= -2 \cdot r_0 + 1 \cdot r_1.$$

Since $r_3 = 1$, the bottom line is that $1 = -2 \cdot r_0 + 1 \cdot r_1$, and discarding the r_1 term while reducing modulo p,

$$-2 \cdot a \equiv 3 \cdot a \equiv 1 \pmod{p},$$

so [3] is the multiplicative inverse of $[a] = [2]$.

✳

We have seen how to do exponentiation quickly by repeated squaring. That is, we have seen how to compute n^k (mod m) with a number of multiplications that is logarithmic in k, and without handling numbers that are much bigger than m. What about logarithms? Whether modular logarithms can be computed quickly is a much more interesting and mysterious question. And important: so-called *discrete logarithms* will play a crucial role when we introduce public-key cryptography in the next chapter.

Let b and m be positive integers, which we can consider to be fixed for the time being. Consider an equation such as

$$y = b^x \quad (\text{mod } m). \tag{30.19}$$

The problem we addressed by means of repeated squaring was to compute y quickly, given x. That is modular exponentiation. But what about the inverse problem: given y, to find an x that makes (30.19) true? Such an x is called a base-b discrete logarithm of y (mod m). For example, since as we showed in (30.13),

$$3^{25} \equiv 3 \quad (\text{mod } 7),$$

25 is a base-3 discrete logarithm of 3, modulo 7. This makes sense as an extension of the standard notion of logarithms—25 is a power to which 3 can be raised in order to get the result 3 (mod 7). Unfortunately, the analogy doesn't help to calculate discrete logarithms. There are no standard math packages, integrals, series expansions, or other mathematical machinery that seem to help. Of course it's possible to plug in one value of x after another and, one at a time, calculate each b^x (mod m) reasonably quickly and compare it to y. But if b, m, and y are big, it may take a long time to stumble across the right x, if one exists at all.

Let's make a comparison between the continuous and discrete logarithms. Suppose we wanted to calculate

$$\log_{123} 2675703636360593169984738701156849484492796490598096370147.$$

With the aid of an infinite-precision mathematical package,[1] this is not hard at all, using the fact that $\log_b a = \frac{\log a}{\log b}$. The answer is 27. (In the quotient, it doesn't matter what base is used for the logarithms as long as it is the same in both the numerator and the denominator.)

On the other hand, what is a discrete base-54321 logarithm of 18789 (mod 70707)? We could try plugging $n = 1, 2, 3, 4, \ldots$ into the expression 54321^n, calculating the result, and reducing modulo 70707, getting 54321, 26517, 57660, 40881, $\ldots$ (mod 70707). But no pattern is apparent, and there is no obvious way to tell when we have tried enough possibilities that we can quit. It turns out that $n = 43210$ works. But there is no known way to find that value that is faster than trying exponents one at a time—though there is also no known proof that this problem is intrinsically time-consuming. And these numbers are relatively small by comparison with the arithmetic capabilities of computers today!

Most of the time, it is a source of frustration and disappointment when we can't find a fast algorithm for a problem. But one set of professionals love such apparently impossibly difficult problems: cryptographers. They exploit hard problems to produce codes that are hard to break. And so we turn to one of the most astonishing developments in twentieth-century mathematics: public key cryptography.

Chapter Summary

- Arithmetic modulo m treats as equivalent all integers that leave the same remainder when they are divided by m.

- Equivalence (mod m) is an equivalence relation on the integers, denoted $\equiv$. The equivalence classes are called the *congruence classes* modulo m. The congruence class of n is written as $[n]$, and the set of m equivalence classes is called $\mathbb{Z}_m$.

- Addition, subtraction, and multiplication of congruence classes (mod m) are well defined. Exponentiation is well defined if the base is a congruence class and the exponent is an integer, but not if the exponent is a congruence class. Division is well defined if m is prime, but not otherwise.

- By repeated squaring, modular exponentiation can be performed in time proportional to the logarithm of the exponent, rather than time proportional to the value of the exponent.

- When m is prime, Euclid's algorithm can be used to find multiplicative inverses in time logarithmic in m.

- Finding discrete logarithms (that is, finding a value of x satisfying the equation $y = b^x \pmod{m}$, given b, m, and y) seems to be intrinsically difficult, but has not been proven to be intrinsically difficult.

Problems

30.1. Prove the equivalence of (30.3)–(30.5).

30.2. Compute the following in $\mathbb{Z}_7$.
   (a) $[5] + [6]$
   (b) $[5] \cdot [6]$
   (c) The additive inverse of $[5]$
   (d) The multiplicative inverse of $[5]$
   (e) $[2]/[5]$

30.3. Write out the complete multiplication tables for $\mathbb{Z}_6$ and $\mathbb{Z}_7$.

30.4. "Fermat's Little Theorem" states that if p is prime and a is a positive integer not divisible by p, then $a^{p-1} \equiv 1 \pmod{p}$.

(a) Prove Fermat's Little Theorem as follows:

- First prove by induction on a that $a^p \equiv a \pmod{p}$ for $0 < a < p$. Start with $a = 1$ and then expand $(a + 1)^p$. You will need to prove along the way that $\binom{p}{i}$ is divisible by p for $0 < i < p$.
- Then extend the result to all positive a not divisible by p.
- Finally, show that it is legal to "cancel" a from each side of the equivalence $a^p \equiv a \pmod{p}$ in order to get $a^{p-1} \equiv 1 \pmod{p}$.

(b) Prove the corollary that if p is prime and a is not divisible by p, then a^{p-2} is a multiplicative inverse of a modulo p.

(c) Calculate $6^{80} \pmod{7}$ and $4^{35} \pmod{11}$ using only pencil and paper.

30.5. (a) Use Euclid's algorithm to find a solution over $\mathbb{Z}$ of the equation $13x + 19y = 1$.

(b) What does the solution to part (a) tell you about the multiplicative inverse of 13 $\pmod{19}$ and of 19 $\pmod{13}$?

30.6. The smallest and largest $(m + 1)$-bit numbers are 2^m and $2^{m+1} - 1$.

(a) Show that if $n = 2^m$, then computing a^n by repeated squaring can be done using m multiplications.

(b) Show that if $n = 2^{m+1} - 1$, then computing a^n by repeated squaring can be done in $2m$ multiplications.

(c) Conclude that the maximum number of multiplications needed to compute a^n, where the binary representation of n has $m + 1$ bits, is $2m = 2\lfloor \lg n \rfloor$.

30.7. Let n be a positive integer, and let $[n]$ denote the congruence class of n modulo some fixed modulus m. Can $[n!]$ be calculated efficiently by determining the congruence class of n modulo m, say $[n] = [a]$, where $0 \le a < m$, and then doing the easier calculation of $[a!]$?

30.8. Consider an equation of the form $c \cdot m + d \cdot n = s$, for fixed integers m, n, and s, for which we seek integer values of the coefficients c and d. (Polynomial equations for which only integer solutions are allowed are called *Diophantine equations*, after the third-century Hellenistic mathematician Diophantus of Alexandria, who studied them.)

(a) Prove that if integers c and d exist such that $c \cdot m + d \cdot n = s$, then s is a multiple of $\gcd(m, n)$.

(b) Prove that if s is a multiple of $\gcd(m, n)$, then there exist integers c and d such that $c \cdot m + d \cdot n = s$, as follows:

- First, prove that such a solution exists if $s = \gcd(m, n)$.
- Then show how the solution for $s = \gcd(m, n)$ can be used to find a solution for $s = k \cdot \gcd(m, n)$ for any integer k.

(c) Prove that if $c \cdot m + d \cdot n = s$ has an integral solution, then it has infinitely many integral solutions. *Hint:* Show that if $c = c_0$, $d = d_0$ is a solution, then for every integer i,

$$c_i = c_0 + \frac{n}{\gcd(m, n)} \cdot i$$

$$d_i = d_0 - \frac{m}{\gcd(m, n)} \cdot i$$

is also a solution.

Chapter 31

Public Key Cryptography

Very few publications have had the impact of Diffie and Hellman's 1976 paper *New Directions in Cryptography*.[1] Almost overnight, secret communications became possible between ordinary people who barely knew each other. No longer were armed guards needed in order to transmit valuable information from place to place. Even the world's most powerful governments couldn't interpret the communications they intercepted. Secure Internet commerce became possible on a massive scale. All this because of the creative application of some simple discrete mathematics.

Cryptography is the art of communicating messages so that only the sender and the recipient know what they say. For thousands of years, kings and generals have been communicating encoded messages about troop movements, treaties, and plots. The route taken by the encoded message is often dangerous, and it is critical that if the encoded message falls into enemy hands, the enemy will not be able to decipher it.

In general terms, encryption transforms an unencrypted message (known as the *plaintext*) into an encrypted message (the *ciphertext*) with the aid of a text string known as the *key*. The recipient of the ciphertext also knows the key and the encoding method, and is therefore able to decipher the ciphertext to retrieve the original message.

Simple methods of encryption have been used since ancient times. The Roman historian Suetonius described the way Julius Caesar encrypted his messages: *If Caesar had anything confidential to say, he wrote it in cipher, that is, by so changing the order of the letters of the alphabet, that not a word could be made out. If anyone wishes to decipher these, and get at their meaning, he must substitute the fourth letter of the alphabet, namely D, for A, and so with the others.* (We encountered these ciphers earlier, in Problem 22.1 on page 241.) The key in this case would be the number 3, the number of positions by which the alphabet was to be shifted. Naturally, if an adversary knew that a so-called Caesar cipher was in use, it would not take long to crack it; there were only 25 possible shifts to try. A cipher is said to be *weak* if it is easy to crack.

Over the centuries, many more sophisticated encryption methods were developed, in which the key might be a secret word or other sequence of

[1] Whitfield Diffie and Martin E. Hellman, "New Directions in Cryptography," *IEEE Transactions on Information Theory* IT-22, no. 6 (November 1976): 644–54.

characters. A *substitution cipher* (also mentioned in Problem 22.1) is a little bit stronger—more difficult to crack—than a Caesar cipher. A substitution cipher relies on a key that is a permutation of the letters A–Z; to encrypt a message, replace each letter of the plaintext by the letter in the corresponding position of the permutation. Cracking a substitution cipher is generally not difficult, because the most frequently occurring letters in the plaintext language are likely to correspond to the most frequently occurring ciphertext letters (see Problem 31.1).

More sophisticated cryptographic methods were developed, in response to which adversaries developed sophisticated methods of analysis. In nearly every case, methods were developed to exploit patterns in the ciphertext to infer the plaintext. A much more powerful cipher is the *one-time pad*, in which the key is the same length as the message, and each character of the plaintext is scrambled by combining it (by an exclusive or of their binary codes, perhaps) with the corresponding character of the key. If the sender and recipient use a one-time pad properly, the encryption is unbreakable, because there is no pattern to the ciphertext.[2]

But all these methods, even the one-time pad, shared the same problem. The sender and the recipient both had to know the same key, and keep it secret. If the key was compromised, the code would become useless. How could both possess the same, secret key?

They could get together, agree on the key, and go their separate ways. If they were planning to communicate over great distances, this presented obvious problems. There was no way to provide a new key except for the parties to travel; and if they wrote the key down before parting, the key might be compromised during their journey.

A courier or some other communication vector could be used for sending the key from one party to the other. But communicating the key in secret requires an answer to the question that was to be solved. If it is possible to communicate the key securely, perhaps the plaintext of the message itself could have been communicated over the same channel. The key may be shorter and therefore easier to hide, but there is no fundamental difference between communicating the key and communicating the message itself.

Figure 31.1 summarizes the problem. Alice wants to send Bob a secret message—in the example, "Retreat at dawn." The encryption algorithm may be extremely ingenious, but if the key is compromised, all is lost. Eve is an eavesdropper—she can hear anything Alice and Bob are saying to each other. But without the secret key, she can't make sense of what she is hearing. How can Alice and Bob agree on a key without any risk that Eve or anyone else will learn it?

In the world of the Internet, "Alice" and "Bob" are computers, executing programs designed to achieve secret communication through an open network. Perhaps "Alice" is you and "Bob" is Amazon or some other online retailer to which you are trying to communicate your password. There is

[2] The "pad" is a set of pages, each with one key on it. Sender and recipient have identical pads, and each discards sheets as they are used to encode or decode one message.

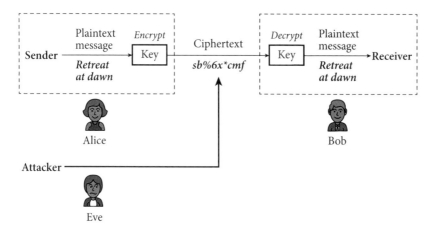

Figure 31.1. The cipher system scenario. Alice encodes her plaintext using a secret key, and sends the ciphertext to Bob. Bob uses the same key to decipher the message and recover the plaintext. Eve, the eavesdropper, overhears or intercepts the ciphertext, but without the secret key can't do anything with it.

no possibility of "Alice" and "Bob" getting together to agree on a secret key—they can communicate only using the Internet itself. Between "Alice" and "Bob" lie hundreds of Internet switching points, cables, and radio links controlled by unknown and untrustworthy parties. "Eve" could be anyone who might "overhear" their communications—between Alice and her WiFi hotspot, between her building and her ISP, or anywhere in the uncharted reaches of the Internet, within national borders and beyond. How can "Alice" and "Bob" possibly use the open Internet to agree on a key so that their Internet communications are secure?

✳

What Diffie and Hellman proposed was a way to implement the following procedure:

1. Alice chooses a secret number a that only Alice knows. (We indicate secret information by writing it in red.)

2. Bob chooses a secret number b that only Bob knows.

3. Alice does a computation on her secret number a and produces a new number, which we will call her public number A. (Information that can be shared through the public channel is written in blue.)

4. Bob does a computation on his secret number b and produces a new number, which we will call his public number B.

5. Alice and Bob exchange their public numbers—Alice sends hers, A, to Bob, and Bob sends his, B, to Alice. Eve and the rest of the world are listening and learn those numbers, which is why we are calling them "public."

6. Alice does a computation based on her secret number a and the public number B she got from Bob. The result of her computation is a new number, which she keeps to herself.

7. Bob does a computation based on his secret number b and the public number A he got from Alice. The result of his computation is a new number, which he keeps to himself.

8. Things have been arranged in such a way that *the number Alice computes from her secret number and Bob's public number is the same as the number Bob computes from his secret number and Alice's public number.* We call this number K the shared key.

9. Eve can't figure out the value of K from Alice and Bob's public numbers or from anything else she is overhearing.

10. Alice and Bob use this key K to encrypt their messages, using some conventional encryption system.

The timeline is laid out in Figure 31.2. Time flows down, and arrows show which information is derived from which other information. For example, Alice derives K from a and B.

The surprise is in steps 8 and 9. How did Alice and Bob wind up with the same key value? And why can't Eve, who learns the public numbers Alice and Bob have shared with each other, use them to figure out their secret numbers and thereby calculate that same key value?

The central idea that makes this scheme work is the notion of a *one-way function*—informally, a function that is easy to compute but hard to "uncompute"; that is, a function for which it is hard to infer the argument from the value. We have already encountered what seems to be a one-way function: modular exponentiation. We saw in Chapter 30 that computing $g^n \bmod p$ can be done quickly by repeated squaring—the number of multiplications increases as $\log n$; that is, as the number of bits in the binary representation of n. But we know of no efficient way, given g, p, and a value x, to find an n such that $g^n \equiv x \bmod p$. (We have suggestively named the modulus p, as it is generally chosen to be prime for the Diffie-Hellman procedure.)

Let's be sure we understand how much faster the modular exponentiation can be done, compared to the exhaustive-search method for finding discrete logarithms. If n is a number of 500 decimal digits, its binary representation is about 1700 bits long, so computing an exponential to that power can be done with two or three thousand multiplications. But searching through all 500-digit exponents to find one that yields the right value could involve 10^{500} modular multiplications, computing each in succession. That is an unfathomably large number—the number of *nanoseconds* since the birth of the universe is less than 10^{27}. (Exhaustive search is not the fastest approach, but no known algorithm is sufficiently better.)

So let's assume that computing discrete logarithms is hard, and see how Alice and Bob proceed.

First, not just Alice and Bob but the whole world agrees on the base g and the modulus p. To emphasize that these parameters are not only

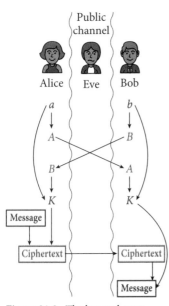

Figure 31.2. The key exchange protocol. Secret information is in red, and public information is in blue. Only blue information passes through the public channel and is available for Eve, the eavesdropper, to learn.

fixed but intentionally published, we will write them in green. Of course, Alice and Bob may not personally know the values of g and p, but their computers do: the programmers who wrote the systems software incorporated these numbers into the computer algorithms. All of the messages that Alice and Bob exchange in developing their shared key, as well as the key itself, will be members of Z_p, that is, numbers in the range from 0 to $p-1$.

Now let's fill in the details in the outline, using the same step numbering.

1. Alice picks a secret number a, chosen at random such that $1 \le a < p-1$. It's important that this number be chosen at random, so Alice invokes her computer's random number generator to do it.

2. Bob picks a secret number b, chosen at random such that $1 \le b < p-1$.

3. Alice calculates $A = g^a \bmod p$. She can do this quickly using modular exponentiation by repeated squaring. This is Alice's *public key*.

4. Bob calculates his public key $B = g^b \bmod p$.

5. Alice sends her public key A to Bob, and Bob sends his public key B to Alice. Eve is listening, and learns both A and B. But neither she nor anyone else can calculate a from A without solving a discrete logarithm problem, because a is a base g discrete logarithm of A modulo p. Similarly, knowing B does not make it easy to learn b without solving a discrete logarithm problem.

6. Alice calculates $B^a \bmod p$, using her secret key and Bob's public key.

7. Bob calculates $A^b \bmod p$, using his secret key and Alice's public key.

8. Alice and Bob have calculated the same value, which we call K:

$$
\begin{aligned}
K \equiv B^a &\equiv (g^b)^a \\
&\equiv g^{ab} \\
&\equiv (g^a)^b \\
&\equiv A^b \qquad (\bmod\ p).
\end{aligned}
\tag{31.1}
$$

9. The key K remains secret; neither Alice nor Bob communicates it to each other or to anyone else. Nobody can figure out what it is without knowing one of the secret keys a or b.

10. Alice and Bob encrypt their communications using some standard encryption algorithm. Each uses K for the encryption key, and can decrypt messages received from the other using the same key.

Equation (31.1) seems like magic, but it's not. The cardinal rule of modular arithmetic means that all the operations can be performed while reducing modulo p at any time. So even though Bob has sent Alice only B, the result of reducing g^b modulo p, that is all Alice needs to calculate the same result that Bob will calculate when he receives A from Alice and computes A^b. Raising g to both the a and b powers can be done in either order, reducing modulo p at any time. Alice and Bob really have found the same value, without ever communicating it to each other!

Current thinking is that to be secure against exhaustive search attacks, p should be a prime number of at least 2048 bits, and g should be chosen such that if we define

$$G = \{g^i \in Z_p : 1 \leq i \leq p - 1\},$$

its size $|G|$ is a large prime, or at least has a large prime factor. If $|G|$ is composite, say with prime factorization $\Pi_i r_i$, then Eve can decompose the problem of calculating a from A into a set of smaller problems: finding $a \bmod r_i$ for each factor r_i. As long as one of the factors r_i is a large prime, this is still intractable. Note that g itself need not be large, as long as it generates a large G.

There is a great deal more to be said about public-key cryptography. Other one-way functions are used in other algorithms. In 1977 Ron Rivest, Adi Shamir, and Len Adleman published a cryptographic system that is secure provided that it is difficult, starting with the product of two large prime numbers, to find those factors.[3] The resulting *RSA cryptosystem* is widely deployed and has important uses beyond enabling two parties to agree publicly on a secret encryption key.

However, neither modular exponentiation, nor multiplication of large primes, nor any other candidate one-way function has been proven to be difficult to invert. It is possible that they are not; in fact, it is possible that no one-way functions exist. It is even possible that some eavesdropper knows how to invert the functions undergirding modern cryptography and is reading all of our banking transactions already. No one considers this likely. It is more likely, however, that protection against brute-force attacks by strongly motivated agents may require longer keys than are currently in use.

One final note. It may be a concern that the secret keys a and b are chosen at random, and we are relying on the vast number of possible keys for the conclusion that Eve will not be able to guess them. But she could get lucky! Why are we not worried about that possibility?

The answer is that we are solving a practical problem, and in practice it is enough to lower the odds of Eve having a lucky guess to be lower than the odds of other forms of failure, such as Bob having a heart attack and never receiving the message, or an asteroid collision destroying the earth and making moot the security of the communication between Alice and Bob. Such eventualities are not impossible, and by making p large enough, we can lower the odds of Eve having a lucky guess to be far smaller than other forms of failure. Make p large enough—and the odds of a lucky guess decrease exponentially with each additional bit in the length of p—and the odds of a lucky guess become insignificant worries.

[3] Clifford Cocks, an English mathematician, had discovered this method in 1973 while working for British intelligence. His discovery was not made public until 1997.

The more immediate risks in public key cryptosystems are not so exotic. The code implementing the cryptographic algorithms may be incorrect because of programmers' errors. Or there may be unanticipated ways (called *sidechannels*) by which an adversary might penetrate the "private" spaces of Alice and Bob (as pictured on the left and right sides of Figure 31.2). Perhaps by listening to the sound of Alice's keystrokes with a parabolic microphone, or by using a sensitive antenna to detect the weak electromagnetic radiation emanating from her processor chip as it carries out the arithmetic in the cryptographic algorithm, an adversary could steal her private information without using the public channel to do it.

Chapter Summary

- *Cryptography* is the art of communicating secret messages so that only the sender and receiver can understand them, even if they fall into other hands.

- The original message is the *plaintext*. It is encoded using a *key*, producing the *ciphertext*.

- A classical encryption method is the *substitution cipher*, in which the key is a permutation of the letters of the alphabet, used to map plaintext letters to ciphertext letters. Substitution ciphers are fairly easily cracked using frequency analysis.

- A *one-time pad* is unbreakable, but clumsy, because a new key is used for every message, and the key has the same length as the message.

- The method of Diffie and Hellman enables the sender and receiver to agree on a secret key via communications over a public channel, such as the Internet.

- Public-key cryptography relies on *one-way functions*, which are easy to compute but hard to invert.

- The crux of the Diffie-Hellman algorithm is that exponentials are easy to calculate in modular arithmetic, but discrete logarithms are hard to compute.

- None of the functions used today in public-key cryptography has been mathematically proven to be a true one-way function; that is, intrinsically difficult to invert.

Problems

31.1. *Frequency analysis* is a way of cracking a substitution cipher by exploiting the fact that, for example, "e" is the most frequent letter in English text. So the most frequent letter in a ciphertext that has been encrypted using a substitu-

[4]Because certain pairs of letters occur with almost the same frequency ("H" and "R," and "U" and "C," for example), different sources give slightly different permutations of this list, depending on the texts that were analyzed to derive the frequency counts. The phrase "ETAOIN SHRDLU" has a special significance. Old typesetting machines arranged the keys in order of letter frequency, with the result that "ETAIONSHRDLU" would be produced if the typesetter ran a finger along one row of keys. This sequence of letters occasionally appeared in newspapers by mistake, but has disappeared with the advent of electronic composition.

tion cipher is likely to be the code for "e." The letters of English text in order of decreasing frequency are[4]

ETAOINSRHDLUCMFYWGPBVKXQJZ.

Although in any given text the frequencies are unlikely to follow exactly this pattern, starting from the assumption that they are in approximately this order will often provide enough clues to reconstruct the plaintext. Use this method to guess the plaintext from which the ciphertext below was generated using a substitution cipher. Spaces and punctuation are as in the plaintext.

STSIJODUHK G PUS. STSIJODUHK JXF DSGI, STSIJODUHK JXF ASS. AX RFYD OX ALSC XFO. ODSJ VFAO BSSL YXRUHK, XHS GQOSI GHX-ODSI.

31.2. It's easier to crack a Caesar cipher than a substitution cipher. Of course there are few enough to try that an exhaustive search might work, but a frequency analysis may provide an even faster approach. Why?

31.3. The one-time pad is a perfect code, except for two problems. One was mentioned already—it is as hard to transmit the key as to transmit the plaintext. A second problem is that a one-time pad is insecure if it is used several times—which users are tempted to do, given how hard it is to distribute the key. What would you do to crack the code if you had multiple ciphertexts that you knew had been encrypted using the same "one-time" pad?

31.4. Let $p = 17$ and $g = 13$. Suppose $a = 3$ and $b = 9$.
 (a) Compute $A = g^a \bmod p$, using a calculator only for addition, subtraction, multiplication, and division.
 (b) Compute $B = g^b \bmod p$.
 (c) Show that $B^a \bmod p = A^b \bmod p$.

31.5. (a) Suppose Eve tries to crack a Diffie-Hellman code by exhaustive search, and she needs to check all keys of length 2048 bits. If she can check one key per nanosecond, how long will it take her?
 (b) What if the possible keys are only 1024 bits in length?

31.6. Work this problem with another student. You two are to derive a shared key. You will use numbers represented by short ASCII strings of capital letters; such a string is converted to a number by replacing each letter with the binary representation of its ASCII code, concatenating the results, and interpreting the final value as a binary number. The modulus p will be the code for HVG, and the value of g will be the code for R. The one-way function is $f(n) = g^n \bmod p$.
 (a) Find the decimal values of p and g, and confirm that p is prime.
 (b) Choose a secret 2-letter word n and compute $x = f(n)$ by repeated squaring. A hand calculator may be useful for reducing modulo p.
 (c) Inform the other student of the value of x, but not the value of n. Wait to receive the other student's number y.
 (d) Calculate $k = y^n \bmod p$. Compare your results.

[5]Named after Blaise de Vigenère (1523–96), though actually discovered by another sixteenth-century cryptologist, Giovan Battista Bellaso.

31.7. After it was understood that substitution ciphers were easy to crack by frequency analysis, a more elaborate system called a Vigenère cypher[5] came into use. A *Vigenère cypher* is a sequence of Caesar ciphers to be used in cyclic order. Suppose, for example, that a message is to be encrypted using 12 Caesar ciphers,

	a	b	c	d	e	f	g	h	i	j	k	l	m	n	o	p	q	r	s	t	u	v	w	x	y	z	
1	t	u	v	w	x	y	z	a	b	c	d	e	f	g	h	i	j	k	l	m	n	o	p	q	r	s	1
2	h	i	j	k	l	m	n	o	p	q	r	s	t	u	v	w	x	y	z	a	b	c	d	e	f	g	2
3	o	p	q	r	s	t	u	v	w	x	y	z	a	b	c	d	e	f	g	h	i	j	k	l	m	n	3
4	m	n	o	p	q	r	s	t	u	v	w	x	y	z	a	b	c	d	e	f	g	h	i	j	k	l	4
5	a	b	c	d	e	f	g	h	i	j	k	l	m	n	o	p	q	r	s	t	u	v	w	x	y	z	5
6	s	t	u	v	w	x	y	z	a	b	c	d	e	f	g	h	i	j	k	l	m	n	o	p	q	r	6
7	b	c	d	e	f	g	h	i	j	k	l	m	n	o	p	q	r	s	t	u	v	w	x	y	z	a	7
8	b	c	d	e	f	g	h	i	j	k	l	m	n	o	p	q	r	s	t	u	v	w	x	y	z	a	8
9	r	s	t	u	v	w	x	y	z	a	b	c	d	e	f	g	h	i	j	k	l	m	n	o	p	q	9
10	y	z	a	b	c	d	e	f	g	h	i	j	k	l	m	n	o	p	q	r	s	t	u	v	w	x	10
11	a	b	c	d	e	f	g	h	i	j	k	l	m	n	o	p	q	r	s	t	u	v	w	x	y	z	11
12	n	o	p	q	r	s	t	u	v	w	x	y	z	a	b	c	d	e	f	g	h	i	j	k	l	m	12

Figure 31.3. The Vigenère encryption table used by industrialist Gordon McKay to encrypt an 1894 letter to his lawyer, Thomas B. Bryan, whose name is the encryption key (in the left column). Courtesy of Harvard University Archives.

say $C_0, \ldots, C_{11}$. To encode a plaintext, use C_0 for the letters in positions 0, 12, 24, and so on, and use C_1 for plaintext letters in positions 1, 13, 25, etc. The key is the sequence of 12 shifts, conveniently presented as a sequence of 12 letters, identifying the encryption of plaintext letter "a" in each of the 12 Caesar ciphers. Figure 31.3 shows an actual Vigenère encryption table, with the key "thomasbbryan" running down the left column. To encrypt a message, the sender would cycle through the twelve rows to encode the successive letters of the plaintext, finding the column headed by the plaintext character; the encrypted character would be at the intersection of the chosen row and column.

The Vigenère cipher was at one point thought to be unbreakable, but it actually can be broken fairly easily. Explain how a Vigenère cypher can be cracked by frequency analysis, if the key is short enough that the encryption can be done by hand.

INDEX